$1.123$$29.10

ELECTRONIC FUNDAMENTALS AND APPLICATIONS: FOR ENGINEERS AND SCIENTISTS

ELECTRONIC FUNDAMENTALS AND APPLICATIONS: FOR ENGINEERS AND SCIENTISTS

JACOB MILLMAN
CHRISTOS C. HALKIAS

Professors of
Electrical Engineering
Columbia University

McGraw-Hill Book Company

New York   St. Louis   San Francisco   Auckland
Düsseldorf   Johannesburg   Kuala Lumpur   London
Mexico   Montreal   New Delhi   Panama   Paris
São Paulo   Singapore   Sydney   Tokyo   Toronto

**ELECTRONIC FUNDAMENTALS AND APPLICATIONS:
FOR ENGINEERS AND SCIENTISTS**

234567890 FGRFGR79

This book was set in 8A by Bi-Comp, Incorporated. The editors were
Kenneth J. Bowman and Madelaine Eichberg; the cover was designed by
Joseph Gillians; the production supervisor was Sam Ratkewitch. New
drawings were done by J & R Services, Inc.
Fairfield Graphics was printer and binder.

Library of Congress Cataloging in Publication Data

Millman, Jacob, date
  Electronic fundamentals and applications.

  Based on the authors' Integrated electronics.
  Includes index.
  1. Integrated circuits. 2. Semiconductors.
  3. Digital electronics. I. Halkias, Christos C., joint author. II. Title.
TK7874.M5248   621.381   75-16234
ISBN 0-07-042310-5

CONTENTS

PREFACE

Modern electronic systems (both digital and analog) consist principally of electrical and mechanical assemblies of solid-state integrated-circuit components (called IC chips, IC packages, or simply ICs), together with input and output transducers (a magnetic or paper tape, a cathode-ray tube, a recorder, a microphone, a loud speaker, etc.). Systems for computing, control, communications, information processing, instrumentation, and laboratory research and development using such integrated electronics pervade most engineering and scientific disciplines. Hence, it is evident that persons engaged in these professions should have a clear understanding and working knowledge of electronic building blocks; which IC chips are commercially available; what circuit functions they perform; their applications; their limitations; and so forth. It is the purpose of this book to supply such information for the engineering student, scientist, or technologist. Specifically, it is intended for chemical, industrial, materials, mechanical, nuclear, civil, and aerospace engineers as well as biomedical majors, chemists, computer scientists, geologists, metallurgists, physicists, and experimental psychologists. It should also benefit engineering aids, technicians, and system managers.

The background of the reader should include simple circuit concepts such as are introduced in a freshman physics course in electricity and magnetism. A knowledge of the material given in a first course in network theory is helpful, but need not be made a prerequisite for this text. A comprehensive summary of the important theorems (Kirchhoff's current and voltage laws, Thévenin's theorem etc.) needed for the analysis of the circuits in this book is given in Appendix C. Sections of this appendix may be assigned simultaneously with the text material whenever deemed necessary. It is assumed that the reader is familiar with elementary calculus, but it is not really essential to an understanding of most of the material contained herein. A knowledge of solid-state physics is not presupposed. In summary, this text is to be used for a one-semester course in electronics to be given normally in the second or third year for engineers, scientists, and technologists (but not for electronic specialists).

The authors' "Integrated Electronics: Analog and Digital Circuits and Systems," McGraw-Hill Book Company, New York, 1972 (abbreviated IE), was written as a three-semester text specifically for electrical engineering students. This present book is a simplification, rearrangement, and condensation of the material in IE and is intended for the user of electronics rather than for the electronic design specialist. The emphasis in the present book is upon the *applications* of integrated circuits in instrumentation and in electronic systems, both digital and analog. The much greater depth and breadth of device physics, network sophistication, hardware detail, and design knowledge required of an electronics engineer is given in IE.

The contents of this text reflect the above-mentioned philosophy. Chapters 1 and 2 give a simple, qualitative, but clear physical picture of a semiconductor and why the junction of two semiconductors results in a rectifier. Chapter 3 exploits the applications of these diodes and indicates how to analyze such circuits. Chapter 4 makes use of the physical concepts already developed to explain the characteristics of the bipolar junction transistor. Chapter 5 shows how to fabricate simultaneously hundreds (or thousands) of diodes, transistors, resistors, and capacitors as an integrated circuit. Chapters 6 through 9 consider in detail digital circuits and digital system applications. These include combinational, sequential, and field-effect transistor LSI digital systems. The analog material in the book (with the exception of the diode applications of Chapter 2) begins with Chapter 10, where the small-signal model of the transistor is introduced. In this chapter low-frequency amplifiers are analyzed and the high-frequency response of amplifiers is given in Chapter 11. Feedback amplifiers are discussed in Chapter 12. The characteristics of the operational amplifier (OP AMP), the basic analog IC, are developed in Chap. 13. Analog systems based upon the OP AMP are studied in Chapters 13 (linear applications) and 14 (nonlinear applications).

The scope of the applications is very broad. Among these are rectifiers (including capacitor filters) clippers, voltage regulators, logic gates (AND, OR, NOT, NAND, DTL, TTL, ECL, etc.) binary adders, decoder/demultiplexers, data selectors/multiplexers, encoders, read-only memories (ROM), digital comparators, parity checkers, shift registers, counters, random-access memories (RAM), low-frequency amplifiers, high-frequency amplifiers, feedback amplifiers, differential amplifiers, OP AMPs, voltage-to-current and current-to-voltage converters, analog computers, active filters, tuned amplifiers, analog comparators, precision ac/dc converters, active detectors, square-wave generators, pulse generators, triangle generators, regenerative comparators, digital-to-analog (D/A) and analog-to-digital (A/D) converters.

For the most part, real (commercially available) device characteristics are employed. In this way the reader may become familiar with the order of magnitude of device parameters, the variability of these parameters within a given type and with a change of temperature, the effect of the inevitable shunt capacitances in circuits, and the effect of input and output resistances and loading on circuit operation. These considerations are of utmost importance to the student or the practicing engineer since the circuits to be designed must function properly and reliably in the physical world rather than under hypothetical or ideal circumstances.

There are 350 homework problems in Appendix D and 340 review questions at the end of each chapter. These will test the students' grasp of the fundamental concepts enunciated in the book and will give them experience in the analysis and use of electronic circuits and systems. In almost all numerical problems realistic parameter values and specifications have been chosen. An answer book is available for students, and a solutions manual may be obtained from the publisher by an instructor who has adopted the text.

Considerable thought was given to the pedagogy of presentation, the explanation of circuit behavior, the use of a consistent system of notation, the care with which diagrams are drawn, the illustrative examples worked out in detail in the text, and the review questions. It is hoped that these will facilitate the use of the book in self-study and that the practicing engineers will find the text useful for keeping up to date in the fast-moving field of electronics. Incidentally, the review questions make excellent quiz problems.

We express our thanks to Mrs. Maria Salgado for her skillful service in typing the manuscript. We also appreciate the assistance given us by J. Waldhuter and B. Wah in connection with the problems.

JACOB MILLMAN
CHRISTOS C. HALKIAS

ELECTRONIC FUNDAMENTALS AND APPLICATIONS: FOR ENGINEERS AND SCIENTISTS

1/ SEMICONDUCTORS

The physical characteristics which allow us to distinguish between an insulator, a semiconductor, and a metal are discussed. The current in a metal is due to the flow of negative charges (*electrons*), whereas the current in a semiconductor results from the movement of both electrons and positive charges (*holes*). A semiconductor may be doped with impurity atoms so that the current is due predominantly either to electrons or to holes. The transport of the charges in a crystal under the influence of an electric field (a *drift* current), and also as a result of a nonuniform concentration gradient (a *diffusion* current), is investigated.

1-1 CHARGED PARTICLES

The charge, or quantity, of negative electricity and the mass of the electron have been found to be 1.60×10^{-19} C (coulomb) and 9.11×10^{-31} kg, respectively. The values of many important physical constants are given in Appendix A, and a list of conversion factors and prefixes is given in Appendix B. Some idea of the number of electrons per second that represents current of the usual order of magnitude is readily possible. For example, since the charge per electron is 1.60×10^{-19} C, the number of electrons per coulomb is the reciprocal of this number, or approximately, 6×10^{18}. Further, since a current of 1 A (ampere) is the flow of 1 C/s, then a current of only 1 pA (1 picoampere, or 10^{-12} A) represents the motion of approximately 6 million electrons per second. Yet a current of 1 pA is so small that considerable difficulty is experienced in attempting to measure it.

The charge of a positive ion is an integral multiple of the charge of the electron, although it is of opposite sign. For the case of singly ionized particles, the charge is equal to that of the electron. For the case of doubly ionized particles, the ionic charge is twice that of the electron.

1

In a semiconductor crystal such as silicon, two electrons are shared by each pair of ionic neighbors. Such a configuration is called a *covalent bond*. Under certain circumstances an electron may be missing from this structure, leaving a "hole" in the bond. These vacancies in the covalent bonds may move from ion to ion in the crystal and constitute a current equivalent to that resulting from the motion of free positive charges. The magnitude of the charge associated with the hole is that of a free electron. This very brief introduction to the concept of a hole as an effective charge carrier is elaborated upon in Sec. 1-5.

1-2   FIELD INTENSITY, POTENTIAL, ENERGY

By definition, *the force* $\mathbf{f}$ *(newtons) on a unit positive charge in an electric field is the electric field intensity* $\mathbf{\varepsilon}$ *at that point.* Newton's second law determines the motion of a particle of charge q (coulombs), mass m (kilograms), moving with a velocity $\mathbf{v}$ (meters per second) in a field $\mathbf{\varepsilon}$ (volts per meter).

$$\mathbf{f} = q\mathbf{\varepsilon} = m\frac{d\mathbf{v}}{dt} \tag{1-1}$$

The mks (meter-kilogram-second) rationalized system of units is found to be most convenient for subsequent studies. Unless otherwise stated, this system of units is employed throughout this book.

Potential   By definition, *the potential V (volts) of point B with respect to point A is the work done* against *the field* in taking a unit positive charge from A to B. This definition is valid for a three-dimensional field. For a one-dimensional problem with A at x_0 and B at an arbitrary distance x, it follows that†

$$V \equiv -\int_{x_0}^{x} \varepsilon \, dx \tag{1-2}$$

where ε now represents the X component of the field. Differentiating Eq. (1-2) gives

$$\varepsilon = -\frac{dV}{dx} \tag{1-3}$$

The minus sign shows that the electric field is directed from the region of higher potential to the region of lower potential. In three dimensions, the electric field equals the negative gradient of the potential.

By definition, *the potential energy U (joules) equals the potential multiplied by the charge q under consideration,* or

$$U \equiv qV \tag{1-4}$$

---

† The symbol $\equiv$ is used to designate "equal to by definition."

If an electron is being considered, q is replaced by $-q$ (where q is the *magnitude* of the electronic charge) and U has the same shape as V but is inverted.

The law of conservation of energy states that the total energy W, which equals the sum of the potential energy U and the kinetic energy $\frac{1}{2}mv^2$, remains constant. Thus, at any point in space,

$$W = U + \tfrac{1}{2}mv^2 = \text{constant} \tag{1-5}$$

As an illustration of this law, consider two parallel electrodes (A and B of Fig. 1-1a) separated a distance d, with B at a negative potential V_d with respect to A. An electron leaves the surface of A with a velocity v_o in the direction toward B. How much speed v will it have if it reaches B?

From the definition, Eq. (1-2), it is clear that only differences of potential have meaning, and hence let us arbitrarily ground A, that is, consider it to be at zero potential. Then the potential at B is $V = -V_d$, and the potential energy is $U = -qV = qV_d$. Equating the total energy at A to that at B gives

$$W = \tfrac{1}{2}mv_o{}^2 = \tfrac{1}{2}mv^2 + qV_d \tag{1-6}$$

This equation indicates that v must be less than v_o, which is obviously correct since the electron is moving in a repelling field. Note that the final speed v attained by the electron in this conservative system is independent of the form of the variation of the field distribution between the plates and depends only upon the magnitude of the potential difference V_d. Also, if the electron is to

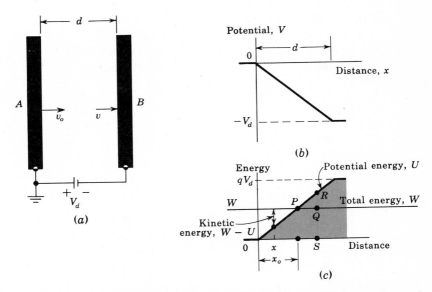

Fig. 1-1 (a) An electron leaves electrode A with an initial speed v_o and moves in a retarding field toward plate B; (b) the potential; (c) the potential-energy barrier between electrodes.

reach electrode B, its initial speed must be large enough so that $\frac{1}{2}mv_o^2 > qV_d$. Otherwise, Eq. (1-6) leads to the impossible result that v is imaginary. We wish to elaborate on these considerations now.

The Concept of a Potential-Energy Barrier For the configuration of Fig. 1-1a with electrodes which are large compared with the separation d, we can draw (Fig. 1-1b) a linear plot of potential V versus distance x (in the interelectrode space). The corresponding potential energy U versus x is indicated in Fig. 1-1c. Since potential is the potential energy per unit charge, curve c is obtained from curve b by multiplying each ordinate by the charge on the electron (a negative number). Since the total energy W of the electron remains constant, it is represented as a horizontal line. The kinetic energy at any distance x equals the difference between the total energy W and the potential energy U at this point. This difference is greatest at O, indicating that the kinetic energy is a maximum when the electron leaves the electrode A. At the point P this difference is zero, which means that no kinetic energy exists, so that the particle is at rest at this point. This distance x_0 is the maximum that the electron can travel from A. At point P (where $x = x_0$) it comes momentarily to rest, and then reverses its motion and returns to A.

Consider a point such as S which is at a greater distance than x_0 from electrode A. Here the total energy QS is less than the potential energy RS, so that the difference, which represents the kinetic energy, is negative. This is an impossible physical condition, however, since negative kinetic energy ($\frac{1}{2}mv^2 < 0$) implies an imaginary velocity. We must conclude that the particle can never advance a distance greater than x_0 from electrode A.

The foregoing analysis leads to the very important conclusion that the shaded portion of Fig. 1-1c can never be penetrated by the electron. Thus, at point P, the particle acts *as if* it had collided with a solid wall, hill, or barrier and the direction of its flight had been altered. *Potential-energy barriers* of this sort play an important role in the analyses of semiconductor devices.

It must be emphasized that the words "collides with" or "rebounds from" a potential "hill" are convenient descriptive phrases and that an actual encounter between two material bodies is not implied.

1-3 THE eV UNIT OF ENERGY

The joule (J) is the unit of energy in the mks system. In some engineering power problems this unit is very small, and a factor of 10^3 or 10^6 is introduced to convert from watts (1 W = 1 J/s) to kilowatts or megawatts, respectively. However, in other problems, the joule is too large a unit, and a factor of 10^{-7} is introduced to convert from joules to ergs. For a discussion of the energies involved in electronic devices, even the erg is much too large a unit. This statement is not to be construed to mean that only minute amounts of energy can be obtained from electron devices. It is true that each electron possesses a tiny amount of energy, but as previously pointed out (Sec. 1-1), an enormous number of electrons are involved even in a small current, so that considerable power may be represented.

A unit of work or energy, called the *electron volt* (eV), is defined as follows:

$$1 \text{ eV} \equiv 1.60 \times 10^{-19} \text{ J}$$

Of course, any type of energy, whether it be electric, mechanical, thermal, etc., may be expressed in electron volts.

The name *electron volt* arises from the fact that, if an electron falls through a potential of one volt, its kinetic energy will increase by the decrease in potential energy, or by

$$qV = (1.60 \times 10^{-19} \text{ C})(1 \text{ V}) = 1.60 \times 10^{-19} \text{ J} = 1 \text{ eV}$$

However, as mentioned above, the electron-volt unit may be used for any type of energy, and is not restricted to problems involving electrons.

A potential-energy barrier of E (electron volts) is equivalent to a potential hill of V (volts) if these quantities are related by

$$qV = 1.60 \times 10^{-19} E \tag{1-7}$$

Note that V and E are *numerically* identical but dimensionally different.

1-4   MOBILITY AND CONDUCTIVITY

In a metal the outer, or valence, electrons of an atom are as much associated with one ion as with another, so that the electron attachment to any individual atom is almost zero. Depending upon the metal, at least one, and sometimes two or three, electrons per atom are free to move throughout the interior of the metal under the action of applied fields.

Figure 1-2 is a two-dimensional schematic picture of the charge distribution within a metal. The shaded regions represent the net positive charge of the nucleus and the tightly bound inner electrons. The black dots represent the outer, or valence, electrons in the atom. It is these electrons that cannot be said to belong to any particular atom; instead, they have completely lost their individuality and can wander freely about from atom to atom in the metal. Thus a metal is visualized as a region containing a periodic three-dimensional array of heavy, tightly bound ions permeated with a swarm of electrons that may move about quite freely. This picture is known as the *electron-gas* description of a metal.

According to the electron-gas theory of a metal, the electrons are in continuous motion, the direction of flight being changed at each collision with the heavy (almost stationary) ions. The average distance between collisions is called the *mean free path*. Since the motion is random, then, on an average,

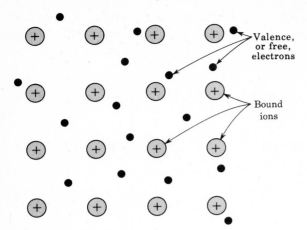

Valence, or free, electrons

Bound ions

Fig. 1-2 Schematic arrangement of the atoms in one plane in a metal, drawn for monovalent atoms. The black dots represent the electron gas, each atom having contributed one electron to this gas.

there will be as many electrons passing through a unit area in the metal in any direction as in the opposite direction in a given time. Hence the average current is zero.

Let us now see how the situation is changed if a constant electric field ε (volts per meter) is applied to the metal. As a result of this electrostatic force, the electrons would be accelerated and the velocity would increase indefinitely with time, were it not for the collisions with the ions. However, at each inelastic collision with an ion, an electron loses energy and changes direction. The probability that an electron moves in a particular direction after a collision is equal to the probability that it travels in the opposite direction after colliding with an ion. Hence, as indicated in Fig. 1-3, the velocity of an electron increases linearly with time between collisions, and (on the average) its velocity is reduced to zero at each collision. A steady-state condition is reached where a finite value of *drift speed v* is attained. This drift velocity is in the direction opposite to that of the electric field. The speed at a time t between collisions is at, where $a = q\varepsilon/m$ is the acceleration. Hence the average speed v is proportional to ε. Thus

$$v = \mu\varepsilon \tag{1-8}$$

where μ (square meters per volt-second) is called the *mobility* of the electrons.

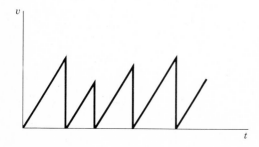

Fig. 1-3 The speed of an electron in a crystal, subjected to an external electric field, increases linearly with time between collisions. Its velocity is effectively reduced to zero at each collision.

According to the foregoing theory, a steady-state drift speed has been superimposed upon the random thermal motion of the electrons. Such a directed flow of electrons constitutes a current. We now calculate the magnitude of the current.

Current Density If N electrons are contained in a length L of conductor (Fig. 1-4), and if it takes an electron a time T sec to travel a distance of L m in the conductor, the total number of electrons passing through any cross section of wire in unit time is N/T. Thus the total charge per second passing any area, which, by definition, is the current in amperes, is

$$I \equiv \frac{Nq}{T} = \frac{Nqv}{L} \qquad (1\text{-}9)$$

because L/T is the average, or *drift*, speed v m/s of the electrons. By definition, the current density, denoted by the symbol J, is the current per unit area of the conducting medium. That is, assuming a uniform current distribution,

$$J \equiv \frac{I}{A} \qquad (1\text{-}10)$$

where J is in amperes per square meter, and A is the cross-sectional area (in meters) of the conductor. This becomes, by Eq. (1-9)

$$J = \frac{Nqv}{LA} \qquad (1\text{-}11)$$

From Fig. 1-4 it is evident that LA is simply the volume containing the N electrons, and so N/LA is the electron concentration n (in electrons per cubic meter). Thus

$$n = \frac{N}{LA} \qquad (1\text{-}12)$$

and Eq. (1-11) reduces to

$$J = nqv = \rho v \qquad (1\text{-}13)$$

where $\rho \equiv nq$ is the charge density, in coulombs per cubic meter, and v is in meters per second.

This derivation is independent of the form of the conducting medium. Consequently, Fig. 1-4 does not necessarily represent a wire conductor. It may represent equally well a portion of a gaseous-discharge tube or a volume

Fig. 1-4 Pertaining to the calculation of current density.

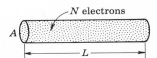

element of a semiconductor.   Furthermore, neither ρ nor v need be constant, but may vary from point to point in space or may vary with time.

Conductivity   From Eqs. (1-13) and (1-8)

$$J = nqv = nq\mu\mathcal{E} = \sigma\mathcal{E} \tag{1-14}$$

where

$$\sigma = nq\mu \tag{1-15}$$

is the *conductivity* of the metal in (ohm-meter)$^{-1}$.   From Eq. (1-14),

$$I = JA = \frac{\sigma AL\mathcal{E}}{L} = \frac{\sigma A V}{L} = \frac{V}{R} \tag{1-16}$$

where $V = L\mathcal{E}$ is the applied voltage across the length L, and R, the resistance of the conductor, is given by

$$R = \frac{L}{\sigma A} \tag{1-17}$$

Equation (1-16) is recognized as Ohm's law, namely, the conduction current is proportional to the applied voltage.   As already mentioned, the energy which the electrons acquire from the applied field is, as a result of collisions, given to the lattice ions.   Hence power is dissipated within the metal by the electrons, and the power density (Joule heat) is given by $J\mathcal{E} = \sigma\mathcal{E}^2$ (watts per cubic meter).

1-5   ELECTRONS AND HOLES IN AN INTRINSIC SEMICONDUCTOR

From Eq. (1-15) we see that the conductivity is proportional to the concentration n of free electrons.   For a good conductor, n is very large ($\sim 10^{28}$ electrons/m^3); for an insulator, n is very small ($\sim 10^7$); and for a semiconductor, n lies between these two values.   The valence electrons in a semiconductor are not free to wander about as they are in a metal, but rather are trapped in a bond between two adjacent ions, as explained below.

The Covalent Bond   Germanium and silicon are the two most important semiconductors used in electronic devices.   The crystal structure of these materials consists of a regular repetition in three dimensions of a unit cell having the form of a tetrahedron with an atom at each vertex.   This structure is illustrated symbolically in two dimensions in Fig. 1-5.   Germanium has a total of 32 electrons in its atomic structure.   Each atom in a germanium crystal

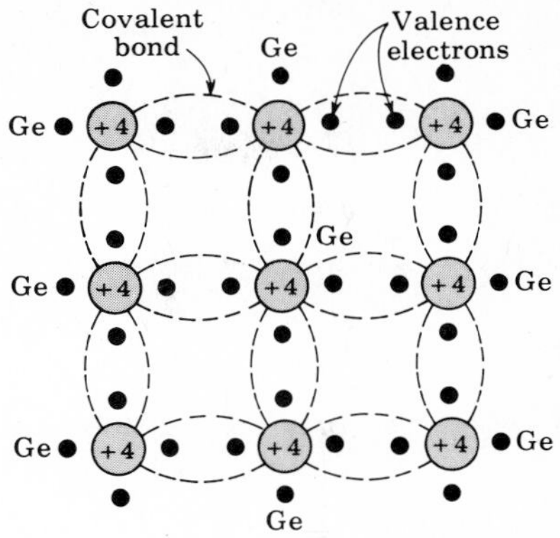

Fig. 1-5 Crystal structure of germanium, illustrated symbolically in two dimensions.

contributes four valence electrons, so that the atom is tetravalent. The inert ionic core of the germanium atom carries a positive charge of $+4$ measured in units of the electronic charge. The binding forces between neighboring atoms result from the fact that each of the valence electrons of a germanium atom is shared by one of its four nearest neighbors. This *electron-pair,* or *covalent, bond* is represented in Fig. 1-5 by the two dashed lines which join each atom to each of its neighbors. The fact that the valence electrons serve to bind one atom to the next also results in the valence electron being tightly bound to the nucleus. Hence, in spite of the availability of four valence electrons, the crystal has a low conductivity.

The Hole At a very low temperature (say 0°K) the ideal structure of Fig. 1-5 is approached, and the crystal behaves as an insulator, since no free carriers of electricity are available. However, at room temperature, some of the covalent bonds will be broken because of the thermal energy supplied to the crystal, and conduction is made possible. This situation is illustrated in Fig. 1-6. Here an electron, which for the far greater period of time forms part of a covalent bond, is pictured as being dislodged, and therefore free to wander in a random fashion throughout the crystal. The energy E_G required to break such a covalent bond is about 0.72 eV for germanium and 1.1 eV for silicon at room temperature. The absence of the electron in the covalent bond is represented by the small circle in Fig. 1-6, and such an incomplete covalent bond is called a *hole.* The importance of the hole is that it may serve as a carrier of electricity comparable in effectiveness with the free electron.

The mechanism by which a hole contributes to the conductivity is qualitatively as follows: When a bond is incomplete so that a hole exists, it is relatively easy for a valence electron in a neighboring atom to leave its covalent bond to fill this hole. An electron moving from a bond to fill a hole leaves a hole in its initial position. Hence the hole effectively moves in the direction opposite to that of the electron. This hole, in its new position, may now be

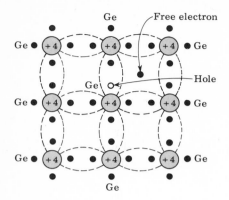

Fig. 1-6 Germanium crystal with a broken covalent bond.

filled by an electron from another covalent bond, and the hole will correspondingly move one more step in the direction opposite to the motion of the electron. Here we have a mechanism for the conduction of electricity which does
not involve *free* electrons. This phenomenon is illustrated schematically in
Fig. 1-7, where a circle with a dot in it represents a completed bond, and an
empty circle designates a hole. Figure 1-7a shows a row of 10 ions, with a
broken bond, or hole, at ion 6. Now imagine that an electron from ion 7 moves
into the hole at ion 6, so that the configuration of Fig. 1-7b results. If we
compare this figure with Fig. 1-7a, it looks as if the hole in (a) has moved
toward the right in (b) (from ion 6 to ion 7). This discussion indicates that
the motion of the hole in one direction actually means the transport of a negative charge an equal distance in the opposite direction. So far as the flow
of electric current is concerned, the hole behaves like a positive charge equal
in magnitude to the electronic charge. We can consider that the holes are
physical entities whose movement constitutes a flow of current. The heuristic
argument that a hole behaves as a *free* positive charge carrier may be justified
by quantum mechanics.

In a pure (*intrinsic*) semiconductor the number of holes is equal to the
number of free electrons. Thermal agitation continues to produce new hole-
electron pairs, whereas other hole-electron pairs disappear as a result of recombination. The hole concentration p must equal the electron concentration n,
so that

$$n = p = n_i \tag{1-18}$$

where n_i is called the *intrinsic concentration*.

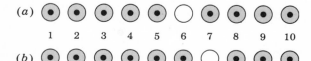

Fig. 1-7 The mechanism by which a hole contributes to the conductivity.

1-6   DONOR AND ACCEPTOR IMPURITIES

If, to intrinsic silicon or germanium, there is added a small percentage of trivalent or pentavalent atoms, a *doped, impure,* or *extrinsic,* semiconductor is formed.

Donors If the dopant has five valence electrons, the crystal structure of Fig. 1-8 is obtained. The impurity atoms will displace some of the germanium atoms in the crystal lattice. Four of the five valence electrons will occupy covalent bonds, and the fifth will be nominally unbound and will be available as a carrier of current. The energy required to detach this fifth electron from the atom is of the order of only 0.01 eV for Ge or 0.05 eV for Si. Suitable pentavalent impurities are antimony, phosphorus, and arsenic. Such impurities donate excess (negative) electron carriers, and are therefore referred to as *donor,* or *n*-type, impurities.

If intrinsic semiconductor material is "doped" with *n*-type impurities, not only does the number of electrons increase, but the number of holes decreases below that which would be available in the intrinsic semiconductor. The reason for the decrease in the number of holes is that the larger number of electrons present increases the rate of recombination of electrons with holes.

Acceptors If a trivalent impurity (boron, gallium, or indium) is added to an intrinsic semiconductor, only three of the covalent bonds can be filled, and the vacancy that exists in the fourth bond constitutes a hole. This situation is illustrated in Fig. 1-9. Such impurities make available positive carriers because they create holes which can accept electrons. These impurities are consequently known as *acceptor,* or *p*-type, impurities. The amount of impurity which must be added to have an appreciable effect on the conductivity is very small. For example, if a donor-type impurity is added to the extent of 1 part in 10^8, the conductivity of germanium at 30°C is multiplied by a factor of 12.

Fig. 1-8 Crystal lattice with a germanium atom displaced by a pentavalent impurity atom.

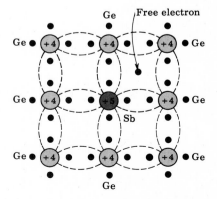

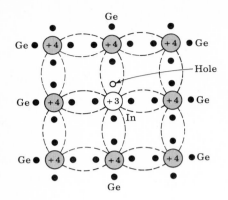

Fig. 1-9 Crystal lattice with a germanium atom displaced by an atom of a trivalent impurity.

The Mass-Action Law We noted above that adding n-type impurities decreases the number of holes. Similarly, doping with p-type impurities decreases the concentration of free electrons below that in the intrinsic semiconductor. A theoretical analysis leads to the result that, under thermal equilibrium, the product of the free negative and positive concentrations is a constant independent of the amount of donor and acceptor impurity doping. This relationship is called the *mass-action law* and is given by

$$np = n_i{}^2 \qquad\qquad (1\text{-}19)$$

The intrinsic concentration n_i is a function of temperature (Sec. 1-8).

We have the important result that the doping of an intrinsic semiconductor not only increases the conductivity, but also serves to produce a conductor in which the electric carriers are either predominantly holes or predominantly electrons. In an n-type semiconductor, the electrons are called the *majority carriers,* and the holes are called the *minority carriers.* In a p-type material, the holes are the majority carriers, and the electrons are the minority carriers.

1-7   CHARGE DENSITIES IN A SEMICONDUCTOR

Equation (1-19), namely, $np = n_i{}^2$, gives one relationship between the electron n and the hole p concentrations. These densities are further interrelated by the law of electrical neutrality, which we shall now state in algebraic form: Let N_D equal the concentration of donor atoms. Since, as mentioned above, these are practically all ionized, N_D positive charges per cubic meter are contributed by the donor ions. Hence the total positive-charge density is $N_D + p$. Similarly, if N_A is the concentration of acceptor ions, these contribute N_A negative charges per cubic meter. The total negative-charge density is $N_A + n$. Since the semiconductor is electrically neutral, the magnitude of the positive-charge density must equal that of the negative concentration, or

$$N_D + p = N_A + n \qquad\qquad (1\text{-}20)$$

Consider an n-type material having $N_A = 0$. Since the number of electrons is much greater than the number of holes in an n-type semiconductor ($n \gg p$), then Eq. (1-20) reduces to

$$n \approx N_D \tag{1-21}$$

In an n-type material the free-electron concentration is approximately equal to the density of donor atoms.

The concentration p of holes *in the n-type semiconductor* is obtained from Eq. (1-19). Thus,

$$p = \frac{n_i^2}{N_D} \tag{1-22}$$

Similarly, *in a p-type semiconductor,*

$$p \approx N_A \tag{1-23}$$

and

$$n = \frac{n_i^2}{N_A} \tag{1-24}$$

It is possible to add donors to a p-type crystal or, conversely, to add acceptors to n-type material. If equal concentrations of donors and acceptors permeate the semiconductor, it remains intrinsic. The hole of the acceptor combines with the conduction electron of the donor to give no additional free carriers. Thus, from Eq. (1-20) with $N_D = N_A$, we observe that $p = n$, and from Eq. (1-19), $n^2 = n_i^2$, or $n = n_i =$ the intrinsic concentration.

An extension of the above argument indicates that if the concentration of donor atoms added to a p-type semiconductor exceeds the acceptor concentration $(N_D > N_A)$, the specimen is changed from a p-type to an n-type semiconductor. [In Eqs. (1-21) and (1-22) N_D should be replaced by $N_D - N_A$.]

Generation and Recombination of Charges In a pure (intrinsic) semiconductor the number of holes is equal to the number of free electrons. Thermal agitation, however, continues to generate g new hole-electron pairs per unit volume per second, while other hole-electron pairs disappear as a result of recombination; in other words, free electrons fall into empty covalent bonds, resulting in the loss of a pair of mobile carriers. On an average, a hole (an electron) will exist for $\tau_p (\tau_n)$ s before recombination. This time is called the *mean lifetime* of the hole and electron, respectively. These parameters are very important in semiconductor devices because they indicate the time required for electron and hole concentrations which have been caused to change to return to their equilibrium concentrations.

1-8   ELECTRICAL PROPERTIES OF Ge AND Si

A fundamental difference between a metal and a semiconductor is that the former is *unipolar* [conducts current by means of charges (electrons) of one sign only], whereas a semiconductor is *bipolar* (contains two charge-carrying "particles" of opposite sign).

Conductivity   One carrier is negative (the free electron), of mobility μ_n, and the other is positive (the hole), of mobility μ_p. These particles move in opposite directions in an electric field $\mathcal{E}$, but since they are of opposite sign, the current of each is in the same direction. Hence the current density J is given by (Sec. 1-4)

$$J = (n\mu_n + p\mu_p)q\mathcal{E} = \sigma\mathcal{E} \tag{1-25}$$

where n = magnitude of free-electron (negative) concentration
p = magnitude of hole (positive) concentration
σ = conductivity

Hence $\sigma = (n\mu_n + p\mu_p)q$ $\tag{1-26}$

For the pure semiconductor, $n = p = n_i$, where n_i is the intrinsic concentration.

Intrinsic Concentration   With increasing temperature, the density of hole-electron pairs increases and, correspondingly, the conductivity increases.

TABLE 1-1   Properties of germanium and silicon†

Property	Ge	Si
Atomic number	32	14
Atomic weight	72.6	28.1
Density, g/cm³	5.32	2.33
Dielectric constant (relative)	16	12
Atoms/cm³	4.4×10^{22}	5.0×10^{22}
E_{GO}, eV, at 0°K	0.785	1.21
E_G, eV, at 300°K	0.72	1.1
n_i at 300°K, cm⁻³	2.5×10^{13}	1.5×10^{10}
Intrinsic resistivity at 300°K, Ω-cm	45	230,000
μ_n, cm²/V-s at 300°K	3,800	1,300
μ_p, cm²/V-s at 300°K	1,800	500
D_n, cm²/s $= \mu_n V_T$	99	34
D_p, cm²/s $= \mu_p V_T$	47	13

† G. L. Pearson and W. H. Brattain, History of Semiconductor Research, *Proc. IRE*, vol. 43, pp. 1794–1806, December, 1955. E. M. Conwell, Properties of Silicon and Germanium, Part II, *Proc. IRE*, vol. 46, no. 6, pp. 1281–1299, June, 1958.

Theoretically it is found that the intrinsic concentration n_i varies with T as

$$n_i{}^2 = A_o T^3 \epsilon^{-E_{GO}/kT} \tag{1-27}$$

where E_{GO} is the energy gap (the energy required to break a covalent bond) at $0°$K in electron volts, k is the Boltzmann constant in $eV/°K$ (Appendix A), and A_o is a constant independent of T. The constants E_{GO}, μ_n, μ_p, and many other important physical quantities for germanium and silicon are given in Table 1-1. Note that germanium has of the order of 10^{22} atoms/cm³, whereas at room temperature $(300°K)$, $n_i \approx 10^{13}$/cm³. Hence only 1 atom in about 10^9 contributes a free electron (and also a hole) to the crystal because of broken covalent bonds. For silicon this ratio is even smaller, about 1 atom in 10^{12}.

The Energy Gap   The energy E_G in a semiconductor depends upon temperature. Experimentally it is found that E_G decreases linearly with T. For germanium and silicon the values of E_G at $0°$K and $300°$K are given in Table 1-1.

The Mobility   This parameter μ varies in a complicated fashion with temperature T and electric field ε.

EXAMPLE   (a) Using Avogadro's number, verify the numerical value given in Table 1-1 for the concentration of atoms in germanium. (b) Find the resistivity of intrinsic germanium at $300°$K. (c) If a donor-type impurity is added to the extent of 1 part in 10^8 germanium atoms, find the resistivity. (d) If germanium were a monovalent metal, find the ratio of its conductivity to that of the n-type semiconductor in part c.

Solution   a. A quantity of any substance equal to its molecular weight in grams is a *mole* of that substance. Further, a mole of any substance contains the same number of molecules as a mole of any other material. This number is called *Avogadro's number* and equals 6.02×10^{23} molecules per mole (Appendix A). Thus, for monatomic germanium (using Table 1-1),

$$\text{Concentration} = 6.02 \times 10^{23} \frac{\text{atoms}}{\text{mole}} \times \frac{1 \text{ mole}}{72.6 \text{ g}} \times \frac{5.32 \text{ g}}{\text{cm}^3} = 4.41 \times 10^{22} \frac{\text{atoms}}{\text{cm}^3}$$

b. From Eq. (1-26), with $n = p = n_i$,

$$\sigma = n_i q(\mu_n + \mu_p) = (2.5 \times 10^{13} \text{ cm}^{-3})(1.60 \times 10^{-19} \text{ C})(3,800 + 1,800) \frac{\text{cm}^2}{\text{V-s}}$$

$$= 0.0224 \ (\Omega\text{-cm})^{-1}$$

$$\text{Resistivity} = \frac{1}{\sigma} = \frac{1}{0.0224} = 44.6 \ \Omega\text{-cm}$$

in agreement with the value in Table 1-1.

c. If there is 1 donor atom per 10^8 germanium atoms, then $N_D = 4.41 \times 10^{14}$ atoms/cm³. From Eq. (1-21) $n \approx N_D$ and from Eq. (1-22)

$$p = \frac{n_i^2}{N_D} = \frac{(2.5 \times 10^{13})^2}{4.41 \times 10^{14}} = 1.42 \times 10^{12} \text{ holes/cm}^3$$

Since $n \gg p$, we can neglect p in calculating the conductivity. From Eq. (1-26)

$$\sigma = nq\mu_n = 4.41 \times 10^{14} \times 1.60 \times 10^{-19} \times 3,800 = 0.268 \ (\Omega\text{-cm})^{-1}$$

The resistivity $= 1/\sigma = 1/0.268 = 3.72 \ \Omega\text{-cm}$.

NOTE: The addition of 1 donor atom in 10^8 germanium atoms has multiplied the conductivity by a factor of $44.6/3.72 = 11.9$.

d. If each atom contributed one free electron to the "metal," then

$$n = 4.41 \times 10^{22} \ \text{electrons/cm}^3$$

and

$$\sigma = nq\mu_n = 4.41 \times 10^{22} \times 1.60 \times 10^{-19} \times 3,800$$
$$= 2.68 \times 10^7 \ (\Omega\text{-cm})^{-1}$$

Hence the conductivity of the "metal" is higher than that of the n-type semiconductor by a factor of

$$\frac{2.68 \times 10^7}{0.268} = 10^8$$

1-9   DIFFUSION

In addition to a conduction current, the transport of charges in a semiconductor may be accounted for by a mechanism called *diffusion*, not ordinarily encountered in metals. The essential features of diffusion are now discussed.

It is possible to have a nonuniform concentration of particles in a semiconductor. As indicated in Fig. 1-10, the concentration p of holes varies with distance x in the semiconductor, and there exists a concentration gradient, dp/dx, in the density of carriers. The existence of a gradient implies that if an imaginary surface (shown dashed) is drawn in the semiconductor, the density of holes immediately on one side of the surface is larger than the density on the other side. The holes are in a random motion as a result of their thermal energy. Accordingly, holes will continue to move back and forth across this surface. We may then expect that, in a given time interval, more holes will cross the surface from the side of greater concentration to the side of smaller concentration than in the reverse direction. This net transport of holes across

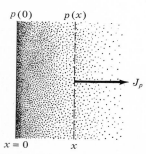

$p(0)$ $p(x)$

J_p

$x = 0$ x

Fig. 1-10   A nonuniform concentration $p(x)$ results in a diffusion current J_p.

the surface constitutes a current in the positive X direction. It should be noted that this net transport of charge is not the result of mutual repulsion among charges of like sign, but is simply the result of a statistical phenomenon. This diffusion is exactly analogous to that which occurs in a neutral gas if a concentration gradient exists in the gaseous container. The diffusion hole-current density J_p (amperes per square meter) is proportional to the concentration gradient, and is given by

$$J_p = -qD_p \frac{dp}{dx} \tag{1-28}$$

where D_p (square meters per second) is called the *diffusion constant* for holes. Since p in Fig. 1-10 decreases with increasing x, then dp/dx is negative and the minus sign in Eq. (1-28) is needed, so that J_p will be positive in the positive X direction. A similar equation exists for diffusion electron-current density [p is replaced by n, and the minus sign is replaced by a plus sign in Eq. (1-28)].

Einstein Relationship Since both diffusion and mobility are statistical thermodynamic phenomena, D and μ are not independent. The relationship between them is given by the Einstein equation

$$\frac{D_p}{\mu_p} = \frac{D_n}{\mu_n} = V_T \tag{1-29}$$

where V_T is the "volt-equivalent of temperature," defined by

$$V_T \equiv \frac{\bar{k}T}{q} = \frac{T}{11,600} \tag{1-30}$$

where $\bar{k}$ is the Boltzmann constant in joules per degree Kelvin. Note the distinction between $\bar{k}$ and k; the latter is the Boltzmann constant in electron volts per degree Kelvin. (Numerical values of $\bar{k}$ and k are given in Appendix A. From Sec. 1-3 it follows that $\bar{k} = 1.60 \times 10^{-19}k$.) At room temperature (300°K), $V_T = 0.026$ V, and $\mu = 39D$. Measured values of μ and computed values of D for silicon and germanium are given in Table 1-1.

Total Current It is possible for both a potential gradient and a concentration gradient to exist simultaneously within a semiconductor. In such a situation the total hole current is the sum of the drift current [Eq. (1-14), with n replaced by p] and the diffusion current [Eq. (1-28)], or

$$J_p = q\mu_p p\varepsilon - qD_p \frac{dp}{dx} \tag{1-31}$$

Similarly, the net electron current is

$$J_n = q\mu_n n\varepsilon + qD_n \frac{dn}{dx} \tag{1-32}$$

1-10   RECAPITULATION

The fundamental principles governing the electrical behavior of semiconductors, discussed in this chapter, are summarized as follows:

1. Two types of mobile charge carriers (positive holes and negative electrons) are available. This bipolar nature of a semiconductor is to be contrasted with the unipolar property of a metal, which possesses only free electrons.

2. A semiconductor may be fabricated with donor (acceptor) impurities; so it contains mobile charges which are primarily electrons (holes).

3. The intrinsic concentration of carriers is a function of temperature. At room temperature, essentially all donors or acceptors are ionized.

4. Current is due to two distinct phenomena:

a. Carriers drift in an electric field (this conduction current is also available in a metal).

b. Carriers diffuse if a concentration gradient exists (a phenomenon which does not take place in a metal).

5. Carriers are continuously being generated (due to thermal creation of hole-electron pairs) and are simultaneously disappearing (due to recombination).

These basic concepts are applied in the next chapter to the study of the p-n junction diode.

REVIEW QUESTIONS

1-1 Define *potential energy* in words and as an equation.

1-2 Define an *electron volt*.

1-3 Give the electron-gas description of a metal.

1-4 (*a*) Define *mobility*. (*b*) Give its dimensions.

1-5 (*a*) Define *conductivity*. (*b*) Give its dimensions.

1-6 Explain why a semiconductor acts as an insulator at 0°K and why its conductivity increases with increasing temperature.

1-7 What is the distinction between an intrinsic and an extrinsic semiconductor?

1-8 Define a *hole* (in a semiconductor).

1-9 Indicate pictorially how a hole contributes to conduction.

1-10 (*a*) Define *intrinsic concentration* of holes. (*b*) What is the relationship between this density and the intrinsic concentration for electrons? (*c*) What do these equal at 0°K?

1-11 Show (in two dimensions) the crystal structure of silicon containing a donor impurity atom.

1-12 Repeat Rev. 1-11 for an acceptor impurity atom.

1-13 Define (*a*) donor, (*b*) *acceptor* impurities.

1-14 A semiconductor is doped with both donors and acceptors of concentrations N_D and N_A, respectively. Write the equation or equations from which to determine the electron and hole concentrations (n and p).

1-15 Define *mean lifetime* of a carrier.

1-16 Explain physically the meaning of the following statement: An electron and a hole recombine and disappear.

1-17 Define the *volt-equivalent of temperature*.

1-18 (*a*) Define *diffusion constant* for holes. (*b*) Give its dimensions.

1-19 Repeat Rev. 1-18 for electrons.

1-20 (*a*) Write the equation for the net electron current in a semiconductor. What is the physical significance of each term? (*b*) How is this equation modified for a metal?

2 / JUNCTION-DIODE CHARACTERISTICS

In this chapter we demonstrate that if a junction is formed between a sample of p-type and one of n-type semiconductor, this combination possesses the properties of a rectifier. The volt-ampere characteristics of such a two-terminal device (called a *junction diode*) is studied. The capacitance across the junction is discussed.

Although the transistor is a triode (three-terminal) semiconductor, it may be considered as one diode biased by the current from a second diode. Hence most of the theory developed here is utilized in Chap. 4 in connection with the study of the transistor.

2-1　THE OPEN–CIRCUITED p-n JUNCTION

If donor impurities are introduced into one side and acceptors into the other side of a single crystal of a semiconductor, a p-n junction is formed. Such a system is illustrated in more schematic detail in Fig. 2-1a. The donor ion is represented by a plus sign because, after this impurity atom "donates" an electron, it becomes a positive ion. The acceptor ion is indicated by a minus sign because, after this atom "accepts" an electron, it becomes a negative ion. Initially, there are nominally only p-type carriers to the left of the junction and only n-type carriers to the right.

Space-Charge Region　Because there is a density gradient across the junction, holes will initially diffuse to the right across the junction, and electrons to the left. We see that the positive holes which neutralized the acceptor ions near the junction in the p-type silicon have disappeared as a result of combination with electrons which have diffused across the junction. Similarly, the neutralizing electrons in

20

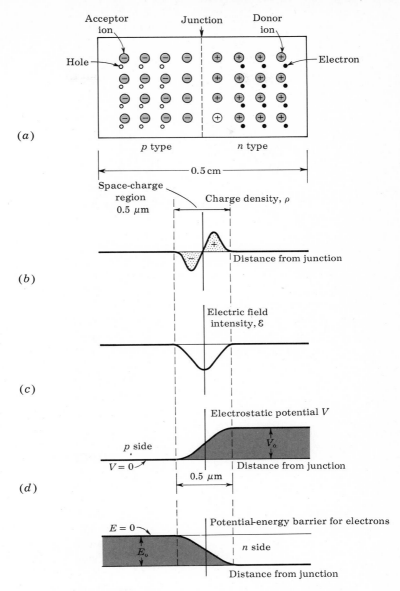

Fig. 2-1   A schematic diagram of a p-n junction, including the charge density, electric field intensity, and potential-energy barrier at the junction.   Since potential energy = potential × charge, the curve in (d) is proportional to the potential energy for a hole (a positive charge) and the curve in (e) is proportional to the negative of that in (d) (an electron is a negative charge). (Not drawn to scale.)

the n-type silicon have combined with holes which have crossed the junction from the p material. The unneutralized ions in the neighborhood of the junction are referred to as *uncovered charges*. The general shape of the charge density ρ (Fig. 2-1b) depends upon how the diode is doped (a step-graded junction is considered in detail in Sec. 2-6). Since the region of the junction is depleted of mobile charges, it is called the *depletion region*, the *space-charge region*, or the *transition region*. The thickness of this region is of the order of the wavelength of visible light (0.5 micron = 0.5 μm). Within this very narrow space-charge layer there are no mobile carriers. To the left of this region the carrier concentration is $p \approx N_A$, and to its right it is $n \approx N_D$.

Electric Field Intensity   The space-charge density ρ is zero at the junction. It is positive to the right and negative to the left of the junction. This distribution constitutes an electrical dipole layer, giving rise to electric lines of flux from right to left, corresponding to negative field intensity $\mathcal{E}$ as depicted in Fig. 2-1c. Equilibrium is established when the field is strong enough to restrain the process of diffusion. Stated alternatively, under steady-state conditions the drift hole (electron) current must be equal and opposite to the diffusion hole (electron) current so that the net hole (electron) current is reduced to zero—as it must be for an open-circuited device. In other words, there is no steady-state movement of charge across the junction.

The field intensity curve is proportional to the integral of the charge density curve. This statement follows from Poisson's equation or Gauss' law. If $\mathcal{E} = 0$ at $x = x_o$, then

$$\mathcal{E} = \int_{x_o}^{x} \frac{\rho}{\epsilon}\, dx \tag{2-1}$$

where ϵ is the permittivity. If ϵ_r is the (relative) dielectric constant and ϵ_o is the permittivity of free space (Appendix A), then $\epsilon = \epsilon_r \epsilon_o$.

Therefore the curve plotted in Fig. 2-1c is the integral of the function drawn in Fig. 2-1b (divided by ϵ).

Potential   The electrostatic-potential variation in the depletion region is shown in Fig. 2-1d, and, from Eq. (1-2), is the negative integral of the function $\mathcal{E}$ of Fig. 2-1c. This variation constitutes a potential-energy barrier (Sec. 1-2) against the further diffusion of holes across the barrier The form of the potential-energy barrier against the flow of electrons from the n side across the junction is shown in Fig. 2-1e. It is similar to that shown in Fig. 2-1d, except that it is inverted, since the charge on an electron is negative. Note the existence, across the depletion layer, of the barrier V_o, called the *contact potential*.

Summary   Under open-circuited conditions the net hole current must be zero. If this statement were not true, the hole density at one end of the semiconductor would continue to increase indefinitely with time, a situation which is obviously physically impossible. Since the concentration of holes in the p side is much greater than that in the n side, a very large hole diffusion current tends to flow across the junction from the p to the n material. Hence an electric field must build up across the junction in such a direction that a hole drift current will tend to flow across the junction from the n to the p side in order to counterbalance the diffusion current. This equilibrium condition of zero resultant hole current allows us to calculate the height of the potential barrier V_o in terms of the donor and acceptor concentrations. The numerical value for V_o is of the order of magnitude of a few tenths of a volt.

2-2   THE p-n JUNCTION AS A RECTIFIER

The essential electrical characteristic of a p-n junction is that it constitutes a rectifier which permits the easy flow of a charge in one direction but restrains the flow in the opposite direction. We consider now, qualitatively, how this diode rectifier action comes about.

Reverse Bias   In Fig. 2-2, a battery is shown connected across the terminals of a p-n junction. The negative terminal of the battery is connected to the p side of the junction, and the positive terminal to the n side. The polarity of connection is such as to cause both the holes in the p type and the electrons in the n type to move away from the junction. Consequently, the region of negative-charge density is spread to the left of the junction (Fig. 2-1b), and the positive-charge-density region is spread to the right. However, this process cannot continue indefinitely, because in order to have a steady flow of holes to the left, these holes must be supplied across the junction from the n-type silicon. And there are very few holes in the n-type side. Hence, nominally, zero current results. Actually, a small current does flow because a small number of hole-electron pairs are generated throughout the crystal as a result of thermal energy. The holes so formed in the n-type silicon will wander over to the junction. A similar remark applies to the electrons thermally generated in the p-type silicon. This small current (microamperes for germanium and nanoamperes for silicon) is the diode *reverse saturation current*, and its magnitude is designated by I_o. This reverse current will increase with increasing temperature [Eq. (2-4)], and hence the back resistance of a

Fig. 2-2   **(a)** A p-n junction biased in the reverse direction.   **(b)** The rectifier symbol is used for the p-n diode.

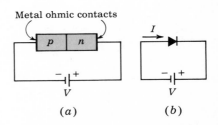

Metal ohmic contacts

(a) (b)

crystal diode decreases with increasing temperature. From the argument presented here, I_o should be independent of the magnitude of the reverse bias.

The mechanism of conduction in the reverse direction may be described alternatively in the following way: When no voltage is applied to the *p-n* diode, the potential barrier across the junction is as shown in Fig. 2-1*d*. When a voltage *V* is applied to the diode in the direction shown in Fig. 2-2, the height of the potential-energy barrier is increased by the amount *qV*. This increase in the barrier height serves to reduce the flow of majority carriers (i.e., holes in *p* type and electrons in *n* type). However, the minority carriers (i.e., electrons in *p* type and holes in *n* type), since they fall down the potential-energy hill, are uninfluenced by the increased height of the barrier. The applied voltage in the direction indicated in Fig. 2-2 is called the *reverse*, or *blocking*, *bias*.

Forward Bias   An external voltage applied with the polarity shown in Fig. 2-3 (opposite to that indicated in Fig. 2-2) is called a *forward* bias. An ideal *p-n* diode has zero ohmic voltage drop across the body of the crystal. For such a diode the height of the potential barrier at the junction will be lowered by the applied forward voltage *V*. The equilibrium initially established between the forces tending to produce diffusion of majority carriers and the restraining influence of the potential-energy barrier at the junction will be disturbed. Hence, for a forward bias, the holes cross the junction from the *p*-type into the *n*-type region, where they constitute an injected minority current. Similarly, the electrons cross the junction in the reverse direction and become a minority current injected into the *p* side. Holes traveling from left to right constitute a current in the *same* direction as electrons moving from right to left. Hence the resultant current crossing the junction is the *sum* of the hole and electron minority currents.

Ohmic Contacts   In Fig. 2-2 (2-3) we show an external reverse (forward) bias applied to a *p-n* diode. We have assumed that the external bias voltage appears directly across the junction and has the effect of raising (lowering) the electrostatic potential across the junction. To justify this assumption we must specify how electric contact is made to the semiconductor from the external bias circuit. In Figs. 2-2 and 2-3 we indicate metal contacts with which

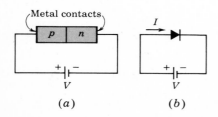

Fig. 2-3   (*a*) A *p-n* junction biased in the forward direction.   (*b*) The rectifier symbol is used for the *p-n* diode.

the homogeneous p-type and n-type materials are provided.   We thus see that
we have introduced two metal-semiconductor junctions, one at each end of
the diode.   We naturally expect a contact potential to develop across these
additional junctions.   However, we shall assume that the metal-semiconductor
contacts shown in Figs. 2-2 and 2-3 have been manufactured in such a way
that they are nonrectifying.   In other words, the contact potential across these
junctions is constant, independent of the direction and magnitude of the current.   A contact of this type is referred to as an *ohmic contact*.

   We are now in a position to justify our assumption that the entire applied
voltage appears as a *change* in the height of the potential barrier.   Inasmuch
as the voltage across the metal-semiconductor ohmic contacts remains constant
and the voltage drop across the bulk of the crystal is neglected, approximately
the entire applied voltage will indeed appear as a change in the height of the
potential barrier at the p-n junction.

   The Short-circuited and Open-circuited p-n Junction   If the voltage V
in Fig. 2-2 or 2-3 were set equal to zero, the p-n junction would be short-circuited.   Under these conditions, as we show below, no current can flow ($I = 0$)
and the electrostatic potential V_o remains unchanged and equal to the value
under open-circuit conditions.   If there were a current ($I \neq 0$), the metal
wire would become heated.   Since there is no external source of energy available, the energy required to heat the metal would have to be supplied by the
p-n bar.   The semiconductor bar, therefore, would have to cool off.   Clearly,
under thermal equilibrium the simultaneous heating of the metal and cooling
of the bar is impossible, and we conclude that $I = 0$.   Since under short-
circuit conditions the sum of the voltages around the closed loop must be zero,
the junction potential V_o must be exactly compensated by the metal-to-semi-
conductor contact potentials at the ohmic contacts.   Since the current is zero,
the wire can be cut without changing the situation, and the voltage drop across
the cut must remain zero.   If in an attempt to measure V_o we connected a
voltmeter across the cut, the voltmeter would read zero voltage.   In other
words, it is not possible to measure contact difference of potential directly
with a voltmeter.

   Large Forward Voltages   Suppose that the forward voltage V in Fig.
2-3 is increased until V approaches V_o.   If V were equal to V_o, the barrier
would disappear and the current could be arbitrarily large, exceeding the rating
of the diode.   As a practical matter we can never reduce the barrier to zero
because, as the current increases without limit, the bulk resistance of the crystal, as well as the resistance of the ohmic contacts, will limit the current.
Therefore it is no longer possible to assume that all the voltage V appears
as a change across the p-n junction.   We conclude that, as the forward voltage
V becomes comparable with V_o, the current through a real p-n diode will be
governed by the ohmic-contact resistances and the crystal bulk resistance.
Thus the volt-ampere characteristic becomes approximately a straight line.

2-3   THE VOLT–AMPERE CHARACTERISTIC

For a p-n junction the current I is related to the voltage V by the equation

$$I = I_o(\epsilon^{V/\eta V_T} - 1) \tag{2-2}$$

A positive value of I means that current flows from the p to the n side. The diode is forward-biased if V is positive, indicating that the p side of the junction is positive with respect to the n side. The symbol η is unity for germanium and is approximately 2 for silicon at rated current.

The symbol V_T stands for the volt equivalent of temperature, and is given by Eq. (1-30), repeated here for convenience:

$$V_T \equiv \frac{T}{11,600} \tag{2-3}$$

At room temperature $(T = 300°\text{K})$, $V_T = 0.026$ V $= 26$ mV.

The form of the volt-ampere characteristic described by Eq. (2-2) is shown in Fig. 2-4a. When the voltage V is positive and several times V_T, the unity in the parentheses of Eq. (2-2) may be neglected. Accordingly, except for a small range in the neighborhood of the origin, the current increases exponentially with voltage. When the diode is reverse-biased and $|V|$ is several times V_T, $I \approx -I_o$. The reverse current is therefore constant, independent of the applied reverse bias. Consequently, I_o is referred to as the *reverse saturation current*.

For the sake of clarity, the current I_o in Fig. 2-4a has been greatly exaggerated in magnitude. Ordinarily, the range of forward currents over which

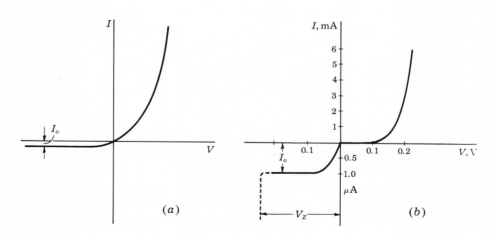

(a)   (b)

Fig. 2-4 (a) The volt-ampere characteristic of an ideal p-n diode. (b) The volt-ampere characteristic for a germanium diode redrawn to show the order of magnitude of currents. Note the expanded scale for reverse currents. The dashed portion indicates breakdown at a voltage V_Z.

a diode is operated is many orders of magnitude larger than the reverse saturation current. To display forward and reverse characteristics conveniently,
it is necessary, as in Fig. 2-4*b*, to use two different current scales. The volt-
ampere characteristic shown in that figure has a forward-current scale in milliamperes and a reverse scale in microamperes.

The dashed portion of the curve of Fig. 2-4*b* indicates that, at a reverse-
biasing voltage V_Z, the diode characteristic exhibits an abrupt and marked
departure from Eq. (2-2). At this critical voltage a large reverse current
flows, and the diode is said to be in the *breakdown* region, discussed in Sec.
2-9.

The Cutin Voltage V_γ Both silicon and germanium diodes are commercially available. A number of differences between these two types are
relevant in circuit design. The difference in volt-ampere characteristics is
brought out in Fig. 2-5. Here are plotted the forward characteristics at room
temperature of a general-purpose germanium switching diode and a general-
purpose silicon diode, the 1N270 and 1N3605, respectively. The diodes have
comparable current ratings. A noteworthy feature in Fig. 2-5 is that there
exists a *cutin, offset, break-point,* or *threshold,* voltage V_γ below which the current is very small (say, less than 1 percent of maximum rated value). Beyond
V_γ the current rises very rapidly. From Fig. 2-5 we see that V_γ is approximately 0.2 V for germanium and 0.6 V for silicon.

Note that the break in the silicon-diode characteristic is offset about 0.4 V
with respect to the break in the germanium-diode characteristic. The reason
for this difference is to be found, in part, in the fact that the reverse saturation
current in a germanium diode is normally larger by a factor of about 1,000
than the reverse saturation current in a silicon diode of comparable ratings.

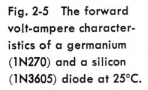

**Fig. 2-5 The forward
volt-ampere character-
istics of a germanium
(1N270) and a silicon
(1N3605) diode at 25°C.**

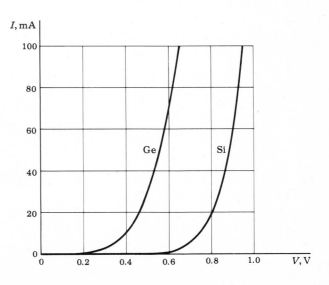

I_o is in the range of microamperes for a germanium diode and nanoamperes for a silicon diode at room temperature.

Since $\eta = 2$ for small currents in silicon, the current increases as $\epsilon^{V/2V_T}$ for the first several tenths of a volt and increases as ϵ^{V/V_T} only at higher voltages. This initial smaller dependence of the current on voltage accounts for the further delay in the rise of the silicon characteristic.

Reverse Saturation Current Many commercially available diodes exhibit an essentially constant value of I_o for negative values of V, as indicated in Fig. 2-4. On the other hand, some diodes show a very pronounced increase in reverse current with increasing reverse voltage.   This variation in I_o results from leakage across the surface of the diode, and also from the additional fact that new charge carriers may be generated by collision in the transition region at the junction.

2-4   THE TEMPERATURE DEPENDENCE OF THE V/I CHARACTERISTIC

The volt-ampere relationship Eq. (2-2) contains the temperature implicitly in the two symbols V_T and I_o.   From experimental data we observe that the reverse saturation current increases approximately 7 percent/°C for both silicon and germanium.   Since $(1.07)^{10} \approx 2.0$, we conclude that *the reverse saturation current approximately doubles for every* 10°C *rise in temperature.*   If $I_o = I_{o1}$ at $T = T_1$, then at a temperature T, I_o is given by

$$I_o(T) = I_{o1} \times 2^{(T-T_1)/10} \tag{2-4}$$

If the temperature is increased at a fixed voltage, the current increases. However, if we now reduce V, then I may be brought back to its previous value. Experimentally it is found that for either silicon or germanium (*at room temperature*)

$$\frac{dV}{dT} \approx -2.5 \text{ mV/°C} \tag{2-5}$$

in order to maintain a constant value of I. It should also be noted that $|dV/dT|$ decreases with increasing T.

2-5   DIODE RESISTANCE

The static resistance R of a diode is defined as the ratio V/I of the voltage to the current.   At any point on the volt-ampere characteristic of the diode (Fig. 2-5), the resistance R is equal to the reciprocal of the slope of a line joining the operating point to the origin.   The static resistance varies widely with V and I and is not a useful parameter.   The rectification property of a diode is indicated on the manufacturer's specification sheet by giving the maximum forward voltage V_F required to attain a given forward current I_F and also the maximum reverse current I_R at a given reverse voltage V_R. Typical values for a silicon planar epitaxial diode are $V_F = 0.8$ V at $I_F = 10$ mA (corresponding to $R_F = 80\ \Omega$) and $I_R = 0.1\ \mu\text{A}$ at $V_R = 50$ (corresponding to $R_R = 500$ M).

For small-signal operation the *dynamic*, or *incremental*, *resistance r* is an important parameter, and is defined as the reciprocal of the slope of the volt-ampere characteristic, $r \equiv dV/dI$. The dynamic resistance is not a constant, but depends upon the operating voltage. For example, for a semiconductor diode, we find from Eq. (2-2) that the dynamic conductance $g \equiv 1/r$ is

$$g \equiv \frac{dI}{dV} = \frac{I_o \epsilon^{V/\eta V_T}}{\eta V_T} = \frac{I + I_o}{\eta V_T} \tag{2-6}$$

For a reverse bias greater than a few tenths of a volt (so that $|V/\eta V_T| \gg 1$), g is extremely small and r is very large. On the other hand, for a forward bias greater than a few tenths of a volt, $I \gg I_o$, and r is given approximately by

$$r \approx \frac{\eta V_T}{I} \tag{2-7}$$

The dynamic resistance varies inversely with current; at room temperature and for $\eta = 1$, $r = 26/I$, where I is in milliamperes and r in ohms. For a forward current of 26 mA, the dynamic resistance is 1 Ω. The ohmic body resistance of the semiconductor may be of the same order of magnitude or even much higher than this value. Although r varies with current, in a small-signal model, it is reasonable to use the parameter r as a constant.

A Piecewise Linear Diode Characteristic A large-signal approximation which often leads to a sufficiently accurate engineering solution is the *piecewise linear* representation. For example, the piecewise linear approximation for a semiconductor diode characteristic is indicated in Fig. 2-6. The break point is not at the origin, and hence V_γ is also called the *offset*, or *threshold, voltage*. The diode behaves like an open circuit if $V < V_\gamma$, and has a constant incre-

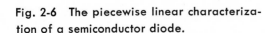

Fig. 2-6 The piecewise linear characterization of a semiconductor diode.

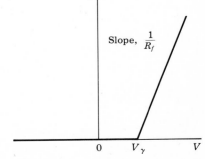

mental resistance $r = dV/dI$ if $V > V_\gamma$. Note that the resistance r (also designated as R_f and called the *forward resistance*) takes on added physical significance even for this large-signal model, whereas the static resistance $R_F = V/I$ is not constant and is not useful.

The numerical values V_γ and R_f to be used depend upon the type of diode and the contemplated voltage and current swings. For example, from Fig. 2-5 we find that, for a current swing from cutoff to 10 mA with a germanium diode, reasonable values are $V_\gamma = 0.2$ V and $R_f = 20$ Ω, and for a silicon diode, $V_\gamma = 0.6$ V and $R_f = 15$ Ω. On the other hand, a better approximation for current swings up to 50 mA leads to the following values; germanium, $V_\gamma = 0.3$ V, $R_f = 6$ Ω; silicon, $V_\gamma = 0.65$ V, $R_f = 5.5$ Ω.

2-6   SPACE–CHARGE, OR TRANSITION, CAPACITANCE C_T

As mentioned in Sec. 2-1, a reverse bias causes majority carriers to move away from the junction, thereby uncovering more immobile charges. Hence the thickness of the space-charge layer at the junction increases with reverse voltage. This increase in uncovered charge with applied voltage may be considered a capacitive effect. We may define an incremental capacitance C_T by

$$C_T \equiv \left| \frac{dQ}{dV} \right| \tag{2-8}$$

where dQ is the increase in charge caused by a change dV in voltage. It follows from this definition that a change in voltage dV in a time dt will result in a current $i = dQ/dt$, given by

$$i = C_T \frac{dV}{dt} \tag{2-9}$$

Therefore a knowledge of C_T is important in considering a diode (or a transistor) as a circuit element. The quantity C_T is referred to as the *transition-region, space-charge, barrier,* or *depletion-region, capacitance.* We now consider C_T quantitatively. As it turns out, this capacitance is not a constant, but depends upon the magnitude of the reverse voltage. It is for this reason that C_T is defined by Eq. (2-8) rather than as the ratio Q/V.

A Step-graded Junction Consider a junction in which there is an abrupt change from acceptor ions on one side to donor ions on the other side. Such a junction is formed experimentally, for example, by placing indium, which is trivalent, against n-type germanium and heating the combination to a high temperature for a short time. Some of the indium dissolves into the germanium to change the germanium from n to p type at the junction. Such a step-graded junction is called an *alloy,* or *fusion, junction.* A step-graded junction is also formed between emitter and base of an integrated transistor (Fig. 5-5). It is not necessary that the concentration N_A of acceptor ions equal the concentration N_D of donor impurities. As a matter of fact, it is often advantageous to have an unsymmetrical junction. Figure 2-7 shows the charge density as a function of distance from an alloy junction in which the acceptor impurity density is assumed to be much larger than the donor concentration. Since the net charge must be zero, then

$$N_A W_p = N_D W_n \tag{2-10}$$

If $N_A \gg N_D$, then $W_p \ll W_n \approx W$. The relationship between field and charge density is given by Eq. (2-1), with $\rho = qN_D$. From this equation it follows that the field intensity varies linearly with the distance x as plotted in Fig. 2-7c and that the potential varies quadratically with x as shown in Fig. 2-7d. These graphs should be compared with the corresponding curves of Fig. 2-1.

At $x = W$, $V = V_j$ = junction, or barrier, potential. It is found that W varies as the square root of V_j. Explicitly,

$$V_j = \frac{qN_D W^2}{2\epsilon} \tag{2-11}$$

We have used the symbol V to represent the potential at any distance x from the junction. Hence, let us introduce V_d as the externally applied diode

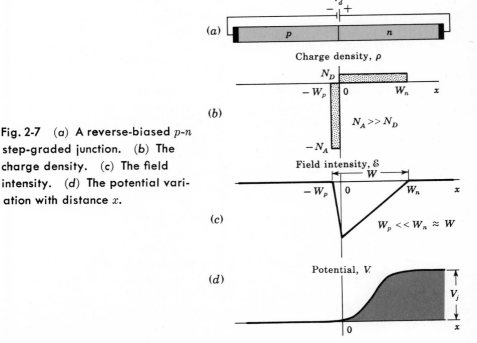

Fig. 2-7 (a) A reverse-biased p-n step-graded junction. (b) The charge density. (c) The field intensity. (d) The potential variation with distance x.

voltage. Since the barrier potential represents a reverse voltage, it is lowered by an applied forward voltage. Thus

$$V_j = V_o - V_d$$

where V_d is a negative number for an applied *reverse* bias and V_o is the contact potential (Fig. 2-1d). This equation confirms our qualitative conclusion that the thickness of the depletion layer increases with applied reverse voltage. We now see that W varies as $V_j^{\frac{1}{2}} = (V_o - V_d)^{\frac{1}{2}}$.

If A is the area of the junction, the charge in the distance W is

$$Q = qN_D WA$$

The transition capacitance C_T, given by Eq. (2-8), is

$$C_T = \left| \frac{dQ}{dV_d} \right| = qN_D A \left| \frac{dW}{dV_j} \right| \tag{2-12}$$

From Eq. (2-11), $|dW/dV_j| = \epsilon/qN_D W$, and hence

$$C_T = \frac{\epsilon A}{W} \tag{2-13}$$

It is interesting to note that this formula is exactly the expression which is obtained for a parallel-plate capacitor of area A (square meters) and plate separation W (meters) containing a material of permittivity ϵ. From Eqs. (2-11) and (2-13) we see that the transition capacitance for a step-graded diode varies inversely with the square root of the junction voltage.

A Linearly Graded Junction A second form of junction is obtained by drawing a single crystal from a melt of germanium whose type is changed during the drawing process by adding first p-type and then n-type impurities. A linearly graded junction is also formed between the collector and base of an integrated transistor (Fig. 5-6). For such a junction the charge density varies gradually (almost linearly), as indicated in Fig. 2-8. If an analysis similar to that given above is carried out for such a junction, Eq. (2-13) is found to be valid where W equals the total width of the space-charge layer. However, it now turns out that W varies as $V_j^{\frac{1}{3}}$ instead of $V_j^{\frac{1}{2}}$.

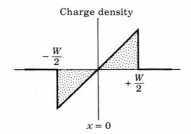

Fig. 2-8 The charge-density variation vs. distance at a linearly graded p-n junction.

Varactor Diodes   We observe from the above equations that the barrier capacitance is not a constant but varies with applied voltage.   The larger the reverse voltage, the larger is the space-charge width W, and hence the smaller the capacitance C_T.   Diodes made especially for applications which are based on the voltage-variable capacitance are called *varactors, varicaps,* or *voltacaps.*

In circuits intended for use with fast waveforms or at high frequencies, it is required that the transition capacitance be as small as possible, for the following reason: a diode is driven to the reverse-biased condition when it is desired to prevent the transmission of a signal.   However, if the barrier capacitance C_T is large enough, the current which is to be restrained by the low conductance of the reverse-biased diode will flow through the capacitor C_T.

2-7   MINORITY–CARRIER STORAGE IN A DIODE

If the voltage across a diode is applied in the forward direction, the potential barrier at the junction is lowered and holes from the p side enter the n region. Similarly, electrons from the n type move into the p side.   Define p_n as *the hole concentration in the n-type semiconductor*.   If the small value of the thermally generated hole concentration is designated by p_{no}, then the *injected*, or *excess*, hole concentration p'_n is defined by $p'_n \equiv p_n - p_{no}$.   As the holes diffuse into the n side, they encounter a plentiful supply of electrons and recombine with them.   Hence, $p_n(x)$ decreases with the distance x into the n material. It is found that the excess hole density falls off exponentially with x:

$$p'_n(x) = p'_n(0)\epsilon^{-x/L_p} = p_n(x) - p_{no} \tag{2-14}$$

where $p'_n(0)$ is the value of the injected minority concentration at the junction $x = 0$.   The parameter L_p is called the *diffusion length for holes* and is related to the diffusion constant D_p and the mean lifetime τ_p by

$$L_p = (D_p\tau_p)^{\frac{1}{2}} \tag{2-15}$$

We see that L_p represents the distance from the junction at which the injected concentration has fallen to $1/\epsilon$ of its value at $x = 0$.   It can be demonstrated that L_p also equals the average distance that an injected hole travels before recombining with an electron.   Hence, L_p is the *mean free path for holes*.

The exponential behavior of the excess minority-carrier density as a function of distance on either side of the junction is shown in Fig. 2-9a. The shaded area under the curve in the n-type (p-type) is proportional to the injected hole (electron) charge.

Charge Storage under Reverse Bias   When an external voltage reverse-biases the junction, the steady-state density of minority carriers is as shown in Fig. 2-9b.   Far from the junction the minority carriers are equal to their

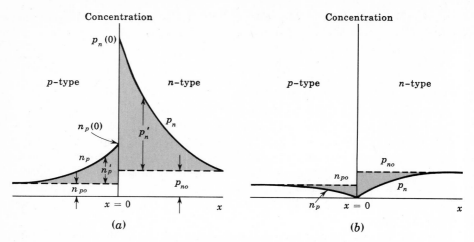

Fig. 2-9 Minority-carrier density distribution as a function of the distance x from a junction. (a) A forward-biased junction; (b) a reverse-biased junction. The excess hole (electron) density $p'_n = p_n - p_{no}$ ($n'_p = n_p - n_{po}$) is positive in (a) and negative in (b). (The transition region is assumed to be so small relative to the diffusion length that it is not indicated in this figure.)

thermal-equilibrium values p_{no} and n_{po}, as is also the situation in Fig. 2-9a. As the minority carriers approach the junction they are rapidly swept across, and the density of minority carriers diminishes to zero at this junction.

Diffusion Currents From Eq. (1-28) it follows that the hole diffusion current $I_p(0)$ crossing the junction under forward bias is proportional to the slope at the origin of the p_n curve in Fig. 2-9a. The corresponding electron current $I_n(0)$ is proportional to the slope at the origin of the n_p curve in Fig. 2-9a. The total diode current I is the sum of these two currents, or

$$I = I_p(0) + I_n(0) \tag{2-16}$$

The reverse saturation hole (electron) current is proportional to the slope at $x = 0$ of the $p_n(n_p)$ curves in Fig. 2-9b. The total reverse saturation current is the sum of these two currents.

Charge Control Description of a Diode For simplicity of discussion we assume that one side of the diode, say, the p material, is so heavily doped in comparison with the n side that the current I is carried across the junction entirely by holes moving from the p to the n side, or $I = I_p(0)$. The excess minority charge Q will then exist only on the n side, and is given by the shaded area in the n region of Fig. 2-9a multiplied by the diode cross section A and the electronic charge q. Integrating Eq. (2-14) we find an equation for Q.

The minority (hole) diffusion current is $I_p = AJ_p$. From Eqs. (1-28) and (2-14) we obtain $I_p(x)$. The hole current I is given by $I_p(x)$ with $x = 0$. In Prob. 2-16 we find that the ratio of Q to I is τ, or

$$I = \frac{Q}{\tau} \tag{2-17}$$

where $\tau \equiv L_p^2/D_p \equiv \tau_p =$ mean lifetime for holes [Eq. (2-15)].

Equation (2-17) is an important relationship, referred to as the *charge-control description of a diode*. It states that the diode current (which consists of holes crossing the junction from the p to the n side) is proportional to the stored charge Q of excess minority carriers. The factor of proportionality is the reciprocal of the decay time constant (the mean lifetime τ) of the minority carriers. Thus, in the steady state, *the current I supplies minority carriers at the rate at which these carriers are disappearing because of the process of recombination.*

The injected charge under reverse bias is given by the shaded area in Fig. 2-9b. This charge is negative since it represents less charge than is available under conditions of thermal equilibrium with no applied voltage. From Eq. (2-17) with Q negative, the diode current I is negative and, of course, equals the reverse saturation current I_o in magnitude. The charge-control characterization of a diode describes the device in terms of the current I and the stored charge Q, whereas the equivalent-circuit characterization uses the current I and the junction voltage V.

2-8 DIFFUSION CAPACITANCE

For a forward bias a capacitance which is much larger than the transition capacitance C_T considered in Sec. 2-6 comes into play. The origin of this larger capacitance lies in the injected charge stored near the junction outside the transition region (Fig. 2-9a). It is convenient to introduce an incremental capacitance, defined as the rate of change of injected charge with voltage, called the *diffusion*, or *storage, capacitance C_D.*

Static Derivation of C_D We now make a quantitative study of C_D. From Eqs. (2-17) and (2-6)

$$C_D \equiv \frac{dQ}{dV} = \tau \frac{dI}{dV} = \tau g = \frac{\tau}{r} \tag{2-18}$$

where the diode incremental conductance $g \equiv dI/dV$. Substituting the expression for the diode incremental resistance $r = 1/g$ given in Eq. (2-7) into Eq. (2-18) yields

$$C_D = \frac{\tau I}{\eta V_T} \tag{2-19}$$

We see that *the diffusion capacitance is proportional to the current I.*

For a reverse bias, g is very small and C_D may be neglected compared with C_T. For a forward current, on the other hand, C_D is usually much larger than C_T. For example, for germanium ($\eta = 1$) at $I = 26$ mA, $g = 1$ ℧, and $C_D = \tau$. If, say, $\tau = 20$ μs, then $C_D = 20$ μF, a value which is about a million times larger than the transition capacitance.

Despite the large value of C_D, the time constant rC_D (which is of importance in circuit applications) may not be excessive because the dynamic forward resistance $r = 1/g$ is small. From Eq. (2-18)

$$rC_D = \tau \tag{2-20}$$

Hence the diode time constant equals the mean lifetime of minority carriers, which lies in the range of nanoseconds to hundreds of microseconds.

2-9   BREAKDOWN DIODES

The reverse-voltage characteristic of a semiconductor diode, including the breakdown region, is redrawn in Fig. 2-10a. Diodes which are designed with adequate power-dissipation capabilities to operate in the breakdown region may be employed as voltage-reference or constant-voltage devices. Such diodes are known as *avalanche, breakdown,* or *Zener diodes.* They are used as voltage regulators (Sec. 3-6) to keep the load voltage essentially constant at the value V_Z, independent of variations in load current or supply voltage. The diode will continue to regulate until the circuit operation requires the diode current to fall to I_{ZK}, in the neighborhood of the knee of the diode volt-ampere curve. The upper limit on diode current is determined by the power-dissipation rating of the diode.

Avalanche Multiplication Two mechanisms of diode breakdown for increasing reverse voltage are recognized. Consider the following situation: A thermally generated carrier (part of the reverse saturation current) falls down the junction barrier and acquires energy from the applied potential.

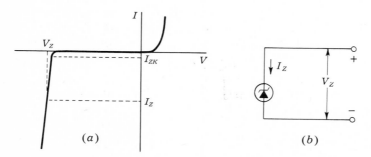

Fig. 2-10   (a) The volt-ampere characteristic of an avalanche, or Zener, diode.   (b) The symbol used for a breakdown diode.

This carrier collides with a crystal ion and imparts sufficient energy to disrupt a covalent bond. In addition to the original carrier, a new electron-hole pair has now been generated. These carriers may also pick up sufficient energy from the applied field, collide with another crystal ion, and create still another electron-hole pair. Thus each new carrier may, in turn, produce additional carriers through collision and the action of disrupting bonds. This cumulative process is referred to as *avalanche multiplication*. It results in large reverse currents, and the diode is said to be in the region of *avalanche breakdown*.

Zener Breakdown Even if the initially available carriers do not acquire sufficient energy to disrupt bonds, it is possible to initiate breakdown through a direct rupture of the bonds. Because of the existence of the electric field at the junction, a sufficiently strong force may be exerted on a bound electron by the field to tear it out of its covalent bond. The new hole-electron pair which is created increases the reverse current. Note that this process, called *Zener breakdown*, does not involve collisions of carriers with the crystal ions (as does avalanche multiplication).

The field intensity ε increases as the impurity concentration increases, for a fixed applied voltage. It is found that Zener breakdown occurs at a field of approximately 2×10^7 V/m. This value is reached at voltages below about 6 V for heavily doped diodes. For lightly doped diodes the breakdown voltage is higher, and avalanche multiplication is the predominant effect. Nevertheless, the term *Zener* is commonly used for the *avalanche*, or *breakdown*, *diode* even at higher voltages. Silicon diodes operated in avalanche breakdown are available with maintaining voltages from several volts to several hundred volts and with power ratings up to 50 W.

Temperature Characteristics A matter of interest in connection with Zener diodes, as with semiconductor devices generally, is their temperature sensitivity. The temperature coefficient is given as the percentage change in reference voltage per centigrade degree change in diode temperature. These data are supplied by the manufacturer. The coefficient may be either positive or negative and will normally be in the range ± 0.1 percent/°C. If the reference voltage is above 6 V, where the physical mechanism involved is avalanche multiplication, the temperature coefficient is positive. However, below 6 V, where true Zener breakdown is involved, the temperature coefficient is negative.

Dynamic Resistance and Capacitance An important characteristic of a breakdown diode is the slope of the diode volt-ampere curve in the operating range. If the reciprocal slope $\Delta V_Z/\Delta I_Z$, called the *dynamic resistance*, is r, then a change ΔI_Z in the operating current of the diode produces a change $\Delta V_Z = r \, \Delta I_Z$ in the operating voltage. Ideally, $r = 0$, corresponding to a volt-ampere curve which, in the breakdown region, is precisely vertical. The variation of r at various currents for a series of avalanche diodes of fixed power-dissipation rating and various voltages show a rather broad minimum in the range 6 to 10 V.   This minimum value of r is of the order of magnitude of a few ohms.   However, for values of V_Z below 6 V or above 10 V, and particularly for small currents ($\sim$1 mA), r may be of the order of hundreds of ohms.

Some manufacturers specify the minimum current I_{ZK} (Fig. 2-10a) below which the diode should not be used.   Since this current is on the knee of the above curve, where the dynamic resistance is large, then for currents lower than I_{ZK} the regulation will be poor.   Some diodes exhibit a very sharp knee even down into the microampere region.

The capacitance across a breakdown diode is the transition capacitance, and hence varies inversely as some power of the voltage.   Since C_T is proportional to the cross-sectional area of the diode, high-power avalanche diodes have very large capacitances.   Values of C_T from 10 to 10,000 pF are common.

Temperature-compensated Reference Diodes   These devices consist of a reverse-biased Zener diode with a positive temperature coefficient, combined in a single package with a forward-biased p-n diode the temperature coefficient of which is negative.   Such a reference diode may have a temperature coefficient of less than 0.001 percent/°C from -55 to $+100$°C, and over a range of current of several milliamperes.   The voltage stability with time of such a reference diode may be comparable with that of a standard cell.

2-10   LIGHT–EMITTING DIODES

Just as it takes energy to generate a hole-electron pair, so energy is released when an electron recombines with a hole.   In silicon and germanium this recombination liberates energy which goes into the crystal as heat.   However, it is found that in other semiconductors, such as gallium arsenide, there is a considerable amount of direct recombination without thermal generation. Under such circumstances when a free electron recombines with a hole, the energy appears in the form of radiation.   Such a p-n diode is called a *light-emitting diode* (*LED*), although the radiation is principally in the infrared. The efficiency of the process of light generation increases with the injected current and with a decrease in temperature.   The light is concentrated near the junction because most of the carriers are to be found within a diffusion length of the junction.

Under certain conditions, the emitted light is coherent (essentially monochromatic).   Such a diode is called an *injection junction laser*.

REVIEW QUESTIONS

2-1   Consider an open-circuited p-n junction.   Sketch curves, as a function of distance across the junction, of space charge, electric field, and potential.

2-2 (*a*) What is the order of magnitude of the space-charge width at a *p-n* junction? (*b*) What does this space charge consist of—electrons, holes, neutral donors, neutral acceptors, ionized donors, ionized acceptors, etc.?

2-3 (*a*) For a reverse-biased diode, does the transition region increase or decrease in width as the applied potential is increased? (*b*) What happens to the junction potential?

2-4 Explain why the *p-n* junction contact potential *cannot* be measured by placing a voltmeter across the diode terminals.

2-5 Explain physically why a *p-n* diode acts as a rectifier.

2-6 (*a*) Write the volt-ampere equation for a *p-n* diode. (*b*) Explain the meaning of each symbol.

2-7 Plot the volt-ampere curves for germanium and silicon to the same scale showing the cutin value for each.

2-8 (*a*) How does the reverse saturation current of a *p-n* diode vary with temperature? (*b*) How does the diode voltage (at constant current) vary with temperature?

2-9 How does the dynamic resistance r of a diode vary with (*a*) current and (*b*) temperature? (*c*) What is the order of magnitude of r for silicon at room temperature and for a dc current of 1 mA?

2-10 (*a*) Sketch the piecewise linear characteristic of a diode. (*b*) What are the approximate cutin voltages for silicon and germanium?

2-11 Consider a step-graded *p-n* junction with equal doping on both sides of the junction ($N_A = N_D$). Sketch the charge density, field intensity, and potential as a function of distance from the junction for a reverse bias.

2-12 (*a*) How does the transition capacitance C_T vary with the depletion-layer width for a step-graded junction? (*b*) With the applied reverse voltage? (*c*) What is the order of magnitude of C_T?

2-13 What is a *varactor diode?*

2-14 Plot the minority-carrier concentration as a function of distance from a *p-n* junction in the *n* side only for (*a*) a forward-biased junction, (*b*) a negatively biased junction. Indicate the excess concentration and note where it is positive and where negative.

2-15 Under steady-state conditions the diode current is proportional to a charge Q. (*a*) What is the physical meaning of the factor of proportionality? (*b*) What charge does Q represent—transition-layer charge, injected minority-carrier charge, majority-carrier charge, etc.?

2-16 (*a*) How does the diffusion capacitance C_D vary with dc diode current? (*b*) What does the product of C_D and the dynamic resistance of a diode equal?

2-17 (*a*) Draw the volt-ampere characteristic of an avalanche diode. (*b*) What is meant by the *knee* of the curve? (*c*) By the dynamic resistance? (*d*) By the temperature coefficient?

2-18 Describe the physical mechanism for avalanche breakdown.

2-19 Describe the physical mechanism for Zener breakdown.

2-20 What is a *light-emitting diode?*

3 / DIODE CIRCUITS

The p-n junction diode is considered as a circuit element. The concept of "load line" is introduced. The piecewise linear diode model is exploited in the following applications: clippers (single-ended and double-ended), voltage regulators, and rectifiers. Capacitor filters are discussed.

3-1   THE DIODE AS A CIRCUIT ELEMENT

The basic diode circuit, indicated in Fig. 3-1, consists of the device in series with a load resistance R_L and an input-signal source v_i. This circuit is now analyzed to find the instantaneous current i and the instantaneous diode voltage v, when the instantaneous input voltage is v_i.

The Load Line   From Kirchhoff's voltage law (KVL),†

$$v = v_i - iR_L \tag{3-1}$$

where R_L is the magnitude of the load resistance. This one equation is not sufficient to determine the two unknowns v and i in this expression. However, a second relation between these two variables is given by the static characteristic of the diode (Fig. 2-5). In Fig. 3-2a is indicated the simultaneous solution of Eq. (3-1) and the diode characteristic. The straight line, which is represented by Eq. (3-1), is called the *load line*. The load line passes through the points $i = 0$, $v = v_i$, and $i = v_i/R_L$, $v = 0$. That is, the intercept with the voltage axis is v_i, and with the current axis is v_i/R_L. The slope of this line is determined, therefore, by R_L; the negative value of the slope is equal to $1/R_L$. The point of intersection A of the load line and the static

---

† Summary of the elementary circuit theory which is used in the analysis of the electronic circuits discussed in this book is given in Appendix C.

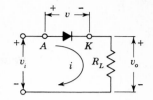

Fig. 3-1   The basic diode circuit.   The anode (the p side) of the diode is marked A, and the cathode (the n side) is labeled K.

curve gives the current i_A that will flow under these conditions.   This construction determines the current in the circuit when the instantaneous input potential is v_i.

A slight complication may arise in drawing the load line if $i = v_i/R_L$ is too large to appear on the printed volt-ampere curve supplied by the manufacturer.   Under such circumstance choose an arbitrary value of current I' which is on the vertical axis of the printed characteristic.   Then the load line is drawn through the point P (Fig. 3-2a), where $i = I'$, $v = v_i - I'R_L$, and through a second point $i = 0$, $v = v_i$.

The Dynamic Characteristic   Consider now that the input voltage is allowed to vary.   Then the above procedure must be repeated for each voltage value.   A plot of current vs. input voltage, called the *dynamic characteristic*, may be obtained as follows: The current i_A is plotted vertically above v_i at point B in Fig. 3-2b.   As v_i changes, the slope of the load line does not vary since R_L is fixed.   Thus, when the applied potential has the value v_i', the corresponding current is $i_{A'}$.   This current is plotted vertically above v_i' at B'.   The resulting curve OBB' that is generated as v_i varies is the dynamic characteristic.

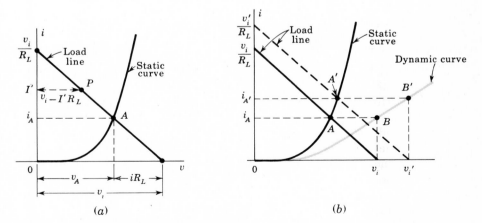

(a)   (b)

Fig. 3-2   (a) The intersection A of the load line with the diode static characteristic gives the current i_A corresponding to an instantaneous input voltage v_i.   (b) The method of constructing the dynamic curve from the static curve and the load line.

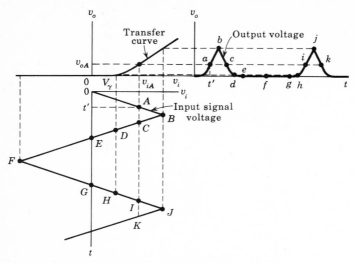

Fig. 3-3 The method of obtaining the output-voltage wave-form from the transfer characteristic for a given input-signal-voltage waveform.

The Transfer Characteristic   The curve which relates the output voltage v_o to the input v_i of any circuit is called the *transfer*, or *transmission*, *character-istic*. Since in Fig. 3-1 $v_o = iR_L$, then for this particular circuit the transfer curve has the same shape as the dynamic characteristic.

It must be emphasized that, regardless of the shape of the static volt-ampere characteristic or the waveform of the input signal, the resultant output waveshape can always be found graphically (at low frequencies) from the transfer curve. This construction is illustrated in Fig. 3-3. The input-signal waveform (not necessarily triangular) is drawn with its time axis vertically downward, so that the voltage axis is horizontal. Suppose that the input voltage has the value v_{iA} indicated by the point A at an instant t'. The corre-sponding output voltage is obtained by drawing a vertical line through A and noting the voltage v_{oA} where this line intersects the transfer curve. This value of v_o is then plotted (a) at an instant of time equal to t'. Similarly, points b, c, d, . . . of the output waveform correspond to points B, C, D, . . . of the input-voltage waveform. Note that $v_o = 0$ for $v_i < V_\gamma$, so that the diode acts as a *clipper* and a portion of the input signal does not appear at the output. Also note the distortion (the deviation from linearity) introduced into the output in the neighborhood of $v_i = V_\gamma$ because of the nonlinearity in the transfer curve in this region.

3-2   THE LOAD–LINE CONCEPT

We now show that the use of the load-line construction allows the graphical analysis of many circuits involving devices which are much more complicated

Fig. 3-4 The output circuit of most devices consists of a supply voltage V in series with a load resistance R_L.

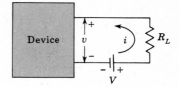

than the p-n diode. The external circuit at the output of almost all devices consists of a dc (constant) supply voltage V in series with a load resistance R_L, as indicated in Fig. 3-4. Since KVL applied to this output circuit yields

$$v = V - iR_L \tag{3-2}$$

we once again have a straight-line relationship between output current i and output (device) voltage v. The load line passes through the point $i = 0$, $v = V$ and has a slope equal to $-1/R_L$ *independently of the device characteristics*. A p-n junction diode or an avalanche diode possesses a single volt-ampere characteristic at a given temperature. However, most other devices must be described by a family of curves.

The volt-ampere characteristics of a transistor are discussed in the following chapter. The output circuit is identical with that in Fig. 3-4, and the graphical analysis begins with the construction of the load line.

3-3   THE PIECEWISE LINEAR DIODE MODEL

If the reverse resistance R_r is included in the diode characteristic of Fig. 2-6, the piecewise linear and continuous volt-ampere characteristic of Fig. 3-5a is obtained. The diode is a *binary* device, in the sense that it can exist in only one of two possible states; that is, the diode is either ON or OFF at a given time. If the voltage applied across the diode exceeds the cutin potential V_γ with the anode A (the p side) more positive than the cathode K (the n side), the diode is forward-biased and is said to be in the ON state. The large-signal model for the ON state is indicated in Fig. 3-5b as a battery V_γ in series with the low forward resistance R_f (of the order of a few tens of ohms or less). For a reverse

Fig. 3-5 (a) The piecewise linear volt-ampere characteristic of a p-n diode. (b) The large-signal model in the ON, or forward, direction (anode A more positive than V_γ with respect to the cathode). (c) The model in the OFF, or reverse, direction ($v < V_\gamma$).

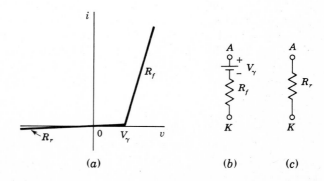

bias ($v < V_\gamma$) the diode is said to be in its OFF state.   The large-signal model
for the OFF state is indicated in Fig. 3-5c as a large reverse resistance R_r (of
the order of several hundred kilohms or more).   Usually R_r is so much larger
than any other resistance in the diode circuit that this reverse resistance may
be considered to be infinite.   We shall henceforth assume that $R_r = \infty$, unless
otherwise stated.

A Simple Application   Consider that in the basic diode circuit of Fig. 3-1
the input is sinusoidal, so that $v_i = V_m \sin \alpha$, where $\alpha = \omega t$, $\omega = 2\pi f$, and f is
the frequency of the input excitation.   Assume that the piecewise linear model
of Fig. 3-5 (with $R_r = \infty$) is valid.   The current in the forward direction
($v_i > V_\gamma$) may then be obtained from the equivalent circuit of Fig. 3-6a.   We
have

$$i = \frac{V_m \sin \alpha - V_\gamma}{R_L + R_f} \tag{3-3}$$

for $v_i = V_m \sin \alpha \geq V_\gamma$ and $i = 0$ for $v_i < V_\gamma$.   This waveform is plotted in
Fig. 3-6b, where the cutin angle ϕ is given by

$$\phi = \arcsin \frac{V_\gamma}{V_m} \tag{3-4}$$

If, for example, $V_m = 2V_\gamma$, then $\phi = 30°$.   For silicon (germanium),

$$V_\gamma = 0.6 \text{ V } (0.2 \text{ V})$$

and hence a cutin angle of 30° is obtained for very small peak sinusoidal
voltages; 1.2 V (0.4 V) for Si (Ge).   On the other hand, if $V_m \geq 10$ V, then
$\phi \leq 3.5°$ (1.2°) for Si (Ge) and the cutin angle may be neglected; the diode
conducts essentially for a full half cycle.   Such a rectifier is considered in more
detail in Sec. 3-7.

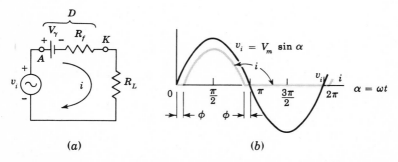

Fig. 3-6   (a) The equivalent circuit of a diode D (in the ON state) in
series with a load resistance R_L and a sinusoidal voltage v_i.   (b)
The input waveform v_i and the rectified current i.

Incidentally, the circuit of Fig. 3-6 may be used to charge a battery from an ac supply line. The battery V_B is placed in series with the diode D, and R_L is adjusted to supply the desired dc (average) charging current. The instantaneous current is given by Eq. (3-3), with V_B added to V_γ.

Analysis of Diode Circuits Using the Piecewise Linear Model Consider a circuit containing several diodes, resistors, supply voltages, and sources of excitation. A general method of analysis of such a circuit consists in assuming (guessing) the state of each diode. For the ON state, replace the diode by a battery V_γ in series with a forward resistance R_f, and for the OFF state replace the diode by the reverse resistance R_r (which can usually be taken as infinite), as indicated in Fig. 3-5b and c. After the diodes have been replaced by these piecewise linear models, the entire circuit is linear and the currents and voltages everywhere can be calculated using Kirchhoff's voltage and current laws. The assumption that a diode is ON can then be verified by observing the sign of the current through it. If the current is in the forward direction (from anode to cathode), the diode is indeed ON and the initial guess is justified. However, if the current is in the reverse direction (from cathode to anode), the assumption that the diode is ON has been proved incorrect. Under this circumstance the analysis must begin again with the diode assumed to be OFF.

Analogous to the above trial-and-error method, we test the assumption that a diode is OFF by finding the voltage across it. If this voltage is either in the reverse direction or in the forward direction but with a voltage less than V_γ, the diode is indeed OFF. However, if the diode voltage is in the forward direction and exceeds V_γ, the diode must be ON and the original assumption is incorrect. In this case the analysis must begin again by assuming the ON state for this diode.

The above method of analysis will be employed in the study of the diode circuits which follows.

3-4   CLIPPING (LIMITING) CIRCUITS

Clipping circuits are used to select for transmission that part of an arbitrary waveform which lies above or below some reference level. Clipping circuits are also referred to as voltage (or current) *limiters, amplitude selectors,* or *slicers.*

In the above sense, Fig. 3-1 is a clipping circuit, and input voltages below V_γ are *not* transmitted to the output, as is evident from the waveforms of Figs. 3-3 and 3-6. Some of the more commonly employed clipping circuits are now to be described.

Consider the circuit of Fig. 3-7a. Using the piecewise linear model, the transfer characteristic of Fig. 3-7b is obtained, as may easily be verified. For example, if D is OFF, the diode voltage $v < V_\gamma$ and $v_i < V_\gamma + V_R$. How-

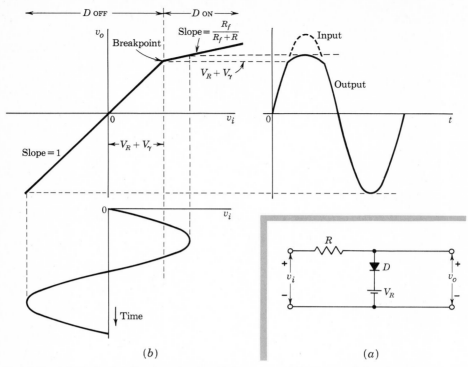

Fig. 3-7 (a) A diode clipping circuit which transmits that part of the waveform more negative than $V_R + V_\gamma$. (b) The piecewise linear transmission characteristic of the circuit. A sinusoidal input and the clipped output are shown.

ever, if D is OFF, the circuit reduces to that in Fig. 3-8a, there is no current in R, and $v_o = v_i$. This argument justifies the linear portion (with unity slope) of the transmission characteristic extending from arbitrary negative values to $v_i = V_R + V_\gamma$. For v_i larger than $V_R + V_\gamma$, the diode conducts, and it behaves as a battery V_γ in series with a resistance R_f, as indicated in Fig. 3-8b. Hence the transfer characteristic is given by

$$v_i \leq V_R + V_\gamma \qquad v_o = v_i$$

$$v_i \geq V_R + V_\gamma \qquad v_o = v_i \frac{R_f}{R + R_f} + (V_R + V_\gamma) \frac{R}{R + R_f} \qquad (3\text{-}5)$$

The second equation above is obtained by superposition, (Sec. C-2) considering v_i as one voltage and $V_R + V_\gamma$ as a second independent source. Equation (3-5) verifies the linear portion of slope $R_f/(R_f + R)$ for $v_i > V_R + V_\gamma$ in the transfer curve. Note that the transmission characteristic is piecewise linear and continuous and has a break point at $V_R + V_\gamma$.

Figure 3-7b shows a sinusoidal input signal of amplitude large enough so that the signal makes excursions past the break point. The corresponding

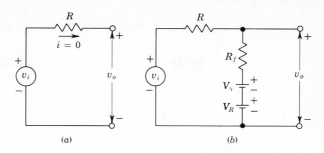

Fig. 3-8 Circuits from
which to obtain the
transfer characteristic of
the clipper of Fig. 3-7.
(a) The diode is OFF.
(b) The diode is ON.

output exhibits a suppression of the positive peak of the signal. If $R_f \ll R$, this suppression will be very pronounced, and the positive excursion of the output will be sharply limited at the voltage $V_R + V_\gamma$. The output will appear as though the positive peak had been "clipped off" or "sliced off." Often it turns out that $V_R \gg V_\gamma$, in which case one may consider that V_R itself is the limiting reference voltage.

In Fig. 3-9a the clipping circuit has been modified in that the diode in Fig. 3-7a has been reversed. The corresponding piecewise linear representa-

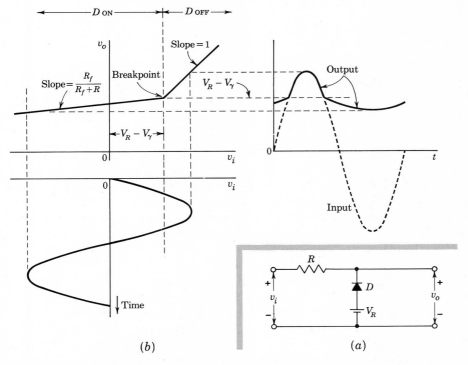

Fig. 3-9 (a) A diode clipping circuit which transmits that part of the waveform more positive than $V_R - V_\gamma$. (b) The piecewise linear transmission character-istic of the circuit. A sinusoidal input and the clipped output are shown.

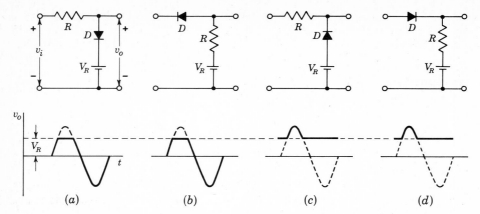

Fig. 3-10   Four diode clipping circuits.   In (a) and (c) the diode appears as a shunt element.   In (b) and (d) the diode appears as a series element.   Under each circuit appears the output waveform (solid) for a sinusoidal input.   The clipped portion of the input is shown dashed.

tion of the transfer characteristic is shown in Fig. 3-9b.   In this circuit, the portion of the waveform more positive than $V_R - V_\gamma$ is transmitted without attenuation, but the less positive portion is greatly suppressed.

In Figs. 3-7b and 3-9b we have assumed R_r arbitrarily large in comparison with R.   If this condition does not apply, the transmission characteristics must be modified.   The portions of these curves which are indicated as having unity slope must instead be considered to have a slope $R_r/(R_r + R)$.

Additional Clipping Circuits   Figures 3-7 and 3-9 appear again in Fig. 3-10, together with variations in which the diodes appear as series elements.   If in each case a sinusoid is applied at the input, the waveforms at the output will appear as shown by the heavy lines.   In these output waveforms we have neglected V_γ in comparison with V_R and also assumed that $R_r \gg R \gg R_f$. In two of these circuits the portion of the waveform transmitted is that part which lies below V_R; in the other two the portion above V_R is transmitted.   In two the diode appears as an element in series with the signal lead; in two it appears as a shunt element.   The use of the diode as a series element has the disadvantage that when the diode is OFF and it is intended that there be no transmission, fast signals or high-frequency waveforms may be transmitted to the output through the diode capacitance.   The use of the diode as a shunt element has the disadvantage that when the diode is open (back-biased) and it is intended that there be transmission, the diode capacitance, together with all other capacitance in shunt with the output terminals, will round sharp edges of input waveforms and attenuate high-frequency signals.   A second disadvantage of the use of the diode as a shunt element is that in such circuits the impedance R_s of the source which supplies V_R must be kept low.   This requirement does not arise in circuits where V_R is in series with R, which is normally large compared with R_s.

3-5   CLIPPING AT TWO INDEPENDENT LEVELS

Diode clippers may be used in pairs to perform double-ended limiting at independent levels. A parallel, a series, or a series-parallel arrangement may be used. A parallel arrangement is shown in Fig. 3-11a. Figure 3-11b shows the piecewise linear and continuous input-output voltage curve for the circuit in Fig. 3-11a. The transfer curve has two break points, one at $v_o = v_i = V_{R1}$ and a second at $v_o = v_i = V_{R2}$, and has the following characteristics (assuming $V_{R2} > V_{R1} \gg V_\gamma$ and $R_f \ll R$):

Input v_i	*Output v_o*	*Diode states*	
$v_i \leq V_{R1}$	$v_o = V_{R1}$	$D1$ ON, $D2$ OFF	
$V_{R1} < v_i < V_{R2}$	$v_o = v_i$	$D1$ OFF, $D2$ OFF	(3-6)
$v_i \geq V_{R2}$	$v_o = V_{R2}$	$D1$ OFF, $D2$ ON	

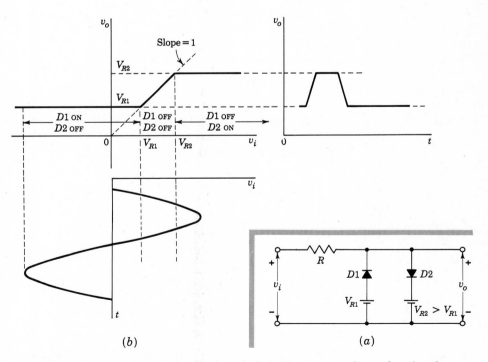

(b)   (a)

Fig. 3-11   (a) A double-diode clipper which limits at two independent levels. (b) The piecewise linear transfer curve for the circuit in (a). The doubly clipped output for a sinusoidal input is shown.

The circuit of Fig. 3-11a is referred to as a *slicer* because the output contains a slice of the input between the two reference levels V_{R1} and V_{R2}.

The circuit is used as a means of converting a sinusoidal waveform into a square wave. In this application, to generate a symmetrical square wave, V_{R1} and V_{R2} are adjusted to be numerically equal but of opposite sign. The transfer characteristic passes through the origin under these conditions, and the waveform is clipped symmetrically top and bottom. If the amplitude of the sinusoidal waveform is very large in comparison with the difference in the reference levels, the output waveform will have been *squared*.

Two avalanche diodes in series opposing, as indicated in Fig. 3-12a, constitute another form of double-ended clipper. If the diodes have identical characteristics, a symmetrical limiter is obtained. If the breakdown (Zener) voltage is V_Z and if the diode cutin voltage is V_γ, then the transfer characteristic of Fig. 3-12b is obtained.

Catching or Clamping Diodes Consider that v_i and R in Fig. 3-11a represent Thévenin's circuit model, Sec. C-2, at the output of a device, such as an amplifier. In other words, R is the output resistance and v_i is the open-circuit output signal. In such a situation $D1$ and $D2$ are called *catching diodes*. The reason for this terminology should be clear from Fig. 3-13, where we see that $D1$ "catches" the output v_o and does not allow it to fall below V_{R1}, whereas $D2$ "catches" v_o and does not permit it to rise above V_{R2} (for $V_\gamma \ll V_{R1}$ and $V_\gamma \ll V_{R2}$).

Generally, whenever a node becomes connected through a low resistance (as through a conducting diode) to some reference voltage V_R, we say that the node has been clamped to V_R, since the voltage at that point in the circuit is unable to depart appreciably from V_R. In this sense the diodes in Fig. 3-13 are called *clamping diodes*.

A circuit for clamping the extremity of a periodic waveform to a reference voltage is considered in Sec. 3-9.

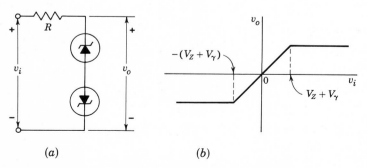

(a)　　　　　　　　　(b)

Fig. 3-12 (a) A double-ended clipper using avalanche diodes; (b) the transfer characteristic.

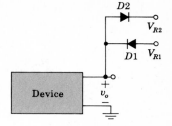

Fig. 3-13 Catching diodes $D1$ and $D2$ limit the output excursion of the device between V_{R1} and V_{R2}.

3-6   A BREAKDOWN DIODE–VOLTAGE REGULATOR

A circuit using an avalanche or Zener diode as a voltage regulator is indicated in Fig. 3-14.   The source V and resistor R are selected so that, initially, the diode is operating in the breakdown region.   Here the diode voltage, which is also the voltage across the load R_L, is V_Z as in Fig. 2-10, and the diode current is I_Z.   The diode will now regulate the load voltage against variations in load current and against variations in supply voltage V because, in the breakdown region, large changes in diode current produce only small changes in diode voltage.   Moreover, as load current or supply voltage changes, the diode current will accommodate itself to these changes to maintain a nearly constant load voltage.   The diode will continue to regulate until the circuit operation requires the diode current to fall to I_{ZK}, in the neighborhood of the knee of the diode volt-ampere curve.   The upper limit on diode current is determined by the power-dissipation rating of the diode.

EXAMPLE   (*a*) The avalanche diode regulates at 50 V over a range of diode currents from 5 to 40 mA.   The supply voltage $V = 200$ V.   Calculate R to allow voltage regulation from a load current $I_L = 0$ up to $I_{L,\text{max}}$, the maximum possible value of I_L.   What is $I_{L,\text{max}}$? (*b*) If R is set as in part *a* and the load current is set at $I_L = 25$ mA, what are the limits between which V may vary without loss of regulation in the circuit?

Solution   *a*. The current from the voltage source V is given by

$$I = \frac{V - V_Z}{R} \qquad\qquad (3\text{-}7)$$

Fig. 3-14   A Zener voltage regulator.

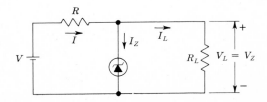

From Kirchhoff's current law (KCL) (Sec. C-1) the diode current is given by

$$I_z = I - I_L \tag{3-8}$$

As the load resistance R_L is varied so that $I_L = V_Z/R_L$ changes, we see from Eq. (3-7) that the current I remains constant. Hence, from Eq. (3-8) the diode current I_Z decreases with an increase in load current. Therefore, the maximum $I_Z = 40$ mA occurs at the minimum $I_L = 0$. From Eq. (3-8), $I = I_Z = 40$ mA. From Eq. (3-7)

$$R = \frac{200 - 50}{40} = 3.75 \text{ K}$$

(NOTE: If the currents in a circuit are expressed in milliamperes and the voltages are given in volts then the resistances are expressed in kilohms.)

Since the minimum Zener current (the value of I_{ZK} in Fig. 2-10a) is 5 mA, then from Eq. (3-8) the maximum load current is

$$I_{L,\text{max}} = I - I_{ZK} = 40 - 5 = 35 \text{ mA}$$

b. At the minimum diode current $I = 5 + 25 = 30$ mA and from Eq. (3-7)

$$V = IR + V_Z = (30)(3.75) + 50 = 162.5 \text{ V}$$

At the maximum Zener current,

$$I = 40 + 25 = 65 \text{ mA} \qquad \text{and} \qquad V = (65)(3.75) + 50 = 293.8 \text{ V}$$

Hence, the source may vary between 162.5 and 293.8 V, and the output voltage will remain constant at 50 V and the load current will be constant at 25 mA.

3-7   RECTIFIERS

Almost all electronic circuits require a dc source of power. For portable low-power systems batteries may be used. More frequently, however, electronic equipment is energized by a *power supply*, a piece of equipment which converts the alternating waveform from the power lines into an essentially direct voltage. The study of ac-to-dc conversion is initiated in this section.

A Half-Wave Rectifier   A device, such as the semiconductor diode, which is capable of converting a sinusoidal input waveform (whose average value is zero) into a unidirectional (though not constant) waveform, with a nonzero average component, is called a *rectifier*. The basic circuit for half-wave rectification is shown in Fig. 3-15. Since in a rectifier circuit the input $v_i = V_m \sin \omega t$ has a peak value V_m which is very large compared with the cutin voltage V_γ of the diode, we assume in the following discussion that $V_\gamma = 0$. (The condition $V_\gamma \neq 0$ is treated in Sec. 3-3, and the current waveform is shown in Fig. 3-6b.) With the diode idealized to be a resistance R_f in the ON state and

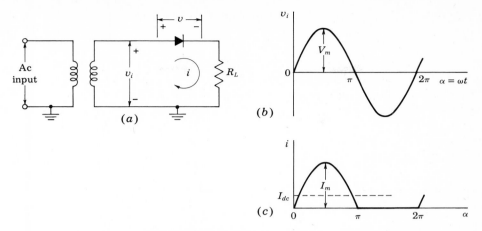

Fig. 3-15   (a) Basic circuit of half-wave rectifier.   (b) Transformer sinusoidal
secondary voltage v_i.   (c) Diode and load current i.

an open circuit in the OFF state, the current i in the diode or load R_L is given by

$$i = I_m \sin \alpha \qquad \text{if } 0 \leq \alpha \leq \pi$$
$$i = 0 \qquad\qquad \text{if } \pi \leq \alpha \leq 2\pi \tag{3-9}$$

where $\alpha \equiv \omega t$ and

$$I_m \equiv \frac{V_m}{R_f + R_L} \tag{3-10}$$

The transformer secondary voltage v_i is shown in Fig. 3-15b, and the rectified
current in Fig. 3-15c. Note that the output current is unidirectional. We
now calculate this nonzero value of the average current.

   *A dc ammeter is constructed so that the needle deflection indicates the
average value of the current passing through it.* By definition, the average
value of a periodic function is given by the area of one cycle of the curve
divided by the base. Expressed mathematically,

$$I_{dc} = \frac{1}{2\pi} \int_0^{2\pi} i \, d\alpha \tag{3-11}$$

For the half-wave circuit under consideration, it follows from Eqs. (3-9) that

$$I_{dc} = \frac{1}{2\pi} \int_0^{\pi} I_m \sin \alpha \, d\alpha = \frac{I_m}{\pi} \tag{3-12}$$

Note that the upper limit of the integral has been changed from 2π to π since
the instantaneous current in the interval from π to 2π is zero and so contributes
nothing to the integral.

The Diode Voltage   The dc output voltage is clearly given as

$$V_{dc} = I_{dc}R_L = \frac{I_m R_L}{\pi} \tag{3-13}$$

However, the reading of a dc voltmeter placed across the diode is *not* given by $I_{dc}R_f$ because the diode cannot be modeled as a constant resistance, but rather it has two values: R_f in the ON state and ∞ in the OFF state.

A *dc voltmeter reads the average value of the voltage across its terminals.* Hence, to obtain V'_{dc} across the diode, the instantaneous voltage must be plotted as in Fig. 3-16 and the average value obtained by integration.   Thus

$$V'_{dc} = \frac{1}{2\pi}\left(\int_0^\pi I_m R_f \sin \alpha \, d\alpha + \int_\pi^{2\pi} V_m \sin \alpha \, d\alpha\right)$$

$$= \frac{1}{\pi}(I_m R_f - V_m) = \frac{1}{\pi}[I_m R_f - I_m(R_f + R_L)]$$

where use has been made of Eq. (3-10).   Hence

$$V'_{dc} = -\frac{I_m R_L}{\pi} \tag{3-14}$$

This result is negative, which means that if the voltmeter is to read upscale, its positive terminal must be connected to the cathode of the diode.   From Eq. (3-13) the dc diode voltage is seen to be equal to the negative of the dc voltage across the load resistor.   This result is evidently correct because the sum of the dc voltages around the complete circuit must add up to zero.

The AC Current (Voltage)   *A root-mean-square ammeter (voltmeter) is constructed so that the needle deflection indicates the effective, or rms, current (voltage).*   Such a "square-law" instrument may be of the thermocouple type. By definition, the effective or rms value squared of a periodic function of time is given by the area of one cycle of the curve, which represents the square of the function, divided by the base.   Expressed mathematically,

$$I_{rms} = \left(\frac{1}{2\pi}\int_0^{2\pi} i^2 \, d\alpha\right)^{\frac{1}{2}} \tag{3-15}$$

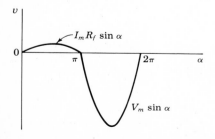

Fig. 3-16   The voltage across the diode in Fig. 3-15.

By use of Eqs. (3-9), it follows that

$$I_{\text{rms}} = \left(\frac{1}{2\pi} \int_0^\pi I_m{}^2 \sin^2 \alpha \; d\alpha \right)^{\frac{1}{2}} = \frac{I_m}{2} \tag{3-16}$$

Applying Eq. (3-15) to the *sinusoidal input voltage*, we obtain

$$V_{\text{rms}} = \frac{V_m}{\sqrt{2}} \tag{3-17}$$

Regulation   The variation of dc output voltage as a function of dc load current is called *regulation*.   The percentage regulation is defined as

$$\% \text{ regulation} \equiv \frac{V_{\text{no load}} - V_{\text{load}}}{V_{\text{load}}} \times 100\% \tag{3-18}$$

where *no load* refers to zero current and *load* indicates the normal load current. For an ideal power supply the output voltage is independent of the load (the output current) and the percentage regulation is zero.

The variation of V_{dc} with I_{dc} for the half-wave rectifier is obtained as follows: From Eqs. (3-12) and (3-10),

$$I_{\text{dc}} = \frac{I_m}{\pi} = \frac{V_m/\pi}{R_f + R_L} \tag{3-19}$$

Solving Eq. (3-19) for $V_{\text{dc}} = I_{\text{dc}} R_L$, we obtain

$$V_{\text{dc}} = \frac{V_m}{\pi} - I_{\text{dc}} R_f \tag{3-20}$$

This result is consistent with the circuit model given in Fig. 3-17 for the dc voltage and current.   Note that the rectifier circuit functions as if it were a constant (open-circuit) voltage source $V = V_m/\pi$ in series with an effective internal resistance (the *output resistance*) $R_o = R_f$.   This model shows that V_{dc} equals V_m/π at no load and that the dc voltage decreases linearly with an increase in dc output current.   In practice, the resistance R_s of the transformer secondary is in series with the diode, and in Eq. (3-20) R_s should be added to R_f.   The best method of estimating the diode resistance is to obtain a regulation plot of V_{dc} versus I_{dc} in the laboratory.   The negative slope of the resulting straight line gives $R_f + R_s$.   Clearly, Fig. 3-17 represents a Thévenin's

Fig. 3-17   The Thévenin's model which gives the dc voltage and current for a power supply.   For the half-wave circuit of Fig. 3-15, $V = V_m/\pi$ and $R_o = R_f$. For the full-wave circuit of Fig. 3-18, $V = 2V_m/\pi$ and $R_o = R_f$.   For the full-wave rectifier with a capacitor filter (Sec. 3-8), $V_o = V_m$ and $R_o = 1/4fC$.

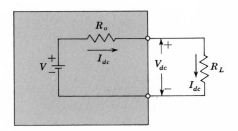

model, and hence a rectifier behaves as a linear circuit with respect to average current and voltage.

A Full-Wave Rectifier   The circuit of a full-wave rectifier is shown in Fig. 3-18a.   This circuit is seen to comprise two half-wave circuits so connected that conduction takes place through one diode during one half of the power cycle and through the other diode during the second half of the cycle.

The current to the load, which is the sum of these two currents, $i = i_1 + i_2$, has the form shown in Fig. 3-18b.   The dc and rms values of the load current and voltage in such a system are readily found to be

$$I_{dc} = \frac{2I_m}{\pi} \qquad I_{rms} = \frac{I_m}{\sqrt{2}} \qquad V_{dc} = \frac{2I_m R_L}{\pi} \tag{3-21}$$

where I_m is given by Eq. (3-10) and V_m is the peak transformer secondary voltage from one end to the center tap.   Note by comparing Eq. (3-21) with Eq. (3-13) that the dc output voltage for the full-wave connection is twice that for the half-wave circuit.

From Eqs. (3-10) and (3-21) we find that the dc output voltage varies with current in the following manner:

$$V_{dc} = \frac{2V_m}{\pi} - I_{dc}R_f \tag{3-22}$$

This expression leads to Thévenin's dc model of Fig. 3-17, except that the internal (open-circuit) supply is $V = 2V_m/\pi$ instead of V_m/π.

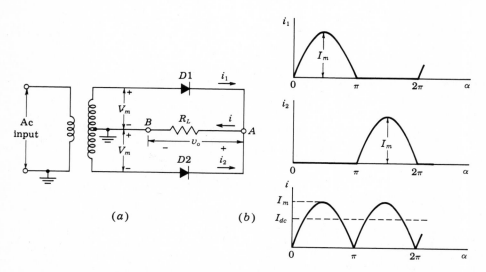

Fig. 3-18   (a) A full-wave rectifier circuit.   (b) The individual diode currents and the load current i.   The output voltage is $v_o = iR_L$.

Peak Inverse Voltage For each rectifier circuit there is a maximum voltage to which the diode can be subjected. This potential is called the *peak inverse voltage* because it occurs during that part of the cycle when the diode is nonconducting. From Fig. 3-15 it is clear that, for the half-wave rectifier, the peak inverse voltage is V_m. We now show that, for a full-wave circuit, twice this value is obtained. At the instant of time when the transformer secondary voltage to midpoint is at its peak value V_m, diode $D1$ is conducting and $D2$ is nonconducting. If we apply KVL around the outside loop and neglect the small voltage drop across $D1$, we obtain $2V_m$ for the peak inverse voltage across $D2$. Note that this result is obtained without reference to the nature of the load, which can be a pure resistance R_L or a combination of R_L and some reactive elements which may be introduced to "filter" the ripple. We conclude that, *in a full-wave circuit, independently of the filter used, the peak inverse voltage across each diode is twice the maximum transformer voltage measured from midpoint to either end.*

3-8 CAPACITOR FILTERS

Filtering is frequently effected by shunting the load with a capacitor. The action of this system depends upon the fact that the capacitor stores energy during the conduction period and delivers this energy to the load during the inverse, or nonconducting, period. In this way, the time during which the current passes through the load is prolonged, and the ripple is considerably decreased. The ripple voltage is defined as the deviation of the load voltage from its average or dc value.

Consider the half-wave capacitive rectifier of Fig. 3-19. Suppose, first that the load resistance $R_L = \infty$. The capacitor will charge to the potential V_m, the transformer maximum value. Further, the capacitor will maintain this potential, for no path exists by which this charge is permitted to leak off, since the diode will not pass a negative current. The diode resistance is infinite in the inverse direction, and no charge can flow during this portion of the cycle. Consequently, the filtering action is perfect, and the capacitor voltage v_o remains constant at its peak value, as is seen in Fig. 3-20.

The voltage v_o across the capacitor is, of course, the same as the voltage across the load resistor, since the two elements are in parallel. The diode

Fig. 3-19 A half-wave capacitor-filtered rectifier.

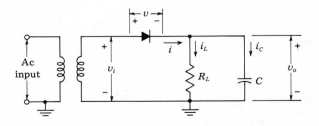

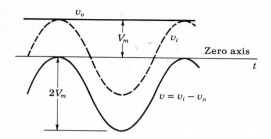

Fig. 3-20 Voltages in a half-wave capacitor-filtered rectifier at no load. The output voltage v_o is a constant, indicating perfect filtering. The diode voltage v is negative for all values of time, and the peak inverse voltage is $2V_m$.

voltage v is given by

$$v = v_i - v_o \qquad\qquad (3\text{-}23)$$

We see from Fig. 3-20 that the diode voltage is always negative and that the peak inverse voltage is twice the transformer maximum. Hence the presence of the capacitor causes the peak inverse voltage to increase from a value equal to the transformer maximum when no capacitor filter is used to a value equal to twice the transformer maximum value when the filter is used.

Suppose, now, that the load resistor R_L is finite. Without the capacitor input filter, the load current and the load voltage during the conduction period will be sinusoidal functions of time. The inclusion of a capacitor in the circuit results in the capacitor charging in step with the applied voltage. Also, the capacitor must discharge through the load resistor, since the diode will prevent a current in the negative direction. Clearly, the diode acts as a switch which permits charge to flow into the capacitor when the transformer voltage exceeds the capacitor voltage, and then acts to disconnect the power source when the transformer voltage falls below that of the capacitor.

Output Voltage Under Load During the time interval when the diode in Fig. 3-19 is conducting, the transformer voltage is impressed directly across the load (assuming that the diode drop can be neglected). Hence, the output voltage is $v_o = V_m \sin \omega t$. During the interval when D is nonconducting, the capacitor discharges through the load with a time constant CR_L. The output waveform in Fig. 3-21 consists of portions of sinusoids (when D is ON) joined

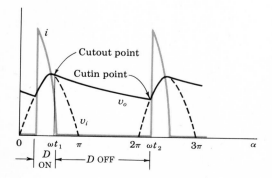

Fig. 3-21 Theoretical sketch of diode current i and output voltage v_o in a half-wave capacitor-filtered rectifier.

to exponential segments (when D is OFF). The point at which the diode starts to conduct is called the *cutin* point t_2, and that at which it stops conducting is called the *cutout point* t_1. These times are indicated in Fig. 3-21.

The cutout time is obtained by finding the expression for the sinusoidal current i in Fig. 3-19 when $v_o = V_m \sin \omega t$. Then the time for which $i = 0$ gives the cutout angle ωt_1. The cutin point t_2 is obtained graphically by finding the time when the exponential portion of v_o in Fig. 3-21 intersects the curve $V_m \sin \omega t$ (in the following cycle). The validity of this statement follows from the fact that at an instant of time greater than t_2, the transformer voltage v_i (the sine curve) is greater than the capacitor voltage v_o (the exponential curve). Since the diode voltage is $v = v_i - v_o$, then v will be positive beyond t_2 and the diode will become conducting. Thus t_2 is the cutin point.

The use of a large capacitance to improve the filtering at a given load R_L is accompanied by a high-peak diode current I_m. For a specified average load current, i becomes more peaked and the conduction period decreases as C is made larger. It is to be emphasized that the use of a capacitor filter may impose serious restrictions on the diode, since the average current may be well within the current rating of the diode, and yet the peak current may be excessive.

Full-Wave Circuit   Consider a full-wave rectifier with a capacitor filter obtained by placing a capacitor C across R_L in Fig. 3-18. The analysis of this circuit requires a simple extension of that just made for the half-wave circuit. If in Fig. 3-21 a dashed half-sinusoid is added between π and 2π, the result is the dashed full-wave voltage in Fig. 3-22. The cutin point now lies between π and 2π, where the exponential portion of v_o intersects this sinusoid. The cutout point is the same as that found for the half-wave rectifier.

Approximate Analysis   It is possible to obtain the dc output voltage for given values of the parameters ω, R_L, C, and V_m from the graphical construction indicated in Fig. 3-22. Such an analysis is involved and tedious. Hence

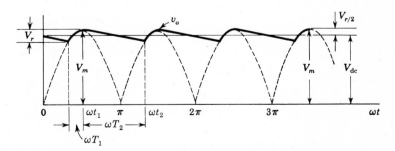

Fig. 3-22   The approximate load-voltage waveform v_o in a full-wave capacitor-filtered rectifier.

we now present an approximate solution which is simple and yet sufficiently accurate for most engineering applications.

We assume that the output-voltage waveform of a full-wave circuit with a capacitor filter may be represented by the approximately piecewise linear curve shown in Fig. 3-22. For large values of C (so that $\omega C R_L \gg 1$) we note that $\omega t_1 \to \pi/2$ and $v_o \to V_m$ at $t = t_1$. Also, with C very large, the exponential decay can be replaced by a linear fall. If the total capacitor discharge voltage (the ripple voltage) is denoted by V_r, then from Fig. 3-22, the average value of the voltage is approximately

$$V_{dc} = V_m - \frac{V_r}{2} \tag{3-24}$$

It is necessary, however, to express V_r as a function of the load current and the capacitance. If T_2 represents the total nonconducting time, the capacitor, when discharging at the constant rate I_{dc}, will lose an amount of charge $I_{dc}T_2$. Hence the change in capacitor voltage is $I_{dc}T_2/C$, or

$$V_r = \frac{I_{dc}T_2}{C} \tag{3-25}$$

The better the filtering action, the smaller will be the conduction time T_1 and the closer T_2 will approach the time of half a cycle. Hence we assume that $T_2 = T/2 = 1/2f$, where f is the fundamental power-line frequency. Then

$$V_r = \frac{I_{dc}}{2fC} \tag{3-26}$$

and from Eq. (3-24),

$$V_{dc} = V_m - \frac{I_{dc}}{4fC} \tag{3-27}$$

This result is consistent with Thévenin's model of Fig. **3-17**, with the open-circuit voltage $V = V_m$ and the effective output resistance $R_o = 1/4fC$.

The ripple is seen to vary directly with the load current I_{dc} and also inversely with the capacitance. Hence, to keep the ripple low and to ensure good regulation, very large capacitances (of the order of tens of microfarads) must be used. The most common type of capacitor for this rectifier application is the electrolytic capacitor. These capacitors are polarized, and care must be taken to insert them into the circuit with the terminal marked $+$ to the positive side of the output.

The desirable features of rectifiers employing capacitor input filters are the small ripple and the high voltage at light load. The no-load voltage is equal, theoretically, to the maximum transformer voltage. The disadvantages of this system are the relatively poor regulation, the high ripple at large load currents, and the peaked currents that the diodes must pass.

An approximate analysis similar to that given above applied to the half-wave circuit shows that the ripple, and also the drop from no load to a given load, are double the values calculated for the full-wave rectifier.

3-9   ADDITIONAL DIODE CIRCUITS

Many applications depend upon the semiconductor diode besides those already considered in this chapter. We mention four others below.

Peak Detector The half-wave capacitor-filtered rectifier circuit of Fig. 3-19 may be used to measure the peak value of an input waveform. Thus, for $R_L = \infty$, the capacitor charges to the maximum value $V_{\max}$ of v_i, the diode becomes nonconducting, and v_o remains at $V_{\max}$ (assuming an ideal capacitor with no leakage resistance shunting C). Refer to Fig. 3-20, where $V_{\max} = V_m =$ the peak value of the input sinusoid. Improved peak detector circuits are given in Sec. 14-3.

In an AM radio the amplitude of the high-frequency wave (called the *carrier*) is varied in accordance with the audio information to be transmitted. This process is called *amplitude modulation,* and such an AM waveform is illustrated in Fig. 3-23. The audio information is contained in the envelope (the locus, shown dashed, of the peak values) of the modulated waveform. The process of extracting the audio signal is called *detection,* or *demodulation.* If the input to Fig. 3-19 is the AM waveform shown in Fig. 3-23, the output v_o is the heavy-weight curve, provided that the time constant $R_L C$ is chosen properly; that is, $R_L C$ must be small enough so that, when the envelope decreases in magnitude, the voltage across C can fall fast enough to keep in step with the envelope, but $R_L C$ must not be so small as to introduce excessive ripple. The order of magnitude of the frequency of an AM radio carrier is 1,000 kHz, and the audio spectrum extends from about 20 Hz to 20 kHz. Hence there should be *at least* 50 cycles of the carrier waveform for each audio cycle. If Fig. 3-23 were drawn more realistically (with a much higher ratio of carrier to audio frequency), then clearly, the ripple amplitude of the demodulated signal would be very much smaller. This low-amplitude high-fre-

Fig. 3-23 An amplitude-modulated wave and the detected audio signal. (For ease of drawing, the carrier waveform is indicated triangular instead of sinusoidal and of much lower frequency than it really is, relative to the audio frequency.)

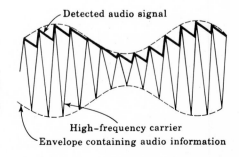

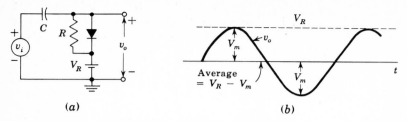

Fig. 3-24 (a) A circuit which clamps to the voltage V_R. (b) The output voltage v_o for a sinusoidal input v_i.

quency ripple in v_o is easily filtered so that the smoothed detected waveform is an excellent reproduction of the audio signal. The capacitor-rectifier circuit of Fig. 3-19 therefore also acts as an *envelope demodulator*.

A Clamping Circuit   A function which must be frequently performed with a periodic waveform is the establishment of the recurrent positive or negative extremity at some constant reference level V_R. Such a clamping circuit is indicated in Fig. 3-24a. Assuming an ideal diode, the drop across the device is zero in the forward direction. Hence the output cannot rise above V_R and is said to be *clamped* to this level. If the input is sinusoidal with a peak value V_m and an average value of zero, then, as indicated in Fig. 3-24b, the output is sinusoidal, with an average value of $V_R - V_m$. This waveform is obtained subject to the following conditions: the diode parameters are $R_f = 0$, $R_r = \infty$, and $V_\gamma = 0$; the source impedance $R_s = 0$; and the time constant RC is much larger than the period of the signal. In practice these restrictions are not completely satisfied and the clamping is not perfect; the output voltage rises slightly above V_R, and the waveshape is a somewhat distorted version of the input.

Digital-Computer Circuits   Since the diode is a binary device existing in either the ON or OFF state for a given interval of time, it is a very usual component in digital-computer applications. Such so-called "logic" circuits are discussed in Chap. 6 in conjunction with transistor binary applications.

REVIEW QUESTIONS

3-1   Explain how to obtain the dynamic characteristic from the static volt-ampere curve of a diode.

3-2   You are given the V-I output characteristic in graphical form of a new device. (a) Sketch the circuit using this device which will require a load-line construction to determine i and v. (b) Is the load line vertical, horizontal, at 135°, or 45° for infinite load resistance? (c) For zero load resistance?

3-3 (*a*) Draw the piecewise linear volt-ampere characteristic of a *p-n* diode. (*b*) What is the circuit model for the ON state? (*c*) The OFF state?

3-4 Consider a circuit consisting of a diode D, a resistance R, and a signal source v_i in series. Define (*a*) static characteristic; (*b*) dynamic characteristic; (*c*) *transfer*, or *transmission*, characteristic. (*d*) What is the correlation between (*b*) and (*c*)?

3-5 In analyzing a circuit containing several diodes by the piecewise linear method, you assume (guess) that certain of the diodes are ON and others are OFF. Explain carefully how you determine whether or not the assumed state of each diode is correct.

3-6 Consider a series circuit consisting of a diode D, a resistance R, a reference battery V_R, and an input signal v_i. The output is taken across R and V_R in series. Draw the transfer characteristic if the anode of D is connected to the positive terminal of the battery. Use the piecewise linear diode model.

3-7 Repeat Rev. 3-6 with the anode of D connected to the negative terminal of the battery.

3-8 If v_i is sinusoidal and D is ideal (with $R_f = 0$, $V_\gamma = 0$, and $R_r = \infty$), find the output waveforms in (*a*) Rev. 3-6 and (*b*) Rev. 3-7.

3-9 Sketch the circuit of a *double-ended clipper* using ideal *p-n* diodes which limit the output between ± 10 V.

3-10 Repeat Rev. 3-9 using avalanche diodes.

3-11 Define in words and as an equation (*a*) dc current I_{dc}; (*b*) dc voltage V_{dc}; (*c*) ac current I_{rms}.

3-12 (*a*) Sketch the circuit for a full-wave rectifier. (*b*) Derive the expression for (1) the dc current; (2) the dc load voltage; (3) the dc diode voltage; (4) the rms load current.

3-13 (*a*) Define *regulation*. (*b*) Derive the regulation equation for a full-wave circuit.

3-14 Draw the Thévenin's model for a full-wave rectifier.

3-15 (*a*) Define *peak inverse voltage*. (*b*) What is the peak inverse voltage for a full-wave circuit using ideal diodes? (*c*) Repeat part *b* for a half-wave rectifier.

3-16 (*a*) Draw the circuit of a half-wave capacitive rectifier. (*b*) At no load draw the steady-state voltage across the capacitor and also across the diode.

3-17 (*a*) Draw the circuit of a full-wave capacitive rectifier. (*b*) Draw the output voltage under load. Indicate over what period of time the diode conducts. Make no calculations. (*c*) Indicate the diode current waveform superimposed upon the output waveform.

3-18 For the circuit of Rev. 3-17, derive the expression for (*a*) the diode current; (*b*) the cutout angle. (*c*) How is the cutin angle found?

3-19 (*a*) Consider a full-wave capacitor rectifier using a large capacitance C. Sketch the approximate output waveform. (*b*) Derive the expression for the peak ripple voltage. (*c*) Derive the Thévenin's model for this rectifier.

3-20 For a full-wave capacitor rectifier circuit, list (*a*) two advantages; (*b*) three disadvantages.

3-21 Describe (*a*) *amplitude modulation* and (*b*) *detection*.

4 / TRANSISTOR CHARACTERISTICS

The physical behavior of a semiconductor triode, called a *bipolar junction transistor* (BJT), is given. The volt-ampere characteristics of this device are studied. It is demonstrated that the transistor is capable of producing amplification. For the transistor operating in either the active region or in saturation, the method of analysis for obtaining the currents and voltages is explained. Typical voltage values are given, for the several possible modes of operation.

4-1 THE JUNCTION TRANSISTOR

A junction transistor consists of a silicon (or germanium) crystal in which a layer of n-type silicon is sandwiched between two layers of p-type silicon. Alternatively, a transistor may consist of a layer of p-type between two layers of n-type material. In the former case the transistor is referred to as a p-n-p transistor, and in the latter case, as an n-p-n transistor. The semiconductor sandwich is extremely small, and is hermetically sealed against moisture inside a metal or plastic case. Manufacturing techniques and constructional details for several transistor types are described in Sec. 4-4.

The two types of transistor are represented in Fig. 4-1a. The representations employed when transistors are used as circuit elements are shown in Fig. 4-1b. The three portions of a transistor are known as *emitter, base,* and *collector.* The arrow on the emitter lead specifies the direction of current flow when the emitter-base junction is biased in the forward direction. In *both* cases, however, the emitter, base, and collector currents, I_E, I_B, and I_C, respectively, are assumed positive when the currents flow *into* the transistor. The symbols V_{EB}, V_{CB}, and V_{CE} are the emitter-base, collector-base, and collector-emitter voltages, respectively. (More specifically, V_{EB} represents the voltage *drop* from emitter to base.)

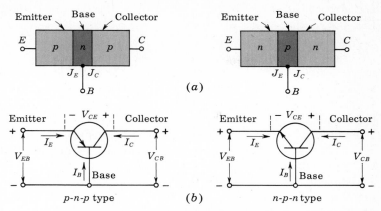

Fig. 4-1   (a) A *p-n-p* and an *n-p-n* transistor. The emitter (collector) junction is $J_E(J_C)$.   (b) Circuit representation of the two
transistor types.

Open-circuited Transistor   If no external biasing voltages are applied,
all transistor currents must be zero.   The potential barriers at the junctions
adjust to the contact difference of potential V_o—given in Fig. 2-1*d* (a few
tenths of a volt)—required so that no free carriers cross each junction.   If
we assume a completely symmetrical junction (emitter and collector regions
having identical physical dimensions and doping concentrations), the barrier
height is identical at the emitter junction J_E and at the collector junction J_C,
as indicated in Fig. 4-2*a*.   The narrow space-charge regions at the junctions
have been neglected.

Under open-circuited conditions, the minority concentration is constant
within each section and is equal to its thermal-equilibrium value, n_{po} in the
p-type emitter and collector regions and p_{no} in the *n*-type base, as shown in
Fig. 4-2*b*.   Since the transistor may be looked upon as a *p-n* diode followed by
an *n-p* diode, much of the theory developed in Chap. 2 for the junction diode
will be used to explain the physical behavior of the transistor, when voltages

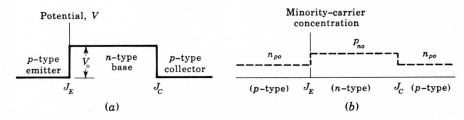

Fig. 4-2   (a) The potential and (b) the minority-carrier density in each section
of an open-circuited symmetrical *p-n-p* transistor.

are applied so as to disturb it from the equilibrium situation pictured in Fig. 4-2.

The Transistor Biased in the Active Region   We may now begin to appreciate the essential feasures of a transistor as an active circuit element by considering the situation depicted in Fig. 4-3a. Here a *p-n-p* transistor is shown with voltage sources which serve to bias the emitter-base junction in the forward direction and the collector-base junction in the reverse direction. The potential variation through the biased transistor is indicated in Fig. 4-3b. The dashed curve applies to the case before the application of external biasing voltages (Fig. 4-2a), and the solid curve to the case after the biasing voltages are applied. The externally applied voltages appear, essentially, across the junctions. Hence, as shown in Fig. 4-3b, the forward biasing of the emitter junction lowers the emitter-base potential barrier by $|V_{EB}|$, whereas the reverse biasing of the collector junction increases the collector-base potential barrier by $|V_{CB}|$. The lowering of the emitter-base barrier permits minority-carrier injection; that is, holes are injected into the base, and electrons are injected into the emitter region. The excess holes diffuse across the *n*-type base,

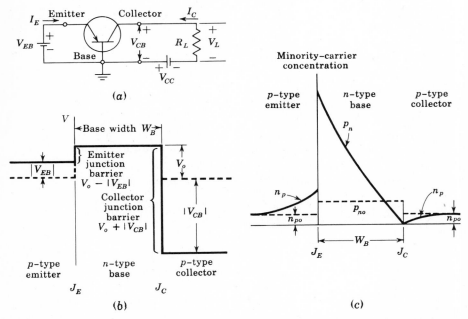

Fig. 4-3   (a) A *p-n-p* transistor biased in the active region (the emitter is forward-biased and the collector is reverse-biased). (b) The potential variation through the transistor. The narrow depletion regions at the junctions are negligibly small. (c) The minority-carrier concentration in each section of the transistor. It is assumed that the emitter is much more heavily doped than the base.

where the electric field intensity $\mathcal{E}$ is zero, to the collector junction. At J_C the field is positive and large ($\mathcal{E} = -dV/dx \gg 0$), and hence holes are accelerated across this junction. In other words, the holes which reach J_C fall down the potential barrier, and are therefore *collected* by the collector. Hence, p_n is reduced to zero at the collector as shown in Fig. 4-3c. Similarly, the reverse collector-junction bias reduces the collector electron density n_p to zero at J_C. The minority-carrier-density curves pictured in Fig. 4-3c should be compared with the corresponding concentration plots for the forward- and reverse-biased p-n junction given in Fig. 2-9.

4-2   TRANSISTOR CURRENT COMPONENTS

In Fig. 4-4 we show the various current components which flow across the forward-biased emitter junction and the reverse-biased collector junction. The emitter current I_E consists of hole current I_{pE} (holes crossing from emitter into base) and electron current I_{nE} (electrons crossing from base into the emitter). In a commercial transistor the doping of the emitter is made much larger than the doping of the base. This feature ensures (in a p-n-p transistor) that the emitter current consists almost entirely of holes. Such a situation is desirable since the current which results from electrons crossing the emitter junction from base to emitter does not contribute carriers which can reach the collector.

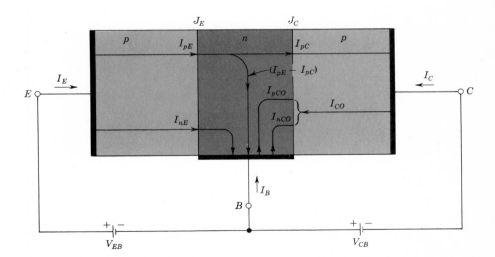

Fig. 4-4 Transistor current components for a forward-biased emitter junction and a reversed-biased collector junction. If a current has a subscript $p(n)$, it consists of holes (electrons) moving in the same (opposite) direction as the arrow indicating the current direction.

The minority current I_{pE} is the hole *diffusion* current into base and its magnitude is proportional to the slope at J_E of the p_n curve [Eq. (1-28)]. Similarly, I_{nE} is the electron *diffusion* current into the emitter, and its magnitude is proportional to the slope at J_E of the n_p curve in Fig. 4-3c. The total emitter current in Fig. 4-4 is the sum of the currents crossing J_E. Hence

$$I_E = I_{pE} + I_{nE} \tag{4-1}$$

All currents in this equation are positive for a *p-n-p* transistor.

Not all the holes crossing the emitter junction J_E reach the collector junction J_C, because some of them recombine with the electrons in the *n*-type base. In Fig. 4-4, let I_{pC} represent the hole current at J_C as a result of holes crossing the base from the emitter. Hence there must be a bulk recombination hole current $I_{pE} - I_{pC}$ leaving the base, as indicated in Fig. 4-4 (actually, electrons enter the base region from the external circuit through the base lead to supply those charges which have been lost by recombination with the holes injected into the base across J_E).

Consider, for the moment, an open-circuited emitter, while the collector junction remains reverse-biased. Then I_C must equal the reverse saturation current I_{CO} of the back-biased diode at J_C. This *reverse* current consists of two components, as shown in Fig. 4-4, I_{nCO} consisting of electrons moving from the p to the n region across J_C and a term, I_{pCO}, resulting from holes crossing from n to p across J_C.

$$-I_{CO} = I_{nCO} + I_{pCO} \tag{4-2}$$

(The minus sign is chosen arbitrarily so that I_C and I_{CO} will have the same sign.) The magnitude of I_{nCO} is proportional to the slope at J_C of the n_p distribution in Fig. 4-3c.

Since $I_E = 0$ under open-circuit conditions, no holes are injected across J_E, and, hence, none can reach J_C from the emitter. Clearly, I_{pCO} results from the small concentration of holes generated thermally within the base.

Now let us return to the situation depicted in Fig. 4-4, where the emitter is forward-biased. Now the total hole current crossing J_C is the sum of I_{pC} and I_{pCO}, and its magnitude is proportional to the slope at J_C of the p_n distribution in Fig. 4-3c. The complete collector current is given by

$$I_C = I_{CO} - I_{pC} = I_{CO} - \alpha I_E \tag{4-3}$$

where α is defined as the fraction of the total emitter current [given in Eq. (4-1)] which represents holes which have traveled from the emitter across the base to the collector. For a *p-n-p* transistor, I_E is positive and both I_C and I_{CO} are negative, which means that the current in the collector lead is in the direction opposite to that indicated by the arrow of I_C in Fig. 4-4. For an *n-p-n* transistor these currents are reversed.

Large-Signal Current Gain α From Eq. (4-3) it follows that α may be defined as the ratio of the negative of the collector-current increment from cutoff ($I_C = I_{CO}$) to the emitter-current change from cutoff ($I_E = 0$), or

$$\alpha \equiv -\frac{I_C - I_{co}}{I_E - 0} \tag{4-4}$$

Alpha is called the *large-signal current gain* of a common-base transistor. Since I_C and I_E have opposite signs (for either a *p-n-p* or an *n-p-n* transistor), then α, as defined, is always positive. Typical numerical values of α lie in the range 0.90 to 0.995. It should be pointed out that α is not a constant, but varies with emitter current I_E, collector voltage V_{CB}, and temperature.

A Generalized Transistor Equation Equation (4-3) is valid only in the *active region*, that is, if the emitter is forward-biased and the collector is reverse-biased. For this mode of operation the collector current is essentially independent of collector voltage and depends only upon the emitter current. Suppose now that we seek to generalize Eq. (4-3) so that it may apply not only when the collector junction is substantially reverse-biased, but also for any voltage across J_C. To achieve this generalization we need only replace I_{co} by the current in a *p-n* diode (that consisting of the base and collector regions). This current is given by the volt-ampere relationship of Eq. (2-2), with I_o replaced by $-I_{co}$ and V by V_C, where the symbol V_C represents the drop across J_C from the *p* to the *n* side. The complete expression for I_C for any V_C and I_E is

$$I_C = -\alpha I_E + I_{co}(1 - \epsilon^{V_C/V_T}) \tag{4-5}$$

Note that if V_C is negative and has a magnitude large compared with V_T, Eq. (4-5) reduces to Eq. (4-3). The physical interpretation of Eq. (4-5) is that the *p-n* junction diode current crossing the collector junction is augmented by the fraction α of the current I_E flowing in the emitter.

4-3 THE TRANSISTOR AS AN AMPLIFIER

A load resistor R_L is in series with the collector supply voltage V_{CC} of Fig. 4-3a. A small voltage change ΔV_i between emitter and base causes a relatively large emitter-current change ΔI_E. We define by the symbol α' that fraction of this current change which is collected and passes through R_L, or $\Delta I_C = \alpha' \, \Delta I_E$. The change in output voltage across the load resistor

$$\Delta V_L = -R_L \, \Delta I_C = -\alpha' R_L \, \Delta I_E \tag{4-6}$$

may be many times the change in input voltage ΔV_i. Under these circumstances, the voltage amplification $A \equiv \Delta V_L/\Delta V_i$ will be greater than unity, and the transistor acts as an amplifier. If the dynamic resistance of the emitter junction is r_e, then $\Delta V_i = r_e \, \Delta I_E$, and

$$A \equiv -\frac{\alpha' R_L \, \Delta I_E}{r_e \, \Delta I_E} = -\frac{\alpha' R_L}{r_e} \tag{4-7}$$

From Eq. (2-7), $r_e = 26/I_E$, where I_E is the quiescent emitter current in milliamperes.   For example, if $r_e = 40 \ \Omega$, $\alpha' = -1$, and $R_L = 3,000 \ \Omega$, $A = +75$. This calculation is oversimplified, but in essence it is correct and gives a physical explanation of why the transistor acts as an amplifier.   The transistor provides power gain as well as voltage or current amplification.   From the foregoing explanation it is clear that current in the low-resistance input circuit is transferred to the high-resistance output circuit.   The word "transistor," which originated as a contraction of "transfer resistor," is based upon the above physical picture of the device.

The Parameter α' The parameter α' introduced above is defined as the ratio of the change in the collector current to the change in the emitter current at constant collector-to-base voltage and is called the *negative of the small-signal short-circuit current transfer ratio, or gain*.   More specifically,

$$\alpha' \equiv \frac{\Delta I_C}{\Delta I_E}\bigg|_{V_{CB}} \tag{4-8}$$

On the assumption that α is independent of I_E, then from Eq. (4-3) it follows that $\alpha' = -\alpha$.

4-4   TRANSISTOR CONSTRUCTION

Four basic techniques have been developed for the manufacture of diodes, transistors, and other semiconductor devices.   Consequently, such devices may be classified into one of the following types: grown, alloy, diffusion, or epitaxial.

Grown Type   The n-p-n grown-junction transistor is illustrated in Fig. 4-5a.   It is made by drawing a single crystal from a melt of silicon or germanium whose impurity concentration is changed during the crystal-drawing operation by adding n- or p-type atoms as required.

Alloy Type   This technique, also called the *fused* construction, is illustrated in Fig. 4-5b for a p-n-p transistor.   The center (base) section is a thin wafer of n-type material.   Two small dots of indium are attached to opposite sides of the wafer, and the whole structure is raised for a short time to a high temperature, above the melting point of indium but below that of germanium. The indium dissolves the germanium beneath it and forms a saturation solution.   On cooling, the germanium contact with the base material recrystallizes, with enough indium concentration to change it from n to p type.   The collector is made larger than the emitter, so that the collector subtends a large angle as viewed from the emitter.   Because of this geometrical arrangement, very little emitter current follows a diffusion path which carries it to the base rather than to the collector.

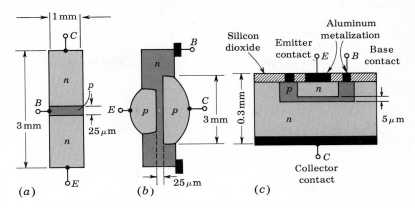

Fig. 4-5 Construction of transistors. (a) Grown, (b) alloy, and (c) diffused planar types. (The dimensions are approximate, and the figures are not drawn to scale.)

Diffusion Type This technique consists of subjecting a semiconductor wafer to gaseous diffusions of both n- and p-type impurities to form both the emitter and the collector junctions. A *planar* silicon transistor of the diffusion type is illustrated in Fig. 4-5c. In this process (described in greater detail in Chap. 5 on integrated-circuit fabrication), the base-collector junction area is determined by a diffusion mask. The emitter is then diffused on the base through a different mask. A thin layer of silicon dioxide is grown over the entire surface and photoetched, so that aluminum contacts can be made for the emitter and base leads (Fig. 4-5c). Because of the passivating action of this oxide layer, most surface problems are avoided and very low leakage currents result. There is also an improvement in the current gain at low currents and in the noise figure.

Epitaxial Type The epitaxial technique (Sec. 5-2) consists of growing a very thin, high-purity, single-crystal layer of silicon or germanium on a heavily doped substrate of the same material. This augmented crystal forms the collector on which the base and emitter may be diffused.

4-5   THE COMMON–BASE CONFIGURATION

In Fig. 4-3a, a p-n-p transistor is shown in a *grounded-base* configuration. This circuit is also referred to as a *common-base*, or CB, configuration, since the base is common to the input and output circuits. For a p-n-p transistor the largest current components are due to holes. Since holes flow from the emitter to the collector and down toward ground out of the base terminal, then, referring to the polarity conventions of Fig. 4-1, we see that I_E is positive, I_C is negative, and I_B is negative. For a forward-biased emitter junction, V_{EB} is positive, and for a reverse-biased collector junction, V_{CB} is negative. For an n-p-n transistor all current and voltage polarities are the negative of those for a p-n-p transistor. In summary, *in the active region the current in the emitter is positive in the direction of the arrow on the emitter lead.* Also, *the sign of I_B is the same as that of I_C and opposite to that of I_E.*

The Output Characteristics From Eq. (4-5) we see that the output (collector) current I_C is completely determined by the input (emitter) current I_E and the output (collector-to-base) voltage $V_{CB} = V_C$. This output relationship is given in Fig. 4-6 for a typical p-n-p germanium transistor and is a plot of collector current I_C versus collector-to-base voltage drop V_{CB}, with emitter current I_E as a parameter. The curves of Fig. 4-6 are known as the *output*, or *collector, static characteristics*.

A qualitative understanding of the form of the output characteristics is not difficult if we consider the fact that the transistor consists of two diodes placed in series "back to back" (with the two cathodes connected together). In the active region the input diode (emitter-to-base) is biased in the forward direction. Note, as in Fig. 4-6, that it is customary to plot along the abscissa and to the right that polarity of V_{CB} which reverse-biases the collector junction even if this polarity is negative. If $I_E = 0$, the collector current is $I_C = I_{CO}$. For other values of I_E, the output-diode reverse current is augmented by the fraction of the input-diode forward current which reaches the collector. Note also that I_C and I_{CO} are negative for a p-n-p transistor and positive for an n-p-n transistor.

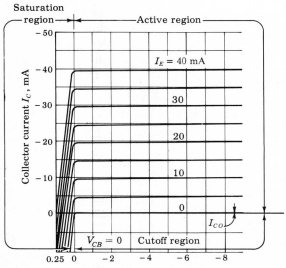

Fig. 4-6 Typical common-base output characteristics of a p-n-p transistor. The cutoff, active, and saturation regions are indicated. Note the expanded voltage scale in the saturation region.

The Input Characteristics These curves are plots of emitter-to-base voltage V_{EB} versus emitter current I_E, with collector-to-base voltage V_{CB} as a parameter. This set of curves is referred to as the *input*, or *emitter*, static characteristics. The input characteristics of Fig. 4-7 represent simply the forward characteristic of the emitter-to-base diode for various collector voltages. A noteworthy feature of the input characteristics is that there exists a *cutin*, *offset*, or *threshold*, voltage V_γ below which the emitter current is very small. In general, V_γ is approximately 0.1 V for germanium transistors (Fig. 4-7) and 0.5 V for silicon.

Active Region In this region *the collector junction is biased in the reverse direction and the emitter junction in the forward direction.* Consider first that the emitter current in Fig. 4-6 is zero. Then the collector current is small and equals the reverse saturation current I_{CO} (microamperes for germanium and nanoamperes for silicon) of the collector junction considered as a diode. Suppose now that a forward emitter current I_E is caused to flow in the emitter circuit. Then a fraction $-\alpha I_E$ of this current will reach the collector, and I_C is therefore given by Eq. (4-3). In the active region, the collector current is essentially independent of collector voltage and depends only upon the emitter current. Because α is less than, but almost equal to, unity, the magnitude of the collector current is (slightly) less than that of the emitter current.

Saturation Region The region to the left of the ordinate, $V_{CB} = 0$, and above the $I_E = 0$ characteristics, in which *both emitter and collector junctions are forward-biased*, is called the *saturation* region. We say that "bottoming" has taken place because the voltage has fallen near the bottom of the characteristic where $V_{CB} \approx 0$. Actually, V_{CB} is slightly positive (for a *p-n-p* transistor) in this region, and this forward biasing of the collector accounts for the large change in collector current with small changes in collector voltage. For a forward bias, I_C increases exponentially with voltage according to the diode relationship [Eq. (2-2)]. A forward bias means that the collector *p* material is made positive with respect to the base *n* side, and hence that hole current flows from the *p* side across the collector junction to the *n* material. This

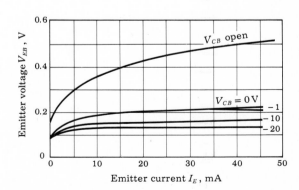

Fig. 4-7 Common-base input characteristics of a typical *p-n-p* germanium junction transistor.

hole flow corresponds to a positive change in collector current. Hence the collector current increases rapidly, and as indicated in Fig. 4-6, I_C may even become positive if the forward bias is sufficiently large.

Cutoff Region The characteristic for $I_E = 0$ passes through the origin, but is otherwise similar to the other characteristics. This characteristic is not coincident with the voltage axis, though the separation is difficult to show because I_{CO} is only a few nanoamperes or microamperes. The region below the $I_E = 0$ characteristic, for which the *emitter and collector junctions are both reverse-biased*, is referred to as the *cutoff* region. The temperature characteristics of I_{CO} are discussed in Sec. 4-7.

4-6   THE COMMON–EMITTER CONFIGURATION

Most transistor circuits have the emitter, rather than the base, as the terminal common to both input and output. Such a *common-emitter* (CE), or *grounded-emitter*, configuration is indicated in Fig. 4-8. In the common-emitter (as in the common-base) configuration, the input current and the output voltage are taken as the independent variables, whereas the input voltage and output current are the dependent variables. Typical output and input characteristic curves for a *p-n-p* junction germanium transistor are given in Figs. 4-9 and 4-10, respectively. In Fig. 4-9 the abscissa is the collector-to-emitter voltage V_{CE}, the ordinate is the collector current I_C, and the curves are given for various values of base current I_B. For a fixed value of I_B, the collector current is not a very sensitive value of V_{CE}. However, the slopes of the curves of Fig. 4-9 are larger than in the common-base characteristics of Fig. 4-6. Observe also that the base current is much smaller than the emitter current.

The locus of all points at which the collector dissipation is 150 mW is indicated in Fig. 4-9 by a solid line $P_C = 150$ mW. This curve is the hyperbola $P_C = V_{CB}I_C \approx V_{CE}I_C =$ constant. To the right of this curve the rated collector dissipation is exceeded. In Fig. 4-9 we have selected $R_L = 500\ \Omega$ and a supply $V_{CC} = 10$ V and have superimposed the corresponding load line on the output characteristics. The method of constructing a load line is identical with that explained in Sec. 3-2 in connection with a diode.

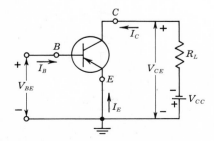

Fig. 4-8 A transistor common-emitter configuration. The symbol V_{CC} is a positive number representing the magnitude of the supply voltage.

Fig. 4-9 Typical common-emitter output characteristics of a *p-n-p* germanium junction transistor. A load line corresponding to $V_{CC} = 10$ V and $R_L = 500$ Ω is superimposed.

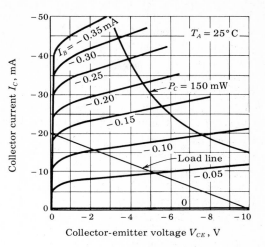

The Input Characteristics   In Fig. 4-10 the abscissa is the base current I_B, the ordinate is the base-to-emitter voltage V_{BE}, and the curves are given for various values of collector-to-emitter voltage V_{CE}.   We observe that, with the collector shorted to the emitter and the emitter forward-biased, the input characteristic is essentially that of a forward-biased diode.   If V_{BE} becomes zero, then I_B will be zero, since under these conditions both emitter and collector junctions will be short-circuited.   In general, increasing $|V_{CE}|$ with constant V_{BE} results in a decreasing recombination base current.   These considerations account for the shape of input characteristics shown in Fig. 4-10.

The input characteristics for silicon transistors are similar in form to those in Fig. 4-10.   The only notable difference in the case of silicon is that the curves break away from zero current in the range 0.5 to 0.6 V, rather than in the range 0.1 to 0.2 V as for germanium.

Fig. 4-10 Typical common-emitter input characteristics of the *p-n-p* germanium junction transistor of Fig. 4-9.

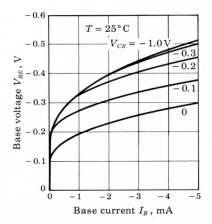

The Output Characteristics   This family of curves may be divided into three regions, just as was done for the CB configuration.   The first of these, the *active region*, is discussed here, and the *cutoff* and *saturation regions* are considered in the next two sections.

In the active region *the collector junction is reverse-biased and the emitter junction is forward-biased*.   In Fig. 4-9 the active region is the area to the right of the ordinate $V_{CE} = $ a few tenths of a volt and above $I_B = 0$.   In this region the transistor output current responds most sensitively to an input signal.   If the transistor is to be used as an amplifying device without appreciable distortion, it must be restricted to operate in this region.

The common-emitter characteristics in the active region are readily understood qualitatively on the basis of our earlier discussion of the common-base configuration.   From Kirchhoff's current law (KCL) applied to Fig. 4-8, the base current is

$$I_B = -(I_C + I_E) \tag{4-9}$$

Combining this equation with Eq. (4-3), we find

$$I_C = \frac{I_{CO}}{1 - \alpha} + \frac{\alpha I_B}{1 - \alpha} \tag{4-10}$$

If we define β by

$$\beta \equiv \frac{\alpha}{1 - \alpha} \tag{4-11}$$

then Eq. (4-10) becomes

$$I_C = (1 + \beta)I_{CO} + \beta I_B \tag{4-12}$$

Note that usually $I_B \gg I_{CO}$, and hence $I_C \approx \beta I_B$ in the active region.

If α were truly constant, then, according to Eq. (4-10), I_C would be independent of V_{CE} and the curves of Fig. 4-9 would be horizontal.   Assume that α increases by only one-half of 1 percent, from 0.98 to 0.985, as $|V_{CE}|$ increases from a few volts to 10 V.   Then the value of β increases from $0.98/(1 - 0.98) = 49$ to $0.985/(1 - 0.985) = 66$,   or   about   34   percent. This numerical example illustrates that a very small change (0.5 percent) in α is reflected in a very large change (34 percent) in the value of β.   It should also be clear that a slight change in α has a large effect on β, and hence upon the common-emitter curves.   Therefore the common-emitter characteristics are normally subject to a wide variation even among transistors of a given type.   This variability is caused by the fact that I_B is the difference between large and nearly equal currents, I_E and I_C.

EXAMPLE   (*a*) Find the transistor currents in the circuit of Fig. 4-11*a*.   A silicon transistor with $\beta = 100$ and $I_{CO} = 20$ $nA = 2 \times 10^{-5}$ mA is under consideration. (*b*) Repeat part *a* if a 2-K emitter resistor is added to the circuit, as in Fig. 4-11*b*.

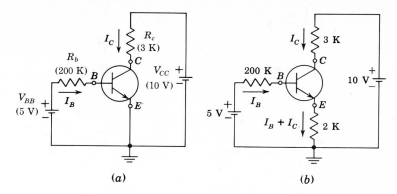

Fig. 4-11 An example illustrating how to determine whether or not a transistor is operating in the active region.

Solution   *a.* Since the base is forward-biased, the transistor is not cut off.   Hence it must be either in its active region or in saturation.   Assume that the transistor operates in the active region.   From KVL applied to the base circuit of Fig. 4-11*a* (with I_B expressed in milliamperes), we have

$$-5 + 200\, I_B + V_{BE} = 0$$

As noted in Table 4-1, a reasonable value for V_{BE} is 0.7 V in the active region. Hence

$$I_B = \frac{5 - 0.7}{200} = 0.0215 \text{ mA}$$

Since $I_{co} \ll I_B$, then $I_C \approx \beta I_B = 2.15$ mA.

We must now justify our assumption that the transistor is in the active region, by verifying that the collector junction is reverse-biased.   From KVL applied to the collector circuit we obtain

$$-10 + 3\, I_C + V_{CB} + V_{BE} = 0$$

or

$$V_{CB} = -(3)(2.15) + 10 - 0.7 = +2.85 \text{ V}$$

For an *n-p-n* device a positive value of V_{CB} represents a reverse-biased collector junction, and hence the transistor is indeed in its active region.

Note that I_B and I_C in the active region are independent of the collector circuit resistance R_C.   Hence, if R_C is increased sufficiently above 3 K, then V_{CB} changes from a positive to a negative value, indicating that the transistor is no longer in its active region.   The method of calculating I_B and I_C when the transistor is in saturation is given in Sec. 4-9.

b. The current in the emitter resistor of Fig. 4-11*b* is

$$I_B + I_C \approx I_B + \beta I_B = 101\, I_B$$

assuming $I_{co} \ll I_B$.   Applying KVL to the base circuit yields

$$-5 + 200 I_B + 0.7 + (2)(101\, I_B) = 0$$

or

$$I_B = 0.0107 \text{ mA} \qquad I_C = 100\, I_B = 1.07 \text{ mA}$$

Note that $I_{CO} = 2 \times 10^{-5}$ mA $\ll I_B$, as assumed.

To check for active circuit operation, we calculate V_{CB}. Thus

$$V_{CB} = -3I_C + 10 - (2)(101 I_B) - 0.7$$

$$= -(3)(1.07) + 10 - (2)(101)(0.0107) - 0.7 = +3.93 \text{ V}$$

Since V_{CB} is positive, this *n-p-n* transistor is in its active region.

4-7   THE CE CUTOFF REGION

We might be inclined to think that cutoff in Fig. 4-9 occurs at the intersection of the load line with the current $I_B = 0$; however, we now find that appreciable collector current may exist under these conditions. From Eqs. (4-9) and (4-10), if $I_B = 0$, then $I_E = -I_C$ and

$$I_C = -I_E = \frac{I_{CO}}{1 - \alpha} \equiv I_{CEO} \tag{4-13}$$

The actual collector current with collector junction reverse-biased and base open-circuited is designated by the symbol I_{CEO}. Since, even in the neighborhood of cutoff, α may be as large as 0.9 for germanium, then $I_C \approx 10I_{CO}$ at zero base current. Accordingly, in order to cut off the transistor, it is not enough to reduce I_B to zero. Instead, it is necessary to reverse-bias the emitter junction slightly. We shall define cutoff as the condition where the collector current is equal to the reverse saturation current I_{CO} and the emitter current is zero. It is found that a reverse-biasing voltage of the order of 0.1 V established across the emitter junction will ordinarily be adequate to cut off a germanium transistor. In silicon, at collector currents of the order of I_{CO}, α is very nearly zero because of recombination in the emitter-junction transition region. Hence, even with $I_B = 0$, we find, from Eq. (4-13), that $I_C = I_{CO} = -I_E$, so that the transistor is still very close to cutoff. We find that, in silicon, cutoff occurs at $V_{BE} \approx 0$ V corresponding to a base short-circuited to the emitter. *In summary, cutoff means that $I_E = 0$, $I_C = I_{CO}$, $I_B = -I_C = -I_{CO}$, and V_{BE} is a reverse voltage whose magnitude is of the order of* 0.1 V *for germanium and* 0 V *for a silicon transistor.*

The Reverse Collector Saturation Current I_{CBO}   The collector current in a physical transistor (a real, nonidealized, or commercial device) when the emitter current is zero is designated by the symbol I_{CBO}. Two factors cooperate to make $|I_{CBO}|$ larger than $|I_{CO}|$. First, there exists a leakage current which flows, not through the junction, but around it and across the surfaces. The leakage current is proportional to the voltage across the junction. The second reason why $|I_{CBO}|$ exceeds $|I_{CO}|$ is that new carriers may be generated by collision in the collector-junction transition region, leading to avalanche multiplication of current and eventual breakdown (Sec. 2-9). But even before breakdown is approached, this *multiplication* component of current may attain considerable proportions.

At 25°C, I_{CBO} for a germanium transistor whose power dissipation is in the range of some hundreds of milliwatts is of the order of microamperes. Under similar conditions a silicon transistor has an I_{CBO} in the range of nano-amperes. The temperature sensitivity of I_{CBO} is the same as that of the reverse saturation current I_o of a *p-n* diode (Sec. 2-4). Specifically, it is found that I_{CBO} approximately doubles for every 10°C increase in temperature for both Ge and Si. However, because of the lower absolute value of I_{CBO} in silicon, these transistors may be used up to about 200°C, whereas germanium transistors are limited to about 100°C.

In addition to the variability of reverse saturation current with tempera-ture, there is also a wide variability of reverse current among samples of a given transistor type. Accordingly, any particular transistor may have an I_{CBO} which differs very considerably from the average characteristic for the type.

Circuit Considerations at Cutoff   Because of temperature effects, ava-lanche multiplication, and the wide variability encountered from sample to sample of a particular transistor type, even silicon may have values of I_{CBO} of the order of many tens of microamperes. Consider the circuit configuration of Fig. 4-12, where V_{BB} represents a biasing voltage intended to keep the tran-sistor cut off. Assume that the transistor is just at the point of cutoff, with $I_E = 0$, so that $I_B = -I_{CBO}$. If we require that at cutoff $V_{BE} \approx -0.1$ V, then the condition of cutoff requires that

$$V_{BE} = -V_{BB} + R_B I_{CBO} \leq -0.1 \text{ V} \qquad (4\text{-}14)$$

As an extreme example consider that R_B is, say, as large as 100 K and that we want to allow for the contingency that I_{CBO} may become as large as 100 μA. Then V_{BB} must be at least 10.1 V. When I_{CBO} is small, the magnitude of the voltage across the base-emitter junction will be 10.1 V. Hence we must use a transistor whose maximum allowable reverse base-to-emitter junction volt-age before breakdown exceeds 10 V. It is with this contingency in mind that a manufacturer supplies a rating for the reverse *breakdown voltage* between emitter and base, represented by the symbol BV_{EBO}. The subscript O indi-cates that BV_{EBO} is measured under the condition that the collector current is

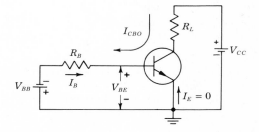

Fig. 4-12   Reverse biasing of the emitter junction to maintain the tran-sistor in cutoff in the presence of the reverse saturation current I_{CBO} through R_B.

zero.   Breakdown voltages BV_{EBO} may be as high as some tens of volts or as low as 0.5 V.   If $BV_{EBO} = 1$ V, then V_{BB} must be chosen to have a maximum value of 1 V.

4-8   THE CE SATURATION REGION

In the saturation region *the collector junction (as well as the emitter junction) is forward-biased by at least the cutin voltage.*   Since the voltage V_{BE} (or V_{BC}) across a forward-biased junction has a magnitude of only a few tenths of a volt, the $V_{CE} = V_{BE} - V_{BC}$ is also only a few tenths of a volt at saturation. Hence, in Fig. 4-9, the saturation region is very close to the zero-voltage axis, where all the curves merge and fall rapidly toward the origin.   A load line has been superimposed on the characteristics of Fig. 4-9 corresponding to a resistance $R_L = 500$ Ω and a supply voltage of 10 V.   We note that in the saturation region the collector current is approximately independent of base current, for given values of V_{CC} and R_L.   Hence we may consider that the onset of saturation takes place at the knee of the transistor curves in Fig. 4-9. Saturation occurs for the given load line at a base current of -0.17 mA, and at this point the collector voltage is too small to be read in Fig. 4-9.   In satu-ration, the collector current is nominally V_{CC}/R_L, and since R_L is small, it may well be necessary to keep V_{CC} correspondingly small in order to stay within the limitations imposed by the transistor on maximum current and dissipation.

We are not able to read the collector-to-emitter saturation voltage, $V_{CE,sat}$, with any precision from the plots of Fig. 4-9.   We refer instead to the char-acteristics shown in Fig. 4-13.   In these characteristics the 0- to -0.5-V region of Fig. 4-9 has been expanded, and we have superimposed the same load line as before, corresponding to $R_L = 500$ Ω.   We observe from Figs. 4-9 and 4-13 that V_{CE} and I_C no longer respond appreciably to base current

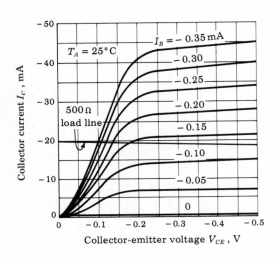

Fig. 4-13 Saturation-region com-mon-emitter characteristics of the type 2N404 germanium transistor. A load line corresponding to V_{CC} = 10 V and R_L = 500 Ω is super-imposed. (Courtesy of Texas In-struments, Inc.)

I_B, after the base current has attained the value -0.15 mA. At this current the transistor enters saturation. For $I_B = -0.15$ mA, $|V_{CE}| \approx 175$ mV. At $I_B = -0.35$ mA, $|V_{CE}|$ has dropped to $|V_{CE}| \approx 100$ mV. Larger magnitudes of I_B will, of course, decrease $|V_{CE}|$ slightly further.

Saturation Resistance For a transistor operating in the saturation region, a quantity of interest is the ratio $V_{CE,\text{sat}}/I_C$. This parameter is called the *common-emitter saturation resistance*, variously abbreviated R_{CS}, R_{CES}, or $R_{CE,\text{sat}}$. To specify R_{CS} properly, we must indicate the operating point at which it was determined. For example, from Fig. 4-13, we find that, at $I_C = -20$ mA and $I_B = -0.35$ mA, $R_{CS} \approx -0.1/(-20 \times 10^{-3}) = 5$ Ω. The usefulness of R_{CS} stems from the fact, as appears in Fig. 4-13, that to the left of the knee each of the plots, for fixed I_B, may be approximated, at least roughly, by a straight line.

The Base-spreading Resistance $r_{bb'}$ Recalling that the base region is very thin (Fig. 4-5), we see that the current which enters the base region across the emitter junction must flow through a long narrow path to reach the base terminal. The cross-sectional area for current flow in the collector (or emitter) is very much larger than in the base. Hence, usually the ohmic resistance of the base is very much larger than that of the collector or emitter [Eq. (1-17)]. The dc ohmic base resistance, designated by $r_{bb'}$, is called the *base-spreading resistance* and is of the order of magnitude of 100 Ω.

The Temperature Coefficient of the Saturation Voltages Since both junctions are forward-biased, a reasonable value for the temperature coefficient of either $V_{BE,\text{sat}}$ or $V_{BC,\text{sat}}$ is -2.5 mV/°C. In saturation the transistor consists of two forward-biased diodes back to back in series opposing. Hence it is to be anticipated that the temperature-induced voltage change in one junction will be canceled by the change in the other junction. We do indeed find such to be the case for $V_{CE,\text{sat}}$, whose temperature coefficient is about one-tenth that of $V_{BE,\text{sat}}$.

The DC Current Gain h_{FE} A transistor parameter of interest is the ratio I_C/I_B, where I_C is the collector current and I_B is the base current. This quantity is designated by β_{dc} or h_{FE}, and is known as the (negative of the) *dc beta*, the *dc forward current transfer ratio*, or the *dc current gain*.

In the saturation region, the parameter h_{FE} is a useful number and one which is usually supplied by the manufacturer when a switching transistor is involved. We know $|I_C|$, which is given approximately by V_{CC}/R_L, and a knowledge of h_{FE} tells us the minimum base current (I_C/h_{FE}) which will be needed to saturate the transistor. For the type 2N404, the variation of h_{FE} with collector current at a low value of V_{CE} is as given in Fig. 4-14. Note the wide spread (a ratio of 3:1) in the value which may be obtained for h_{FE} even for a transistor of a particular type. Commercially available transistors have

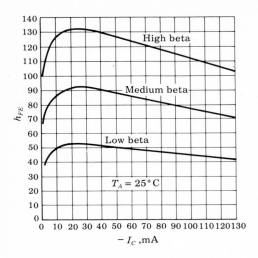

Fig. 4-14   Plots of dc current gain h_{FE} (at $V_{CE} = -0.25$ V) versus collector current for three samples of the type 2N404 germanium transistor.   (Courtesy of General Electric Company.)

values of h_{FE} that cover the range from 10 to 150 at collector currents as small as 5 mA and as large as 30 A.

4-9   TYPICAL TRANSISTOR—JUNCTION VOLTAGE VALUES

The characteristics plotted in Fig. 4-15 of output current I_C as a function of input voltage V_{BE} for n-p-n germanium and silicon transistors are quite instructive and indicate the several regions of operation for a CE transistor circuit.   The numerical values indicated are typical values obtained experimentally or from theoretical equations.   Let us examine the various portions of the transfer curves of Fig. 4-15.

The Cutoff Region   *Cutoff* is defined, as in Sec. 4-7, to mean $I_E = 0$ and $I_C = I_{CO}$, and it is found that a *reverse* bias $V_{BE,\text{cutoff}} = 0.1$ V (0 V) will cut off a germanium (silicon) transistor.

What happens if a larger reverse voltage than $V_{BE,\text{cutoff}}$ is applied?   It turns out that if V_E is reverse-biased and much larger than V_T, that the collector current falls slightly below I_{CO} and that the emitter current *reverses* but remains small in magnitude (less than I_{CO}).

Short-circuited Base   Suppose that, instead of reverse-biasing the emitter junction, we connect the base to the emitter so that $V_E = V_{BE} = 0$.   As indicated in Fig. 4-15, $I_C \equiv I_{CES}$ does not increase greatly over its cutoff value I_{CO}.

Open-circuited Base   If instead of a shorted base we allow the base to "float" so that $I_B = 0$, we obtain the $I_C \equiv I_{CEO}$ given in Eq. (4-13).   At low currents $\alpha \approx 0.9$ (0) for Ge (Si), and hence $I_C \approx 10 I_{CO} (I_{CO})$ for Ge (Si).

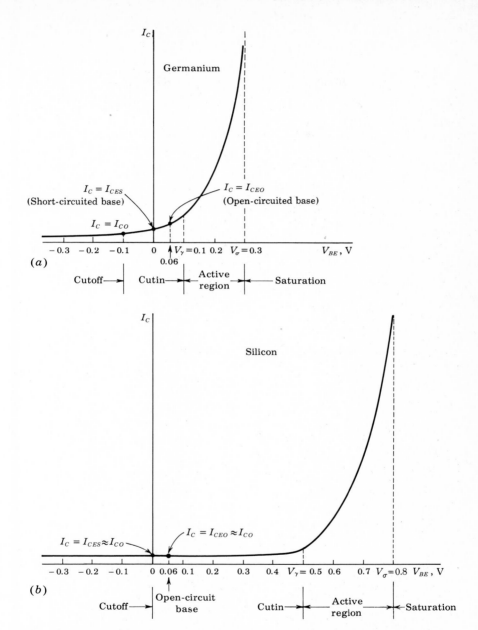

Fig. 4-15   Plots of collector current against base-to-emitter voltage for (a) germanium and (b) silicon n-p-n transistors.   (I_C is not drawn to scale.)

The values of V_{BE} calculated for this open-base condition ($I_C = -I_E$) are a few tens of millivolts of *forward* bias, as indicated in Fig. 4-15.

The Cutin Voltage The volt-ampere characteristic between base and emitter at constant collector-to-emitter voltage (Fig. 4-10) is not unlike the volt-ampere characteristic of a simple junction diode. When the emitter junction is reverse-biased, the base current is very small, being of the order of nanoamperes or microamperes, for silicon and germanium, respectively. When the emitter junction is forward-biased, again, as in the simple diode, no appreciable base current flows until the emitter junction has been forward-biased to the extent where $|V_{BE}| \geq |V_\gamma|$, where V_γ is called the *cutin voltage*. Since the collector current is nominally proportional to the base current, no appreciable collector current will flow until an appreciable base current flows. Therefore a plot of collector current against base-to-emitter voltage will exhibit a cutin voltage, just as does the simple diode.

In principle, a transistor is in its active region whenever the base-to-emitter voltage is on the forward-biasing side of the cutoff voltage, which occurs at a reverse voltage of 0.1 V for germanium and 0 V for silicon. In effect, however, a transistor enters its active region when $V_{BE} > V_\gamma$.

We may estimate the cutin voltage V_γ by assuming that $V_{BE} = V_\gamma$ when the collector current reaches, say, 1 percent of the maximum (saturation) current in the CE circuit of Fig. 4-8. Typical values of V_γ are 0.1 V for germanium and 0.5 V for silicon.

Figure 4-16 shows plots, for several temperatures, of the collector current as a function of the base-to-emitter voltage at constant collector-to-emitter voltage for a typical silicon transistor. We see that a value for V_γ of the order of 0.5 V at room temperature is entirely reasonable. The temperature dependence results from the temperature coefficient of the emitter-junction diode. Therefore the lateral shift of the plots with change in temperature and the change with temperature of the cutin voltage V_γ are approximately -2.5 mV/°C [Eq. (2-5)].

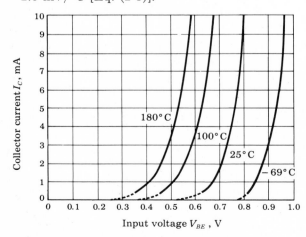

Fig. 4-16 Plot of collector current against base-to-emitter voltage for various temperatures for the type 2N337 silicon transistor. (Courtesy of Transitron Electronic Corporation.)

TABLE 4-1   Typical n-p-n transistor-junction voltages at 25°C

	$V_{CE,\text{sat}}$	$V_{BE,\text{sat}} \equiv V_\sigma$	$V_{BE,\text{active}}$	$V_{BE}\dagger,\text{cutin}} \equiv V_\gamma$	$V_{BE,\text{cutoff}}$
Si	0.2	0.8	0.7	0.5	0.0
Ge	0.1	0.3	0.2	0.1	−0.1

† The temperature variation of these voltages is discussed in Secs. 4-8 and 4-9.

Saturation Voltages   Manufacturers specify saturation values of input and output voltages in a number of different ways, in addition to supplying characteristic curves such as Figs. 4-10 and 4-13. For example, they may specify R_{CS} for several values of I_B or they may supply curves of $V_{CE,\text{sat}}$ and $V_{BE,\text{sat}}$ as functions of I_B and I_C. The saturation voltages depend not only on the operating point, but also on the semiconductor material (germanium or silicon) and on the type of transistor construction. Typical values of saturation voltages are indicated in Table 4-1.

Summary   The voltages referred to above and indicated in Fig. 4-15 are summarized in Table 4-1. The entries in the table are appropriate for an n-p-n transistor. For a p-n-p transistor the signs of all entries should be reversed. Observe that the total range of V_{BE} between cutin and saturation is rather small, being only 0.3 V. The voltage $V_{BE,\text{active}}$ has been located somewhat arbitrarily, but nonetheless reasonably, near the midpoint of the active region in Fig. 4-15.

Of course, particular cases will depart from the estimates of Table 4-1. But it is unlikely that the numbers will be found in error by more than 0.1 V.

EXAMPLE   (a) The circuits of Fig. 4-11a and b are modified by changing the base-circuit resistance from 200 to 50 K (as indicated in Fig. 4-17). If $h_{FE} = 100$, determine whether or not the silicon transistor is in saturation and find I_B and I_C. (b) Repeat with the 2-K emitter resistance added.

Solution   Assume that the transistor is in saturation. Using the values $V_{BE,\text{sat}}$ and $V_{CE,\text{sat}}$ in Table 4-1, the circuit of Fig. 4-17a is obtained. Applying KVL to the base circuit gives

$$-5 + 50\, I_B + 0.8 = 0$$

or

$$I_B = \frac{4.2}{50} = 0.084 \text{ mA}$$

Applying KVL to the collector circuit yields

$$-10 + 3\, I_C + 0.2 = 0$$

or

$$I_C = \frac{9.8}{3} = 3.27 \text{ mA}$$

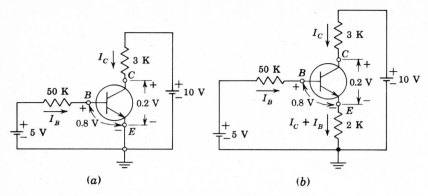

Fig. 4-17 An example illustrating how to determine whether or not a transistor is operating in the saturation region.

The minimum value of base current required for saturation is

$$(I_B)_{\min} = \frac{I_C}{h_{FE}} = \frac{3.27}{100} = 0.033 \text{ mA}$$

Since $I_B = 0.084 > I_{B,\min} = 0.033$ mA, we have verified that the transistor is in saturation.

b. If the 2-K emitter resistance is added, the circuit becomes that in Fig. 4-17b. Assume that the transistor is in saturation. Applying KVL to the base and collector circuits, we obtain

$$-5 + 50 \, I_B + 0.8 + 2(I_C + I_B) = 0$$
$$-10 + 3 \, I_C + 0.2 + 2(I_C + I_B) = 0$$

If these simultaneous equations are solved for I_C and I_B, we obtain

$$I_C = 1.95 \text{ mA} \qquad I_B = 0.0055 \text{ mA}$$

Since $(I_B)_{\min} = I_C/h_{FE} = 0.0195$ mA $> I_B = 0.0055$, the transistor is *not* in saturation. Hence the device must be operating in the active region. Proceeding exactly as we did for the circuit of Fig. 4-11b (but with the 200 K replaced by 50 K), we obtain

$$I_C = 1.71 \text{ mA} \qquad I_B = 0.0171 \text{ mA} = 17 \text{ } \mu\text{A} \qquad V_{CB} = 0.72 \text{ V}$$

4-10   COMMON–EMITTER CURRENT GAIN

Three different definitions of current gain appear in the literature. The interrelationships between these are now to be found.

Large-Signal Current Gain β We define β in terms of α by Eq. (4-11). From Eq. (4-12), with I_{CO} replaced by I_{CBO}, we find

$$\beta = \frac{I_C - I_{CBO}}{I_B - (-I_{CBO})} \tag{4-15}$$

In Sec. 4-7 we define *cutoff* to mean that $I_E = 0$, $I_C = I_{CBO}$, and $I_B = -I_{CBO}$. Consequently, Eq. (4-15) gives the ratio of the collector-current increment to the base-current change from cutoff to I_B, and hence β *represents the* (negative of the) *large-signal current gain of a common-emitter transistor.*   This parameter is of primary importance in connection with the biasing and stability of transistor circuits, as discussed in Chap. 10.

DC Current Gain h_{FE} In Sec. 4-8 we define the dc current gain by

$$\beta_{\text{dc}} \equiv \frac{I_C}{I_B} \equiv h_{FE} \tag{4-16}$$

In that section it is noted that h_{FE} is most useful in connection with determining whether or not a transistor is in saturation.   In general, the base current (and hence the collector current) is large compared with I_{CBO}.   Under these conditions the large-signal and the dc betas are approximately equal; then $h_{FE} \approx \beta$.

Small-Signal Current Gain h_{fe} We define β' as the ratio of a collector-current increment ΔI_C for a small base-current change ΔI_B (at a given quiescent operating point, at a fixed collector-to-emitter voltage V_{CE}), or

$$\beta' \equiv \left. \frac{\partial I_C}{\partial I_B} \right|_{V_{CE}} = h_{fe} \tag{4-17}$$

Clearly, β' is (the negative of) the *small-signal* current gain.   If β were independent of current, we see from Eq. (4-16) that $\beta' = \beta \approx h_{FE}$.   However, Fig. 4-14 indicates that β is a function of current.   Over most of the wide current range in Fig. 4-14, h_{fe} differs from h_{FE} by less than 20 percent.

It should be emphasized that $h_{fe} \approx h_{FE}$ is valid in the active region only. From Fig. 4-13 we see that $h_{fe} \rightarrow 0$ in the saturation region because $\Delta I_C \rightarrow 0$ for a small increment ΔI_B.

REVIEW QUESTIONS

4-1   Draw the circuit symbol for a *p-n-p* transistor and indicate the reference directions for the three currents and the reference polarities for the three voltages.

4-2   Repeat Rev. 4-1 for an *n-p-n* transistor.

4-3   For a *p-n-p* transistor biased in the active region, plot (in each region *E*, *B*, and *C*) (*a*) the potential variation; (*b*) the minority-carrier concentration.   (*c*) Explain the shapes of the plots in (*a*) and (*b*).

4-4 (a) For a p-n-p transistor biased in the active region, indicate the various electron and hole current components crossing each junction and entering (or leaving) the base terminal. (b) Which of the currents is proportional to the gradient of p_n at J_E and J_C, respectively? (c) Repeat part b with p_n replaced by n_p. (d) What is the physical origin of the several current components crossing the base terminal?

4-5 (a) From the currents indicated in Rev. 4-4 obtain an expression for the collector current I_C. Define each symbol in this equation. (b) Generalize the equation for I_C in part a so that it is valid even if the transistor is not operating in its active region.

4-6 (a) Define the *current gain* α in words and as an equation. (b) Repeat part a for the parameter α'.

4-7 Describe the fabrication of an alloy transistor.

4-8 For a p-n-p transistor in the active region, what is the sign (positive or negative) of I_E, I_C, I_B, V_{CB}, and V_{EB}?

4-9 Repeat Rev. 4-8 for an n-p-n transistor.

4-10 (a) Sketch a family of CB output characteristics for a transistor. (b) Indicate the active, cutoff, and saturation regions. (c) Explain the shapes of the curves qualitatively.

4-11 (a) Sketch a family of CB input characteristics for a transistor. (b) Explain the shapes of the curves qualitatively.

4-12 Define the following regions in a transistor: (a) active; (b) saturation; (c) cutoff.

4-13 (a) Draw the circuit of a transistor in the CE configuration. (b) Sketch the output characteristics. (c) Indicate the active, saturation, and cutoff regions.

4-14 (a) Sketch a family of CE input characteristics. (b) Explain the shape of these curves qualitatively.

4-15 (a) Derive the expression for I_C versus I_B for a CE transistor configuration in the active region. (b) For $I_B = 0$, what is I_C?

4-16 (a) What is the order of magnitude of the reverse collector saturation current I_{CBO} for a silicon transistor? (b) How does I_{CBO} vary with temperature?

4-17 Repeat Rev. 4-16 for a germanium transistor.

4-18 Why does I_{CBO} differ from I_{CO}?

4-19 (a) Define *saturation resistance* for a CE transistor. (b) Give its order of magnitude.

4-20 (a) Define *base-spreading resistance* for a transistor. (b) Give its order of magnitude.

4-21 What is the order of magnitude of the temperature coefficients of $V_{BE,\text{sat}}$, $V_{BC,\text{sat}}$, and $V_{CE,\text{sat}}$?

4-22 (a) Define h_{FE}. (b) Plot h_{FE} versus I_C.

4-23 (a) Give the order of magnitude of V_{BE} at cutoff for a silicon transistor. (b) Repeat part a for a germanium transistor. (c) Repeat parts a and b for the cutin voltage.

4-24 Is $|V_{BE,\text{sat}}|$ greater or less than $|V_{CE,\text{sat}}|$? Explain.

4-25 (a) What is the range in volts for V_{BE} between cutin and saturation for a silicon transistor? (b) Repeat part a for a germanium transistor.

4-26 What is the collector current relative to I_{CO} in a silicon transistor (a) if the base is shorted to the emitter? (b) If the base floats? (c) Repeat parts a and b for a germanium transistor.

4-27 Consider a transistor circuit with resistors R_b, R_c, and R_e in the base, collector, and emitter legs, respectively. The biasing voltages are V_{BB} and V_{CC} in base and collector circuits, respectively. (a) Outline the method for finding the quiescent currents, assuming that the transistor operates in the active region. (b) How do you test to see if your assumption is correct?

4-28 Repeat Rev. 4-27, assuming that the transistor is in saturation.

4-29 For a CE transistor define (in words and symbols) (a) β; (b) $\beta_{dc} = h_{FE}$; (c) $\beta' = h_{fe}$.

5 / INTEGRATED CIRCUITS: FABRICATION AND CHARACTERISTICS

An integrated circuit consists of a single-crystal chip of silicon, typically 50 by 50 mils in cross section, containing both active and passive elements and their interconnections. Such circuits are produced by the same processes used to fabricate individual transistors and diodes. These processes include epitaxial growth, masked impurity diffusion, oxide growth, and oxide etching, using photolithography for pattern definition. A method of batch processing is employed which offers excellent repeatability and is adaptable to the production of large numbers of integrated circuits at low cost. In this chapter we describe the basic processes involved in fabricating an integrated circuit.

5-1   INTEGRATED–CIRCUIT TECHNOLOGY

The fabrication of integrated circuits is based on materials, processes, and design principles which constitute a highly developed semiconductor (planar-diffusion) technology. The basic structure of an integrated circuit is shown in Fig. 5-1b, and consists of four distinct layers of material. The bottom layer ① (6 mils thick) is p-type silicon and serves as a *substrate* upon which the integrated circuit is to be built. The second layer ② is thin (typically 25 μm = 1 mil) n-type material which is grown as a single-crystal extension of the substrate. This process is called *epitaxial growth*. All active and passive components are built within the thin n-type layer using a series of diffusion steps. These components are transistors, diodes, capacitors, and resistors, and they are made by diffusing p-type and n-type impurities.   The most complicated component fabricated is the transistor, and all other elements are constructed with one or more of the processes required to make a transistor.   In the fabrication of all the above elements it is necessary to distribute impurities in certain precisely defined regions within the second (n-type) layer.   The selective diffusion of impurities is accomplished by using SiO_2 as a barrier which protects portions of the wafer against impurity penetration.   Thus the third layer of material ③ is silicon dioxide, and it also provides protection of the semiconductor surface against contamination. In the regions where diffusion is to take place, the SiO_2 layer is etched away, leaving the rest of the wafer protected against diffusion.   To permit selective etching, the SiO_2 layer must be subjected to a photolithographic process, described in Sec. 5-4.   Finally, a fourth metallic (aluminum) layer ④ is added to supply the necessary interconnections between components.

The p-type substrate which is required as a foundation for the integrated circuit is obtained by growing an ingot (1 to 2 in. in diameter and about 10 in. long) from a silicon melt with a predetermined number of impurities.   The crystal ingot is subsequently sliced into round wafers approximately 6 mils thick, and one side of each wafer is lapped and polished to eliminate surface imperfections.

We are now in a position to appreciate some of the significant advantages

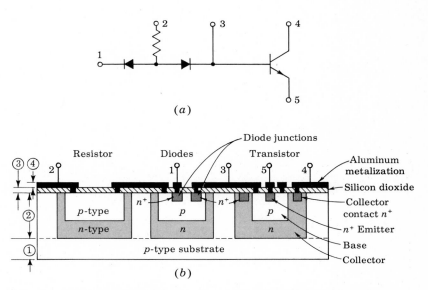

(a)

(b)

Fig. 5-1   (a) A circuit containing a resistor, two diodes, and a transistor.   (b) Cross-sectional view of the circuit in (a) when transformed into a monolithic form (not drawn to scale).   The four layers are ① substrate, ② n-type crystal containing the integrated circuit, ③ silicon dioxide, and ④ aluminum metalization.   (Not drawn to scale.)

of the integrated-circuit technology.   Let us consider a 1-in.-square wafer divided into 400 chips of surface area 50 by 50 mils.   We demonstrate in this chapter that a reasonable area under which a component (say, a transistor) is fabricated is 50 mils^2.   Hence each chip (each integrated circuit) contains 50 separate components, and there are $50 \times 400 = 20,000$ components/in.^{2} on each wafer.

If we process 10 wafers in a batch, we can manufacture 4,000 integrated circuits simultaneously, and these contain 200,000 components.   Some of the chips will contain faults due to imperfections in the manufacturing process, but the *yield* (the percentage of fault-free chips per wafer) is extremely large.

The following advantages are offered by integrated-circuit technology as compared   with   discrete   components   interconnected   by   conventional techniques:

1. Low cost (due to the large quantities processed).
2. Small size.
3. High reliability.   (All components are fabricated simultaneously, and there are no soldered joints.)
4. Improved performance.   (Because of the low cost, more complex circuitry may be used to obtain better functional characteristics.)

In the next sections we examine the processes required to fabricate an integrated circuit.

5-2   BASIC MONOLITHIC INTEGRATED CIRCUITS

We now examine in some detail the various techniques and processes required to obtain the circuit of Fig. 5-1a in an integrated form, as shown in Fig. 5-1b. This configuration is called a monolithic integrated circuit because it is formed on a single silicon chip.   The word "monolithic" is derived from the Greek *monos*, meaning "single," and *lithos*, meaning "stone."   Thus a monolithic circuit is built into a single stone, or single crystal.

In this section we describe qualitatively a complete epitaxial-diffused fabrication process for integrated circuits.   The logic circuit of Fig. 5-1a is chosen for discussion because it contains typical components: a resistor, diodes, and a transistor.   These elements (and also capacitors with small values of capacitances) are the components encountered in integrated circuits.   The monolithic circuit is formed by the steps indicated in Fig. 5-2 and described below.

Step 1.   Epitaxial Growth   An n-type epitaxial layer, typically 25 μm thick, is grown onto a p-type substrate which has a resistivity of typically 10 Ω-cm, corresponding to $N_A = 1.4 \times 10^{15}$ atoms/cm^3. The resistivity of the n-type epitaxial layer can be chosen independently of that of the substrate.

Values of 0.1 to 0.5 Ω-cm are chosen for the n-type layer. After polishing and cleaning, a thin layer (0.5 μm = 5000 Å) of oxide, SiO_2, is formed over the entire wafer, as shown in Fig. 5-2a. The SiO_2 is grown by exposing the epitaxial layer to an oxygen atmosphere while being heated to about 1000°C. Silicon dioxide has the fundamental property of preventing the diffusion of impurities through it. Use of this property is made in the following steps.

Step 2.   Isolation Diffusion   In Fig. 5-2b the wafer is shown with the oxide removed in four different places on the surface. This removal is accomplished by means of a photolithographic etching process described in Sec. 5-4. The remaining SiO_2 serves as a mask for the diffusion of acceptor impurities (in this case, boron). The wafer is now subjected to the so-called *isolation diffusion*, which takes place at the temperature and for the time interval required for the p-type impurities to penetrate the n-type epitaxial layer and reach the p-type substrate. We thus leave the shaded n-type regions in Fig. 5-2b. These sections are called *isolation islands*, or *isolated regions*, because they are separated by two back-to-back p-n junctions. Their purpose is to allow electrical isolation between different circuit components. For example, it will become apparent later in this section that a different isolation region must be used for the collector of each separate transistor. The p-type substrate must always be held at a negative potential with respect to the isolation islands in order that the p-n junctions be reverse-biased. If these diodes were to become forward-biased in an operating circuit, then, of course, the isolation would be lost.

It should be noted that the concentration of acceptor atoms ($N_A \approx 5 \times 10^{20}$ cm^{-3}) in the region between isolation islands will generally be much higher (and hence indicated as p^+) than in the p-type substrate. The reason for this higher density is to prevent the depletion region of the reverse-biased isolation-to-substrate junction from extending into p^+-type material (Sec. 2-6) and possibly connecting two isolation islands.

Parasitic Capacitance   It is now important to consider that these isolation regions, or junctions, are connected by a significant barrier, or transition capacitance C_{Ts}, to the p-type substrate, which capacitance can affect the operation of the circuit. Since C_{Ts} is an undesirable by-product of the isolation process, it is called the *parasitic capacitance*.

The parasitic capacitance is the sum of two components, the capacitance C_1 from the bottom of the n-type region to the substrate (Fig. 5-2b) and C_2 from the sidewalls of the isolation islands to the p^+ region. The bottom component, C_1, results from an essentially step junction due to the epitaxial growth (Sec. 5-3), and hence varies inversely as the square root of the voltage V between the isolation region and the substrate (Sec. 2-6). The sidewall capacitance C_2 is associated with a diffused graded junction, and it varies as $V^{-\frac{1}{3}}$. For this component the junction area is equal to the perimeter of the isolation region times the thickness y of the epitaxial n-type layer. The total capacitance is of the order of a few picofarads.

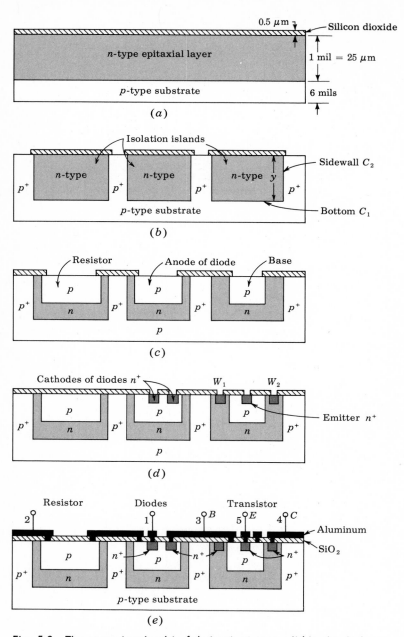

Fig. 5-2 The steps involved in fabricating a monolithic circuit (not drawn to scale).   (a) Epitaxial growth; (b) isolation diffusion; (c) base diffusion; (d) emitter diffusion; (e) aluminum metalization.

Step 3.   Base Diffusion   During this process a new layer of oxide is formed over the wafer, and the photolithographic process is used again to create the pattern of openings shown in Fig. 5-2c.   The p-type impurities (boron) are diffused through these openings.   In this way are formed the transistor base regions as well as resistors, the anode of diodes, and junction capacitors (if any).   It is important to control the depth of this diffusion so that it is shallow and does not penetrate to the substrate.   The resistivity of the base layer will generally be much higher than that of the isolation regions.

Step 4.   Emitter Diffusion   A layer of oxide is again formed over the entire surface, and the masking and etching processes are used again to open windows in the p-type regions, as shown in Fig. 5-2d.   Through these openings are diffused n-type impurities (phosphorus) for the formation of transistor emitters, the cathode regions for diodes, and junction capacitors.

Additional windows (such as W_1 and W_2 in Fig. 5-2d) are often made into the n regions to which a lead is to be connected, using aluminum as the ohmic contact, or interconnecting metal.   During the diffusion of phosphorus a heavy concentration (called n^+) is formed at the points where contact with aluminum is to be made.   Aluminum is a p-type impurity in silicon, and a large concentration of phosphorus prevents the formation of a p-n junction when the aluminum is alloyed to form an ohmic contact.

Step 5.   Aluminum Metalization   All p-n junctions and resistors for the circuit of Fig. 5-1a have been formed in the preceding steps.   It is now necessary to interconnect the various components of the integrated circuit as dictated by the desired circuit.   To make these connections, a fourth set of windows is opened into a newly formed SiO_2 layer, as shown in Fig. 5-2e, at the points where contact is to be made.   The interconnections are made first, using vacuum deposition of a thin even coating of aluminum over the entire wafer. The photoresist technique is now applied to etch away all undesired aluminum areas, leaving the desired pattern of interconnections shown in Fig. 5-2e between resistors, diodes, and transistors.

In production a large number (several hundred) of identical circuits such as that of Fig. 5-1a are manufactured simultaneously on a single wafer.   After the metalization process has been completed, the wafer is scribed with a diamond-tipped tool and separated into individual chips.   Each chip is then mounted on a ceramic wafer and is attached to a suitable header.   The package leads are connected to the integrated circuit by stitch bonding of a 1-mil aluminum or gold wire from the terminal pad on the circuit to the package lead.

Summary   In this section the epitaxial-diffused method of fabricating integrated circuits is described.   We have encountered the following processes:

1. Crystal growth of a substrate
2. Epitaxy
3. Silicon dioxide growth
4. Photoetching
5. Diffusion
6. Vacuum evaporation of aluminum

Using these techniques, it is possible to produce the following elements on the same chip: transistors, diodes, resistors, capacitors, and aluminum interconnections.

5-3   EPITAXIAL GROWTH

The epitaxial process produces a thin film of single-crystal silicon from the gas phase upon an existing crystal wafer of the same material. The epitaxial layer may be either p-type or n-type. The growth of an epitaxial film of pure silicon is obtained from the hydrogen reduction of silicon tetrachloride at 1200°C. The substrate wafers are placed in an oven containing these chemicals, and the crystal continues to grow by capturing the silicon atoms released in the chemical reaction. Since it is required to produce epitaxial films of specific impurity concentrations, it is necessary to introduce impurities such as phosphine for n-type doping or biborane for p-type doping into the silicon tetrachloride–hydrogen gas stream. The junction formed in this manner is an approximately abrupt step p-n junction similar to that shown in Fig. 2-7.

5-4   MASKING AND ETCHING

The monolithic technique described in Sec. 5-2 requires the selective removal of the SiO$_2$ to form openings through which impurities may be diffused. The photoetching method used for this removal is illustrated in Fig. 5-3. During the photolithographic process the wafer is coated with a uniform film of a photosensitive emulsion (such as the Kodak *photoresist* KPR). A large black-and-white layout of the desired pattern of openings is made and then reduced photographically. This negative, or stencil, of the required dimensions is placed as a mask over the photoresist, as shown in Fig. 5-3a. By exposing the KPR to ultraviolet light through the mask, the photoresist becomes polymerized under the transparent regions of the stencil. The mask is now removed, and the wafer is "developed" by using a chemical (such as trichloroethylene) which dissolves the unexposed (unpolymerized) portions of the photoresist film and leaves the surface pattern as shown in Fig. 5-3b. The emulsion which was not removed in development is now *fixed*, or *cured*, so that it becomes resistant to the corrosive etches used next. The chip is im-

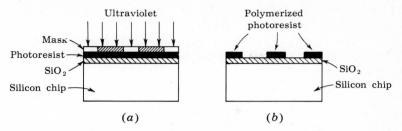

Fig. 5-3   Photoetching technique.   (a) Masking and exposure to
ultraviolet radiation.   (b) The photoresist after development.

mersed in an etching solution of hydrofluoric acid, which removes the oxide
from the areas through which dopants are to be diffused.   Those portions of
the SiO₂ which are protected by the photoresist are unaffected by the acid.
After etching and diffusion of impurities, the resist mask is removed (stripped)
with a chemical solvent (hot H_2SO_4) and by means of a mechanical abrasion
process.

5-5   DIFFUSION OF IMPURITIES

The most important process in the fabrication of integrated circuits is
the diffusion of impurities into the silicon chip.   Reasonable diffusion times
(say, 2 hours) require high diffusion temperatures ($\sim 1000°C$).   Therefore a
high-temperature diffusion furnace, having a closely controlled temperature
over the length (20 in.) of the hot zone of the furnace, is standard equipment
in a facility for the fabrication of integrated circuits.   Impurity sources used

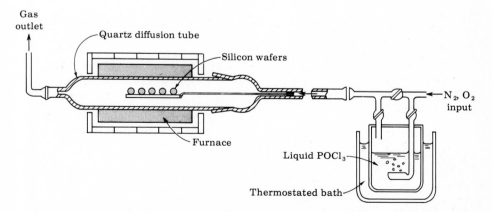

Fig. 5-4   Schematic representation of typical apparatus for POCl₃ diffusion.
(Courtesy of Motorola, Inc.)

in connection with diffusion furnaces can be gases, liquids, or solids. For example, $POCl_3$, which is a liquid, is often used as a source of phosphorus. Figure 5-4 shows the apparatus used for $POCl_3$ diffusion. In this apparatus a carrier gas (mixture of nitrogen and oxygen) bubbles through the liquid-diffusant source and carries the diffusant atoms to the silicon wafers.

5-6   TRANSISTORS FOR MONOLITHIC CIRCUITS

A planar transistor made for monolithic integrated circuits, using epitaxy and diffusion, is shown in Fig. 5-5a. Here the collector is electrically separated from the substrate by the reverse-biased isolation diodes. Since the anode of the isolation diode covers the back of the entire wafer, it is necessary to make the collector contact on the top, as shown in Fig. 5-5a. It is now clear that the isolation diode of the integrated transistor has two undesirable effects: it adds a parasitic shunt capacitance to the collector and a leakage current path. In addition, the necessity for a top connection for the collector increases the collector-current path and thus increases the collector resistance and $V_{CE,\text{sat}}$. All these undesirable effects are absent from the discrete epitaxial transistor shown in Fig. 5-5b. What is then the advantage of the monolithic transistor? A significant improvement in performance arises from the fact that integrated transistors are located physically close together and their electrical characteristics are closely matched. For example, integrated transistors spaced within 30 mils (0.03 in.) have V_{BE} matching of better than 5 mV with less than 10 μV/°C drift and an h_{FE} match of ± 10 percent. These matched transistors make excellent difference amplifiers (Sec. 12-12).

The electrical characteristics of a transistor depend on the size and geometry of the transistor, doping levels, diffusion schedules, and the basic silicon

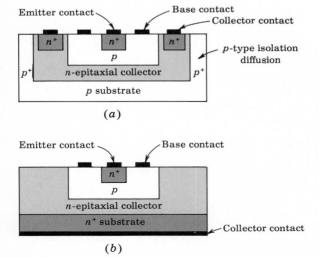

(a)

(b)

Fig. 5-5 Comparison of cross sections of (a) a monolithic integrated circuit transistor with (b) a discrete planar epitaxial transistor. [For a top view of the transistor in (a) see Fig. 5-7.]

material.   Of all these factors the size and geometry offer the greatest flexibility for design.   The doping levels and diffusion schedules are determined by the standard processing schedule used for the desired transistors in the integrated circuit.

Impurity Profiles for Integrated Transistors   Figure 5-6 shows a typical impurity profile for a monolithic integrated circuit transistor.   The background, or epitaxial-collector, concentration N_{BC} is shown as a dashed line in Fig. 5-6.   The concentration N of boron is high (5×10^{18} atoms/cm³) at the surface and falls off with distance into the silicon as indicated in Fig. 5-6. At that distance $x = x_j$, at which N equals the concentration N_{BC}, the net impurity density is zero. For $x < x_j$, the net impurity concentration is positive, and for $x > x_j$, it is negative.   Hence x_j represents the distance from the surface at which the collector junction is formed.   For the transistor whose impurity profile is indicated in Fig. 5-6, $x_j = 2.7 \ \mu m$.

The emitter diffusion (phosphorus) starts from a much higher surface concentration (close to the solid solubility) of about 10^{21} atoms/cm³, and is diffused to a depth of 2 μm, where the emitter junction is formed.   This junction corresponds to the intersection of the base and emitter distributions of impurities.   We now see that the base thickness for this monolithic transistor is 0.7 μm.   The emitter-to-base junction is usually treated as a step-graded junction, whereas the base-to-collector junction is considered a linearly graded junction.

Monolithic Transistor Layout   The physical size of a transistor determines the parasitic isolation capacitance as well as the junction capacitance.

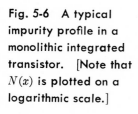

Fig. 5-6 A typical impurity profile in a monolithic integrated transistor.   [Note that $N(x)$ is plotted on a logarithmic scale.]

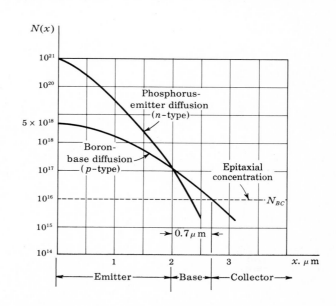

It is therefore necessary to use small-geometry transistors if the integrated circuit is designed to operate at high frequencies or high switching speeds. The geometry of a typical monolithic transistor is shown in Fig. 5-7. The emitter rectangle measures 1 by 1.5 mils, and is diffused into a 2.5- by 4.0-mil base region. Contact to the base is made through two metalized stripes on either side of the emitter. The rectangular metalized area forms the ohmic contact to the collector region. The rectangular collector contact of this transistor reduces the saturation resistance. The substrate in this structure is located about 1 mil below the surface. Since diffusion proceeds in three dimensions, it is clear that the *lateral-diffusion* distance will also be 1 mil. The dashed rectangle in Fig. 5-7 represents the substrate area and is 6.5 by 8 mils.

Lateral *p-n-p* **Transistor** The standard integrated-circuit transistor is an *n-p-n* type, as we have already emphasized. In some applications it is required to have both *n-p-n* and *p-n-p* transistors on the same chip. The lateral *p-n-p* structure shown in Fig. 5-8 is the most common form of the integrated *p-n-p* transistor. This *p-n-p* uses the standard diffusion techniques as the *n-p-n*, but the last *n* diffusion (used for the *n-p-n* transistor) is eliminated.

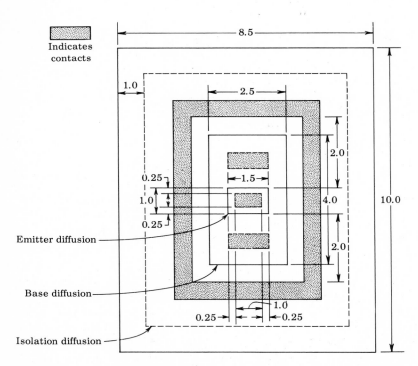

Fig. 5-7 A typical double-base stripe geometry of an integrated-circuit transistor. Dimensions are in mils. (For a side view of the transistor see Fig. 5-5.) (Courtesy of Motorola Monitor.)

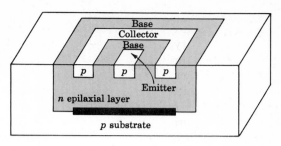

Fig 5-8 A *p-n-p* lateral transistor.

While the p base for the n-p-n transistor is made, the two adjacent p regions are diffused for the emitter and collector of the p-n-p transistor shown in Fig. 5-8. Note that the current flows *laterally* from emitter to collector. Because of inaccuracies in masking, and because, also, of lateral diffusion, the base width between emitter and collector is large (about 1 mil compared with 1 μm for an n-p-n base). Hence the current gain of the p-n-p transistor is very low (0.5 to 5) instead of 50 to 300 for the n-p-n device. Since the base-p resistivity of the n-p-n transistor is relatively high, the collector and emitter resistances of the p-n-p device are high.

Vertical p-n-p Transistor This transistor uses the substrate for the p collector; the n epitaxial layer for the base; and the p base of the standard n-p-n transistor as the emitter of this p-n-p device. We have already emphasized that the substrate must be connected to the most negative potential in the circuit. Hence a vertical p-n-p transistor can be used only if its collector is at a fixed negative voltage. Such a configuration is called an *emitter follower*, and is discussed in Sec. 10-4.

5-7   MONOLITHIC DIODES

The diodes utilized in integrated circuits are made by using transistor structures in one of five possible connections (Prob. 5-4). The three most popular diode structures are shown in Fig. 5-9. They are obtained from a transistor structure by using the emitter-base diode, with the collector short-circuited to the base (*a*); the emitter-base diode, with the collector open (*b*); and the collector-base diode, with the emitter open-circuited (or not fabricated at all) (*c*). The choice of the diode type used depends upon the application and circuit performance desired. Collector-base diodes have the higher collector-base voltage-breakdown rating of the collector junction ($\sim$12 V minimum).

The emitter-base diffusion is very popular for the fabrication of diodes provided that the reverse-voltage requirement of the circuit does not exceed the lower base-emitter breakdown voltage ($\sim$7 V). Common-anode arrays can easily be made with the emitter-base diffusion by using a multiple-emitter transistor within a single isolation area, as shown in Fig. 5-10. The collector

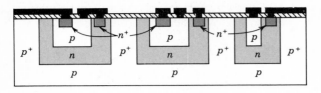

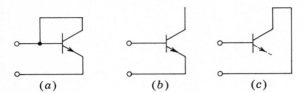

Fig. 5-9 Cross section of various diode structures. (a) Emitter-base diode with collector shorted to base; (b) emitter-base diode with collector open; (c) collector-base diode (no emitter diffusion).

may be either open or shorted to the base. The diode pair in Fig. 5-1 is constructed in this manner, with the collector floating (open). The diode-connected transistor (emitter-base diode with collector short-circuited to the base) provides the highest conduction for a given forward voltage.

5-8 INTEGRATED RESISTORS

A resistor in a monolithic integrated circuit is very often obtained by utilizing the bulk resistivity of one of the diffused areas. The p-type base diffusion is most commonly used, although the n-type emitter diffusion is also employed. Since these diffusion layers are very thin, it is convenient to define a quantity known as the *sheet resistance* R_S.

Sheet Resistance If, in Fig. 5-11, the width w equals the length l, we have a square l by l of material with resistivity ρ, thickness y, and cross-sectional area $A = ly$. The resistance of this conductor (in ohms per square) is

$$R_S = \frac{\rho l}{ly} = \frac{\rho}{y} \tag{5-1}$$

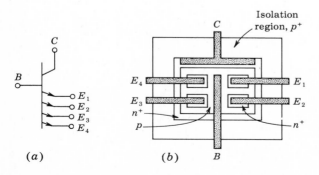

Fig. 5-10 A multiple-emitter n-p-n transistor. (a) Schematic, (b) monolithic surface pattern. If the base is connected to the collector, the result is a multiple-cathode diode structure with a common anode.

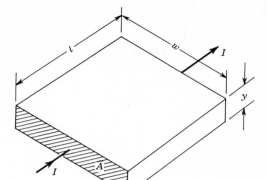

Fig. 5-11   Pertaining to sheet resistance, ohms per square.

Note that R_s is independent of the size of the square.   Typically, the sheet resistance of the base and emitter diffusions whose profiles are given in Fig. 5-6 is 200 Ω/square and 2.2 Ω/square, respectively.

The construction of a base-diffused resistor is shown in Fig. 5-1 and is repeated in Fig. 5-12a.   A top view of this resistor is shown in Fig. 5-12b. The resistance value may be computed from

$$R = \frac{\rho l}{yw} = R_s \frac{l}{w} \tag{5-2}$$

where l and w are the length and width of the diffused area, as shown in the top view.   For example, a base-diffused-resistor stripe 1 mil wide and 10 mils long contains 10 (1 by 1 mil) squares, and its value is $10 \times 200 = 2,000\ \Omega$. Empirical corrections for the end contacts are usually included in calculations of R.

Resistance Values   Since the sheet resistance of the base and emitter diffusions is fixed, the only variables available for diffused-resistor design are

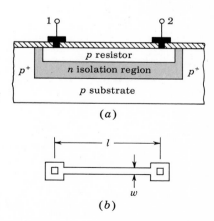

Fig. 5-12   A monolithic resistor.   (a) Cross-sectional view; (b) top view.

stripe length and stripe width. Stripe widths of less than 1 mil (0.001 in.) are not normally used because a line-width variation of 0.0001 in. due to mask drawing error or mask misalignment or photographic-resolution error can result in 10 percent resistor-tolerance error.

The range of values obtainable with diffused resistors is limited by the size of the area required by the resistor. Practical range of resistance is 20 Ω to 30 K for a base-diffused resistor and 10 Ω to 1 K for an emitter-diffused resistor. The tolerance which results from profile variations and surface geometry errors is as high as ± 10 percent of the nominal value at 25°C, with ratio tolerance of ± 1 percent. For this reason the design of integrated circuits should, if possible, emphasize resistance ratios rather than absolute values. The temperature coefficient for these heavily doped resistors is positive and is $+0.06$ percent/°C from -55 to 0°C and $+0.20$ percent/°C from 0 to 125°C.

Equivalent Circuit A model of the diffused resistor is shown in Fig. 5-13, where the parasitic capacitances of the base-isolation (C_1) and isolation-substrate (C_2) junctions are included. In addition, it can be seen that a parasitic p-n-p transistor exists, with the substrate as collector, the isolation n-type region as base, and the resistor p-type material as the emitter. The collector is reverse-biased because the p-type substrate is at the most negative potential. It is also necessary that the emitter be reverse-biased to keep the parasitic transistor at cutoff. This condition is maintained by placing all resistors in the same isolation region and connecting the n-type isolation region surrounding the resistors to the *most positive* voltage present in the circuit. Typical values of h_{fe} for this parasitic transistor range from 0.5 to 5.

Thin-Film Resistors A technique of vapor thin-film deposition can also be used to fabricate resistors for integrated circuits. The metal (usually Nichrome NiCr) film is deposited (to a thickness of less than 1 μm) on the silicon dioxide layer, and masked etching is used to produce the desired geometry. The metal resistor is then covered by an insulating layer, and apertures for the ohmic contacts are opened through this insulating layer. Typical sheet-resistance values for Nichrome thin-film resistors are 40 to 400 Ω/square, resulting in resistance values from about 20 Ω to 50 K.

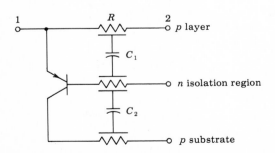

Fig. 5-13 The equivalent circuit of a diffused resistor.

5-9   INTEGRATED CAPACITORS AND INDUCTORS

Capacitors in integrated circuits may be obtained by utilizing the transition capacitance of a reverse-biased p-n junction or by a thin-film technique.

Junction Capacitors   A cross-sectional view of a junction capacitor is shown in Fig. 5-14a. The capacitor is formed by the reverse-biased junction J_2, which separates the epitaxial n-type layer from the upper p-type diffusion areas. An additional junction J_1 appears between the n-type epitaxial plane and the substrate, and a parasitic capacitance C_1 is associated with this reverse-biased junction. The equivalent circuit of the junction capacitor is shown in Fig. 5-14b, where the desired capacitance C_2 should be as large as possible relative to C_1. The value of C_2 depends on the junction area and impurity concentration. Since this junction is essentially linearly graded, C_2 varies as $V^{-\frac{1}{3}}$. The series resistance R (10 to 50 Ω) represents the resistance of the n-type layer.

It is clear that the substrate must be at the most negative voltage so as to mimimize C_1 and isolate the capacitor from other elements by keeping junction J_1 reverse-biased. It should also be pointed out that the junction capacitor C_2 is polarized since the p-n junction J_2 must always be reverse-biased.

Thin-Film Capacitors   A metal-oxide-semiconductor (MOS) nonpolarized capacitor is indicated in Fig. 5-15a. This structure is a parallel-plate capacitor with SiO_2 as the dielectric. A surface thin film of metal (aluminum) is the top plate. The bottom plate consists of the heavily doped n^+ region that is formed during the emitter diffusion. A typical value for capacitance is 0.4 pF/mil^2 for an oxide thickness of 500 Å, and the capacitance varies inversely with the thickness.

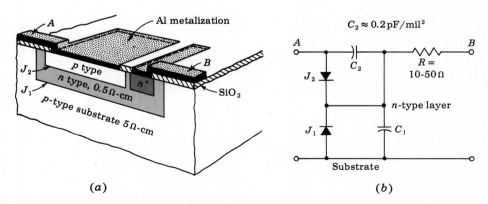

Fig. 5-14 (a) **Junction monolithic capacitor.** (b) **Equivalent circuit.** **(Courtesy of Motorola, Inc.)**

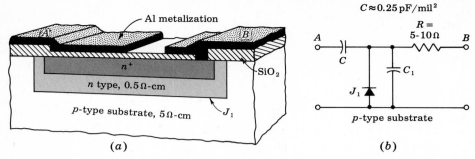

Fig. 5-15 An MOS capacitor. (a) The structure; (b) the equivalent circuit.

The equivalent circuit of the MOS capacitor is shown in Fig. 5-15b, where C_1 denotes the parasitic capacitance of the collector-substrate junction J_1, and R is the small series resistance of the n^+ region.

Inductors   No practical inductance values have been obtained on silicon substrates using semiconductor or thin-film techniques. Therefore their use is avoided in circuit design wherever possible. If an inductor is required, a discrete component is connected externally to the integrated circuit.

Characteristics of Integrated Components   Based upon our discussion of integrated-circuit technology, we can summarize the significant characteristics of integrated circuits (in addition to the advantages listed in Sec. 5-1).

1. A restricted range of values exists for resistors and capacitors. Typically, $10\ \Omega \le R \le 30\ \text{K}$ and $C \le 200\ \text{pF}$.

2. Poor tolerances are obtained in fabricating resistors and capacitors of specific magnitudes. For example, ± 20 percent of absolute values is typical. Resistance ratio tolerance can be specified to ± 1 percent because all resistors are made at the same time using the same techniques.

3. Components have high-temperature coefficients and may also be voltage-sensitive.

4. High-frequency response is limited by parasitic capacitances.

5. The technology is very costly for small-quantity production.

6. No practical inductors or transormers can be integrated.

5-10   THE METAL–SEMICONDUCTOR CONTACT

Two types of metal-semiconductor junctions are possible, *ohmic* and *rectifying*. The former is the type of contact desired when a lead is to be attached to a semiconductor. On the other hand, the rectifying contact results in a metal-semiconductor diode (called a *Schottky barrier*), with volt-ampere characteristics very similar to those of a *p-n* diode. The metal-semiconductor diode was investigated many years ago, but until the late 1960s commercial Schottky diodes were not available because of problems encountered in their manufacture.   It has turned out that most of the fabrication difficulties are due to surface effects; by employing the surface-passivated integrated-circuit techniques described in this chapter, it is possible to construct almost ideal metal-semiconductor diodes very economically.

As mentioned in Sec. 5-2 (step 4), aluminum acts as a p-type impurity when in contact with silicon.   If Al is to be attached as a lead to n-type Si, an ohmic contact is desired and the formation of a p-n junction must be prevented.   It is for this reason that n^+ diffusions are made in the n regions near the surface where the Al is deposited (Fig. 5-2d).   On the other hand, if the n^+ diffusion is omitted and the Al is deposited directly upon the n-type Si, an equivalent p-n structure is formed, resulting in an excellent metal-semiconductor diode.   In Fig. 5-16 contact 1 is a Schottky barrier, whereas contact 2 is an ohmic (nonrectifying) contact, and a metal-semiconductor diode exists between these two terminals, with the anode at contact 1.   Note that the fabrication of a Schottky diode is actually simpler than that of a p-n diode, which requires an extra (p-type) diffusion.

The external volt-ampere characteristic of a metal-semiconductor diode is essentially the same as that of a p-n junction, but the physical mechanisms involved are more complicated.   Note that in the forward direction electrons from the n-type Si cross the junction into the metal, where electrons are plentiful.   In this sense, this is a majority-carrier device, whereas minority carriers account for a p-n diode characteristic.   There is a delay in switching a p-n diode from ON to OFF (called the *storage* time) because the minority carriers stored at the junction must first be removed.   Schottky diodes have a negligible storage time t_s because the current is carried predominantly by majority carriers.   (Electrons from the n side enter the aluminum and become indistinguishable from the electrons in the metal, and hence are not "stored" near the junction.)

It should be mentioned that the voltage drop across a Schottky diode is much less than that of a p-n diode for the same forward current.   Thus, a cutin voltage of about 0.3 V is reasonable for a metal-semiconductor diode

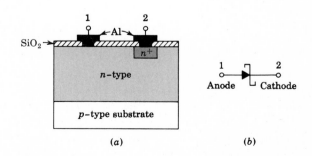

Fig. 5-16   (a) A Schottky diode formed by IC techniques.   The aluminum and the lightly doped n region form a rectifying contact 1, whereas the metal and the heavily doped n^+ region form an ohmic contact 2. (b) The symbol for this metal-semiconductor diode.

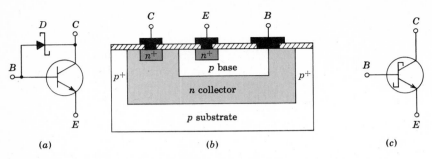

Fig. 5-17 (a) A transistor with a Schottky-diode clamp between base and collector to prevent saturation. (b) The cross section of a mono-lithic IC equivalent to the diode-transistor combination in (a). (c) The Schottky transistor symbol, which is an abbreviation for that shown in (a).

as against 0.6 V for a p-n barrier. Hence the former is closer to the ideal diode than the latter.

The Schottky Transistor To reduce the propagation-delay time in a logic gate, it is desirable to eliminate storage time in all transistors. In other words, a transistor must be prevented from entering saturation. This condition can be achieved, as indicated in Fig. 5-17a, by using a Schottky diode as a clamp between the base and collector. If an attempt is made to saturate this transistor by increasing the base current, the collector voltage drops, D conducts, and the base-to-collector voltage is limited to about 0.4 V. Since the collector junction is forward-biased by less than the cutin voltage (≈ 0.5 V), the transistor does *not* enter saturation (Sec. 4-8).

With no additional processing steps, the Schottky clamping diode can be fabricated at the same time that the transistor is constructed. As indicated in Fig. 5-17b, the aluminum metalization for the base lead is allowed to make contact also with the n-type collector region (but without an intervening n^+ section). This simple procedure forms a metal-semiconductor diode between base and collector. The device in Fig. 5-17b is equivalent to the circuit of Fig. 5-17a. This is referred to as a *Schottky transistor*, and is represented by the symbol in Fig. 5-17c.

REVIEW QUESTIONS

5-1 What are the four advantages of integrated circuits?

5-2 List the five steps involved in fabricating a monolithic integrated circuit (IC), assuming you already have a substrate.

5-3 List the five basic processes involved in the fabrication of an IC, assuming you already have a substrate.

5-4 Describe *epitaxial growth.*

5-5 (a) Describe the *photoetching process.* (b) How many masks are required to complete an IC? List the function performed by each mask.

5-6 (a) Describe the *diffusion process.* (b) What is meant by an *impurity profile?*

5-7 (a) How is the surface layer of SiO_2 formed? (b) How thick is this layer? (c) What are the reasons for forming the SiO_2 layers?

5-8 Explain how isolation between components is obtained in an IC.

5-9 How are the components interconnected in an IC?

5-10 Explain what is meant by *parasitic capacitance* in an IC.

5-11 Give the order of magnitude of (a) the substrate thickness; (b) the epitaxial thickness; (c) the base thickness; (d) the diffusion time; (e) the diffusion temperature; (f) the surface area of a transistor; (g) the chip size.

5-12 Sketch the cross section of an IC transistor.

5-13 Sketch the cross section of a discrete planar epitaxial transistor.

5-14 List the advantages and disadvantages of an IC vs. a discrete transistor.

5-15 Describe a *lateral p-n-p* transistor. Why is its current gain low?

5-16 Describe a *vertical p-n-p* transistor. Why is it of limited use?

5-17 (a) How are IC diodes fabricated? (b) Sketch the cross sections of two types of emitter-base diodes.

5-18 Sketch the top view of a multiple-emitter transistor. Show the isolation, collector, base, and emitter regions.

5-19 (a) Define *sheet resistance R_S.* (b) What is the order of magnitude of R_S for the base region and also for the emitter region? (c) Sketch the cross section of an IC resistor. (d) What are the orders of magnitude of the smallest and the largest values of an IC resistance?

5-20 (a) Sketch the equivalent circuit of a base-diffused resistor, showing all parasitic elements. (b) What must be done (externally) to minimize the effect of the parasitic elements?

5-21 Describe a *thin-film resistor.*

5-22 (a) Sketch the cross section of a junction capacitor. (b) Draw the equivalent circuit, showing all parasitic elements.

5-23 Repeat Rev. 5-22 for an MOS capacitor.

5-24 (a) What are the two basic distinctions between a junction and an MOS capacitor? (b) What is the order of magnitude of the capacitance per square mil? (c) What is the order of magnitude of the maximum value of C?

5-25 (a) To what voltage is the substrate connected? Why? (b) To what voltage is the isolation island containing the resistors connected? Why? (c) Can several transistors be placed in the same isolation island? Explain.

5-26 How is an aluminum contact made with n-type silicon so that it is (a) ohmic; (b) rectifying?

5-27 Why is storage time eliminated in a metal-semiconductor diode?

5-28 What is a *Schottky transistor?* Why is storage time eliminated in such a transistor?

5-29 Sketch the cross section of an IC Schottky transistor.

6 / DIGITAL CIRCUITS

Even in a large-scale digital system, such as a computer, or a data-processing, control, or digital-communication system, there are only a few basic operations which must be performed. These operations, to be sure, may be repeated very many times. The four circuits most commonly employed in such systems are known as the OR, AND, NOT, and FLIP-FLOP. These are called *logic* gates, or circuits, because they are used to implement Boolean algebraic equations (as we shall soon demonstrate). This algebra was invented by G. Boole in the middle of the nineteenth century as a system for the mathematical analysis of logic.

This chapter discusses in detail the first three basic logic circuits mentioned above. These basic gates are combined into FLIP-FLOPS and other digital-system building blocks in Chaps. 7, 8, and 9.

6-1   DIGITAL (BINARY) OPERATION OF A SYSTEM

A digital system functions in a binary manner. It employs devices which exist only in two possible states. A transistor is allowed to operate at cutoff or in saturation, but not in its active region. A node may be at a high voltage of, say, 4 ± 1 V or at a low voltage of, say, 0.2 ± 0.2 V, but no other values are allowed. Various designations are used for these two quantized states, and the most common are listed in Table 6-1. In logic, a statement is characterized as *true* or *false*, and this is the first binary classification listed in the table. A switch may be *closed* or *open*, which is the notation under 9, etc. Binary arithmetic and mathematical manipulation of switching or logic functions are best carried out with classification 3, which involves two symbols, 0 (zero) and 1 (one).

TABLE 6-1   Binary-state terminology

	1	2	3	4	5	6	7	8	9	10	11
One of the states.....	True	High	1	Up	Pulse	Excited	Off	Hot	Closed	North	Yes
The other state.......	False	Low	0	Down	No pulse	Non-excited	On	Cold	Open	South	No

The binary system of representing numbers will now be explained by making reference to the familiar *decimal system*. In the latter the base is 10 (ten), and ten numerals, 0, 1, 2, 3, . . . , 9, are required to express an arbitrary number. To write numbers larger than **9**, we assign a meaning to the *position* of a numeral in an array of numerals. For example, the number 1,264 (one thousand two hundred sixty four) has the meaning

$$1{,}264 \equiv 1 \times 10^3 + 2 \times 10^2 + 6 \times 10^1 + 4 \times 10^0$$

Thus the individual digits in a number represent the coefficients in an expansion of the number in powers of 10. The digit which is farthest to the right is the coefficient of the zeroth power, the next is the coefficient of the first power, and so on.

In the *binary system* of representation the base is **2**, and only the two numerals 0 and 1 are required to represent a number. The numerals 0 and 1 have the same meaning as in the decimal system, but a different interpretation is placed on the position occupied by a digit. In the binary system the individual digits represent the coefficients of powers of *two* rather than *ten* as in the decimal system. For example, the decimal number 19 is written in the binary representation as 10011 since

$$10011 \equiv 1 \times 2^4 + 0 \times 2^3 + 0 \times 2^2 + 1 \times 2^1 + 1 \times 2^0$$
$$= \quad 16 \quad + \quad 0 \quad + \quad 0 \quad + \quad 2 \quad + \quad 1 \quad = 19$$

A short list of equivalent numbers in decimal and binary notation is given in Table 6-2.

A general method for converting from a decimal to a binary number is indicated in Table 6-3. The procedure is the following. Place the decimal number (in this illustration, 19) on the extreme right. Next divide by 2 and place the quotient (9) to the left and indicate the remainder (1) directly below it. Repeat this process (for the next column $9 \div 2 = 4$ and a remainder of 1) until a quotient of 0 is obtained. The array of 1's and 0's in the second row is the binary representation of the original decimal number. In this example, decimal 19 = 10011 binary.

A binary digit (a 1 or a 0) is called a *bit*. A group of bits having a significance is a *byte, word,* or *code*. For example, to represent the 10 numerals

TABLE 6-2   Equivalent numbers in decimal and binary notation

Decimal notation	Binary notation	Decimal notation	Binary notation
0	00000	11	01011
1	00001	12	01100
2	00010	13	01101
3	00011	14	01110
4	00100	15	01111
5	00101	16	10000
6	00110	17	10001
7	00111	18	10010
8	01000	19	10011
9	01001	20	10100
10	01010	21	10101

(0, 1, 2, . . . , 9) and the 26 letters of the English alphabet would require 36 different combinations of 1's and 0's. Since $2^5 < 36 < 2^6$, then a minimum of 6 bits per byte are required in order to accommodate all the alphanumeric characters. In this sense a bit is sometimes referred to as a *character* and a group of one or more characters as a *word*.

Logic Systems   In a *dc*, or *level-logic*, system a bit is implemented as one of two voltage levels. If, as in Fig. 6-1a, the more positive voltage is the 1 level and the other is the 0 level, the system is said to employ dc *positive* logic. On the other hand, a dc *negative*-logic system, as in Fig. 6-1b, is one which designates the more negative voltage state of the bit as the 1 level and the more positive as the 0 level. It should be emphasized that the absolute values of the two voltages are of no significance in these definitions. In partic-

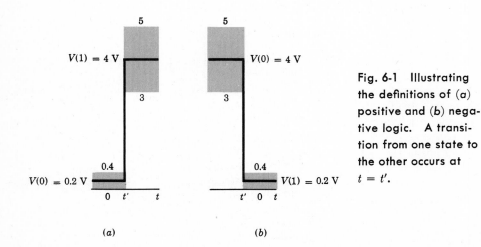

Fig. 6-1   Illustrating the definitions of (a) positive and (b) negative logic. A transition from one state to the other occurs at $t = t'$.

(a)   (b)

TABLE 6-3   Decimal-to-binary conversion

Divide by 2..........	0	1	2	4	9	19 decimal
Remainder..........	1	0	0	1	1	Binary

ular, the 0 state need not represent a zero voltage level (although in some systems it might).

The parameters of a physical device (for example, $V_{CE,\text{sat}}$ of a transistor) are not identical from sample to sample, and they also vary with temperature. Furthermore, ripple or voltage spikes may exist in the power supply or ground leads, and other sources of unwanted signals, called *noise*, may be present in the circuit. For these reasons the digital levels are not specified precisely, but as indicated by the shaded regions in Fig. 6-1, each state is defined by a voltage range about a designated level, such as 4 ± 1 V and 0.2 ± 0.2 V.

In a *dynamic,* or *pulse-logic* system a bit is recognized by the presence or absence of a pulse. A 1 signifies the existence of a positive pulse in a dynamic positive-logic system; a negative pulse denotes a 1 in a dynamic negative-logic system. In either system a 0 at a particular input (or output) at a given instant of time designates that no pulse is present at that particular moment.

6-2   THE OR GATE

An OR gate has two or more inputs and a single output, and it operates in accordance with the following definition: *The output of an* OR *assumes the 1 state if one or more inputs assume the 1 state.* The n inputs to a logic circuit will be designated by $A, B, \ldots, N$ and the ouput by Y. It is to be understood that each of these symbols may assume one of two possible values, either 0 or 1. A standard symbol for the OR circuit is given in Fig. 6-2a, together with the Boolean expression for this gate. The equation is to be read "Y equals A *or* B *or* $\cdots$ *or* N." Instead of defining a logical operation in words, an alternative method is to give a *truth table* which contains a tabulation of all possible input values and their corresponding outputs. It should be clear that the two-input truth table of Fig. 6-2b is equivalent to the above definition of the OR operation.

In a *diode-logic* (DL) system the logical gates are implemented by using diodes. A diode OR for negative logic is shown in Fig. 6-3. The generator source resistance is designated by R_s. We consider first the case where the supply voltage V_R has a value equal to the voltage $V(0)$ of the 0 state for dc logic.

If all inputs are in the 0 state, the voltage across each diode is $V(0) - V(0) = 0$. Since, in order for a diode to conduct, it must be for-

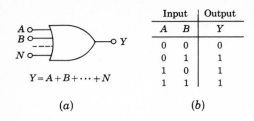

| Input | | Output |
A	B	Y
0	0	0
0	1	1
1	0	1
1	1	1

$Y = A + B + \cdots + N$

(a)   (b)

Fig. 6-2 (a) The standard symbol for an OR gate and its Boolean expression; (b) the truth table for a two-input OR gate.

ward-biased by at least the cutin voltage V_γ (Fig. 3-5), none of the diodes conducts. Hence the output voltage is $v_o = V(0)$, and Y is in the 0 state.

If now input A is changed to the 1 state, which for negative logic is at the potential $V(1)$, less positive than the 0 state, then $D1$ will conduct. Usually R is chosen much larger than $R_s + R_f$, where R_f is the forward resistance of the diode. Under this restriction,

$$v_o \approx V(1) + V_\gamma \tag{6-1}$$

Hence the output voltage exceeds the more negative level $V(1)$ by V_γ (approximately 0.2 V for germanium or 0.6 V for silicon). Furthermore, the step in output voltage is *smaller* by V_γ than the change in input voltage.

From now on, unless explicitly stated otherwise, we shall assume $R \gg R_s$ and ideal diodes with $R_f = 0$ and $V_\gamma = 0$. The output, for input A excited, is then $v_o = V(1)$, and the circuit has performed the following logic: if $A = 1$, $B = 0, \ldots, N = 0$, then $Y = 1$, which is consistent with the OR operation.

For the above excitation, the output is at $V(1)$, and each diode, except $D1$, is back-biased. Hence the presence of signal sources at $B, C, \ldots, N$ does not result in an additional load on generator A. Since the OR configuration minimizes the interaction of the sources on one another, this gate is sometimes referred to as a *buffer* circuit. Since it allows several independent sources to be applied at a given node, it is also called a (nonlinear) *mixing* gate.

If two or more inputs are in the 1 state, the diodes connected to these inputs conduct and all other diodes remain reverse-biased. The output is $V(1)$, and again the OR function is satisfied. If for any reason the level $V(1)$ is not identical for all inputs, *the most negative value of $V(1)$ (for negative logic) appears at the output*, and all diodes except one are nonconducting.

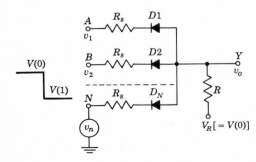

Fig. 6-3 A diode OR circuit for negative logic. [It is also possible to choose the supply voltage such that $V_R > V(0)$, but that arrangement has the disadvantage of drawing standby current when all inputs are in the 0 state.]

A positive-logic OR gate uses the same configuration as that in Fig. 6-3, except that all diodes must be reversed. *The output now is equal to the most positive level $V(1)$* [or more precisely is smaller than the most positive value of $V(1)$ by V_γ]. If a dynamic logic system is under consideration, *the output-pulse magnitude is* (approximately) *equal to the largest input pulse* (regardless of whether the system uses positive or negative logic).

A second mode of operation of the OR circuit of Fig. 6-3 is possible if V_R is set equal to a voltage more positive than $V(0)$ by at least V_γ. For this condition *all diodes conduct in the 0 state*, and $v_o \approx V(0)$ if $R \gg R_s + R_f$. If one or more inputs are excited, then the diode connected to the most negative $V(1)$ conducts, the output equals this value of $V(1)$, and all other diodes are back-biased. Clearly, the OR function has been satisfied.

A third mode of operation of the circuit of Fig. 6-3 results if we select $V_R < V(0)$. This arrangement has the disadvantage that the output will not respond until the input falls enough to overcome the initial reverse bias of the diodes.

Boolean Identities If it is remembered that A, B, and C can take on only the value 0 or 1, the following equations from Boolean algebra pertaining to the OR (+) operation are easily verified:

$$A + B + C = (A + B) + C = A + (B + C) \tag{6-2}$$

$$A + B = B + A \tag{6-3}$$

$$A + A = A \tag{6-4}$$

$$A + 1 = 1 \tag{6-5}$$

$$A + 0 = A \tag{6-6}$$

These equations may be justified by referring to the definition of the OR operation, to a truth table, or to the action of the OR circuits discussed above.

6-3   THE AND GATE

An AND gate has two or more inputs and a single output, and it operates in accordance with the following definition: *The output of an AND assumes the 1 state if and only if all the inputs assume the 1 state.* A symbol for the AND circuit is given in Fig. 6-4a, together with the Boolean expression for this

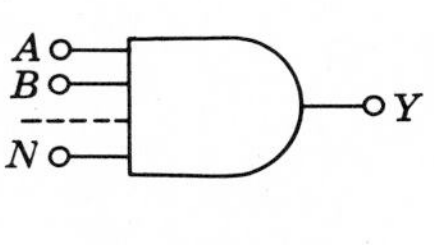

Fig. 6-4 (a) The standard symbol for an AND gate and its Boolean expression; (b) the truth table for a two-input AND gate.

$Y = AB \cdots N$

(a)

Input		Output
A	B	Y
0	0	0
0	1	0
1	0	0
1	1	1

(b)

gate. The equation is to be read "Y equals A *and* B *and* . . . *and* N." [Some-times a dot ($\cdot$) or a cross ($\times$) is placed between symbols to indicate the AND operation.] It may be verified that the two-input truth table of Fig. 6-4b is consistent with the above definition of the AND operation.

A diode-logic (DL) configuration for a negative AND gate is given in Fig. 6-5a. To understand the operation of the circuit, assume initially that all source resistances R_s are zero and that the diodes are ideal. If *any* input is at the 0 level $V(0)$, the diode connected to this input conducts and the output is clamped at the voltage $V(0)$, or $Y = 0$. However, if *all* inputs are at the 1 level $V(1)$, then all diodes are reverse-biased and $v_o = V(1)$, or $Y = 1$. Clearly, the AND operation has been implemented. The AND gate is also called a *coincidence circuit*.

A positive-logic AND gate uses the same configuration as that in Fig. 6-5a, except that all diodes are reversed. This circuit is indicated in Fig. 6-5b and should be compared with Fig. 6-3. It is to be noted that the symbol $V(0)$ in Fig. 6-3 designates the same voltage as $V(1)$ in Fig. 6-5b because each represents the upper binary level. Similarly, $V(1)$ in Fig. 6-3 equals $V(0)$ in Fig. 6-5b, since both represent the lower binary level. Hence these two circuits are identical, and we conclude that *a negative OR gate is the same circuit as a positive AND gate*. This result is not restricted to diode logic, and by using Boolean algebra, we show in Sec. 6-7 that it is valid independently of the hardware used to implement the circuit.

In Fig. 6-5b it is possible to choose V_R to be more positive than $V(1)$. If this condition is met, all diodes will conduct upon a coincidence (all inputs in the 1 state) and the output will be clamped to $V(1)$. The output impedance is low in this mode of operation, being equal to $(R_s + R_f)/n$ in parallel with R. On the other hand, if $V_R = V(1)$, then all diodes are cut off at a coincidence, and the output impedance is high (equal to R). If for any reason not all inputs have the same upper level $V(1)$, then the output of the positive AND gate of Fig. 6-5b will equal $V(1)_{min}$, the *least* positive value of $V(1)$. Note that the diode connected to $V(1)_{min}$ conducts, clamping the output to this minimum value of $V(1)$ and maintaining all other diodes in the reverse-

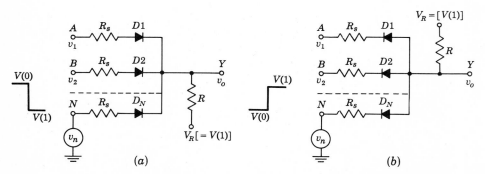

Fig. 6-5 A diode-logic AND circuit for (a) negative logic and (b) positive logic.

biased condition. If, on the other hand, V_R is smaller than all inputs $V(1)$, then all diodes will be cut off upon coincidence and the output will rise to the voltage V_R. Similarly, if the inputs are pulses, then *the output pulse will have an amplitude equal to the smallest input amplitude* [provided that V_R is greater than $V(1)_{min}$].

Boolean Identities Since A, B, and C can have only the value 0 or 1, the following expressions involving the AND operation may be verified:

$$ABC = (AB)C = A(BC) \tag{6-7}$$

$$AB = BA \tag{6-8}$$

$$AA = A \tag{6-9}$$

$$A1 = A \tag{6-10}$$

$$A0 = 0 \tag{6-11}$$

$$A(B + C) = AB + AC \tag{6-12}$$

These equations may be proved by reference to the definition of the AND operation, to a truth table, or to the behavior of the AND circuits discussed above. Also, by using Eqs. (6-10), (6-12), and (6-5), it can be shown that

$$A + AB = A \tag{6-13}$$

Similarly, if follows from Eqs. (6-12), (6-9), and (6-5) that

$$A + BC = (A + B)(A + C) \tag{6-14}$$

We shall have occasion to refer to the last two equations later.

6-4 THE NOT, OR INVERTER, CIRCUIT

The NOT circuit has a single input and a single output and performs the operation of *logic negation* in accordance with the following definition: *The out-*

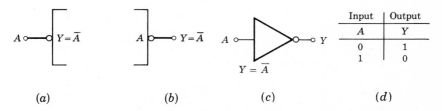

Input	Output
A	Y
0	1
1	0

$A \circ\!\!-\!\!\boxed{}\!\!-\!\!\circ Y = \bar{A}$

$A \circ\!\!-\!\!\boxed{}\!\!-\!\!\circ Y = \bar{A}$

$A \circ\!\!-\!\!\triangleright\!\!-\!\!\circ Y$

$Y = \bar{A}$

(a)   (b)   (c)   (d)

Fig. 6-6 Logic negation at (a) the input and (b) the output of a logic block; (c) a symbol often used for a NOT gate and the Boolean equation; (d) the truth table.

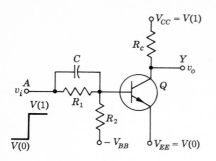

Fig. 6-7 An INVERTER for positive logic. A similar circuit using a p-n-p transistor is used for a negative-logic NOT circuit.

put of a NOT *circuit takes on the* 1 *state if and only if the input does* not *take on the* 1 *state.* The standard to indicate a *logic negation* is a small circle drawn at the point where a signal line joins a logic symbol. Negation at the input of a logic block is indicated in Fig. 6-6a and at the output in Fig. 6-6b. The symbol for a NOT gate and the Boolean expression for negation are given in Fig. 6-6c. The equation is to be read "Y equals NOT A" or "Y is the complement of A." [Sometimes a prime (′) is used instead of the bar (‾) to indicate the NOT operation.] The truth table is given in Fig. 6-6d.

A circuit which accomplishes a logic negation is called a NOT circuit, or, since it inverts the sense of the output with respect to the input, it is also known as an *inverter.* The output of an INVERTER is relatively more positive if and only if the input is relatively less positive. In a truly binary system only two levels $V(0)$ and $V(1)$ are recognized, and the output, as well as the input, of an inverter must operate between these two voltages. When the input is at $V(0)$, the output must be at $V(1)$, and vice versa. Ideally, then, a NOT circuit inverts a signal while preserving its shape and the binary levels between which the signal operates.

The transistor circuit of Fig. 6-7 implements an inverter for positive logic having a 0 state of $V(0) = V_{EE}$ and a 1 state of $V(1) = V_{CC}$. If the input is low, $v_i = V(0)$, then the parameters are chosen so that the Q is OFF, and hence $v_o = V_{CC} = V(1)$. On the other hand, if the input is high, $v_i = V(1)$, then the circuit parameters are picked so that Q is in saturation and then $v_o = V_{EE} = V(0)$, if we neglect the collector-to-emitter saturation voltage $V_{CE,\text{sat}}$. A detailed calculation of quiescent conditions is made in the following example.

EXAMPLE If the silicon transistor in Fig. 6-8 has a minimum value of h_{FE} of 30, find the output levels for input levels of 0 and 12 V.

Solution For $v_i = V(0) = 0$ the open-circuited base voltage V_B is

$$V_B = -12 \times \frac{15}{100 + 15} = -1.56 \text{ V}$$

Since a bias of about 0 V is adequate to cut off a silicon emitter junction (Table 4-1, page 85), then Q is indeed cut off. Hence $v_o = 12$ V for $v_i = 0$.

Fig. 6-8 An inverter calculation.

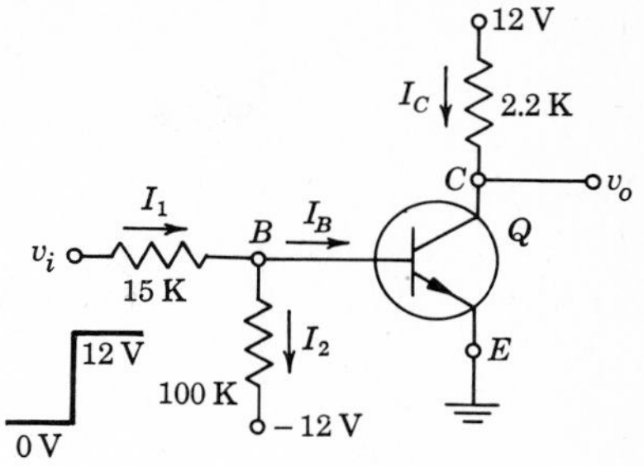

For $v_i = V(1) = 12$ V let us verify the assumption that Q is in saturation. The minimum base current required for saturation is

$$(I_B)_{\min} = \frac{I_C}{h_{FE}}$$

It is usually sufficiently accurate to use the approximate values for the saturation junction voltages given in Table 4-1, which for silicon are $V_{BE,\text{sat}} = 0.8$ V and $V_{CE,\text{sat}} = 0.2$ V. With these values

$$I_C = \frac{12 - 0.2}{2.2} = 5.36 \text{ mA} \qquad (I_B)_{\min} = \frac{5.36}{30} = 0.18 \text{ mA}$$

$$I_1 = \frac{12 - 0.8}{15} = 0.75 \text{ mA} \qquad I_2 = \frac{0.8 - (-12)}{100} = 0.13 \text{ mA}$$

and

$$I_B = I_1 - I_2 = 0.75 - 0.13 = 0.62 \text{ mA}$$

Since this value exceeds $(I_B)_{\min}$, Q is indeed in saturation and the drop across the transistor is $V_{CE,\text{sat}}$. Hence $v_o = 0.2$ V for $v_i = 12$ V, and the circuit has performed the NOT operation.

If the input to the inverter is obtained from the output of a similar gate, the input levels are $V(0) = V_{CE,\text{sat}} = 0.2$ V and $V(1) = 12$ V. The corresponding output levels are 12 and 0.2 V, respectively.

The capacitor C across R_1 in Fig. 6-7 is added to improve the transient response of the inverter. This capacitor aids in the removal of the minority-carrier charge stored in the base when the signal changes abruptly between logic states. The order of magnitude of C is 100 pF, but its exact value depends upon the transistor.

Transistor Limitations   There are certain transistor characteristics as well as certain circuit features which must particularly be taken into account in designing transistor inverters.

1. *The back-bias emitter-junction voltage* V_{EB}. This voltage must not exceed the emitter-to-base breakdown voltage BV_{EBO} specified by the manu-

facturer. For the type 2N914, $BV_{EBO} = 5$ V, and for the 2N1304, $BV_{EBO} = 25$ V. However, for some (diffused-base) transistors, BV_{EBO} may be quite small (less than 1 V).

2. *The dc current gain* h_{FE}. Since h_{FE} decreases with decreasing temperature, the circuit must be designed so that at the lowest expected temperature the transistor will remain in saturation. The maximum value of R_1 is determined principally by this condition.

3. *The reverse collector saturation current* I_{CBO}. Since $|I_{CBO}|$ increases about 7 percent/°C (doubles every 10°C for either germanium or silicon), we cannot continue to neglect the effect of I_{CBO} at high temperatures. At cutoff the emitter current is zero and the base current is I_{CBO} (in a direction opposite to that indicated as I_B in Fig. 6-8). Let us calculate the value of I_{CBO} which just brings the transistor to the point of cutoff. If we assume, as in Table 4-1, that at cutoff, $V_{BE} = 0$ V, then $I_1 = 0$ and the drop across the 100-K resistor is

$$100 \, I_{CBO} = 12 \text{ V} \quad \text{or} \quad I_{CBO} = 0.12 \text{ mA}$$

The ambient temperature at which $I_{CBO} = 0.12$ mA $= 120 \, \mu$A is the maximum temperature at which the inverter will operate satisfactorily. A silicon transistor can be operated at temperatures in excess of 185°C.

Boolean Identities From the basic definition of the NOT, AND, and OR connectives we can verify the following Boolean identities:

$$\bar{\bar{A}} = A \tag{6-15}$$

$$\bar{A} + A = 1 \tag{6-16}$$

$$\bar{A}A = 0 \tag{6-17}$$

$$A + \bar{A}B = A + B \tag{6-18}$$

EXAMPLE Verify Eq. (6-18).

Solution Since $B + 1 = 1$ and $A1 = A$, then

$$A + \bar{A}B = A(B + 1) + \bar{A}B = AB + A + \bar{A}B = (A + \bar{A})B + A = B + A$$

where use is made of Eq. (6-16).

6-5 THE INHIBIT (ENABLE) OPERATION

A NOT circuit preceding one terminal (S) of an AND gate acts as an *inhibitor*. This modified AND circuit implements the logical statement. *If* $A = 1$, $B = 1, \ldots, M = 1$, *then* $Y = 1$ *provided that* $S = 0$. *However, if* $S = 1$, *then the coincidence of* $A, B, \ldots, M$ *is inhibited, and* $Y = 0$. Such a configuration is also called an *anticoincidence* circuit. The logical block symbol is drawn in Fig. 6-9a, together with its Boolean equation. The equation is

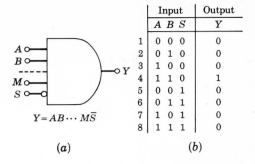

Fig. 6-9 (a) The logic block and Boolean expression for an AND with an enable terminal S. (b) The truth table for $Y = AB\bar{S}$. The column on the left numbers the eight possible input combinations.

	Input			Output
	A	B	S	Y
1	0	0	0	0
2	0	1	0	0
3	1	0	0	0
4	1	1	0	1
5	0	0	1	0
6	0	1	1	0
7	1	0	1	0
8	1	1	1	0

$Y = AB \cdots M\bar{S}$

(a)   (b)

to be read "Y equals A and B and . . . and M and not S." The truth table for a three-input AND gate with one inhibitor terminal (S) is given in Fig. 6-9b.

The terminal S is also called a *strobe* or an *enable input*. The enabling bit $S = 0$ allows the gate to perform its AND logic, whereas the inhibiting bit $S = 1$ causes the output to remain at $Y = 0$, independently of the values of the input bits.

6-6   THE EXCLUSIVE OR CIRCUIT

An EXCLUSIVE OR gate obeys the definition: *The output of a two-input* EXCLUSIVE OR *assumes the 1 state if one and only one input assumes the 1 state.* The standard symbol for an EXCLUSIVE OR is given in Fig. 6-10a and the truth table in Fig. 6-10b. The circuit of Sec. 6-2 is referred to as an INCLUSIVE OR if it is desired to distinguish it from the EXCLUSIVE OR.

The above definition is equivalent to the statement: "If $A = 1$ or $B = 1$ but not simultaneously, then $Y = 1$." In Boolean notation,

$$Y = (A + B)(\overline{AB}) \qquad (6\text{-}19)$$

This function is implemented in logic diagram form in Fig. 6-11a.

A second logic statement equivalent to the definition of the EXCLUSIVE OR is the following: "If $A = 1$ and $B = 0$, or if $B = 1$ and $A = 0$, then $Y = 1$." The Boolean expression is

$$Y = A\bar{B} + B\bar{A} \qquad (6\text{-}20)$$

The block diagram which satisfies this logic is indicated in Fig. 6-11b.

Fig. 6-10 (a) The standard symbol for an EXCLUSIVE OR gate and its Boolean expression; (b) the truth table.

$Y = A \oplus B$

(a)

Input		Output
A	B	Y
0	0	0
0	1	1
1	0	1
1	1	0

(b)

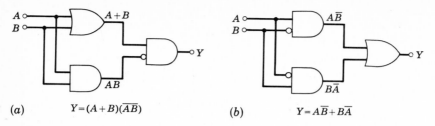

(a) $Y = (A+B)(\overline{AB})$ (b) $Y = A\overline{B} + B\overline{A}$

Fig. 6-11 Two logic block diagrams for the EXCLUSIVE OR **gate.**

An EXCLUSIVE OR is employed within the arithmetic section of a computer. Another application is as an *inequality comparator, matching circuit,* or *detector* because, as can be seen from the truth table, $Y = 1$ only if $A \neq B$. This property is used to check for the inequality of two bits. If bit A is not identical with bit B, then an output is obtained. Equivalently, "If A and B are both 1 or if A and B are both 0, then no output is obtained, and $Y = 0$." This latter statement may be put into Boolean form as

$$Y = \overline{AB + \overline{A}\overline{B}} \tag{6-21}$$

This equation leads to a third implementation for the EXCLUSIVE OR block, which is indicated by the logic diagram of Fig. 6-12a. An *equality detector* gives an output $Z = 1$ if A and B are both 1 or if A and B are both 0, and hence

$$Z = \bar{Y} = AB + \bar{A}\bar{B} \tag{6-22}$$

where use was made of Eq. (6-15). If the output Z is desired, the negation in Fig. 6-12a may be omitted or an additional inverter may be cascaded with the output of the EXCLUSIVE OR.

A fourth possibility for this gate is

$$Y = (A + B)(\bar{A} + \bar{B}) \tag{6-23}$$

which may be verified from the definition or from the truth table. This logic is depicted in Fig. 6-12b.

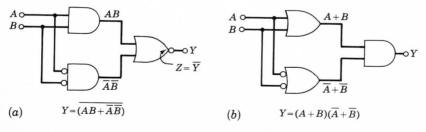

(a) $Y = \overline{(AB + \overline{A}\overline{B})}$ (b) $Y = (A+B)(\overline{A}+\overline{B})$

Fig. 6-12 Two additional logic block diagrams for the EXCLUSIVE OR **gate.**

We have demonstrated that there often are several ways to implement a logical circuit. In practice one of these may be realized more advantageously than the others. Boolean algebra is sometimes employed for manipulating a logic equation so as to transform it into a form which is better from the point of view of implementation in hardware. In the next section we shall verify through the use of Boolean algebra that the four expressions given above for the EXCLUSIVE OR are equivalent.

Two-Level Logic Digital design often calls for several gates (AND, OR, or combinations of those) feeding into an OR (or AND) gate. Such a combination is known as *two-level (or two-wide) logic*. The EXCLUSIVE OR circuits of Figs. 6-11 and 6-12 are examples of two-level logic. In the discussion of digital systems in Chap. 7 it is found that the most useful logic array consists of several ANDs which feed an OR which is followed by a NOT gate. This cascade of gates is called an AND-OR-INVERT (AOI) configuration. The detailed circuit topology for an AOI is given in Fig. 7-2.

6-7   DE MORGAN'S LAWS

The statement "If and only if all inputs are true (1), then the output is true (1)" is logically equivalent to the statement "If at least one input is false (0), then the output is false (0)." In Boolean notation this equivalence is written

$$ABC \cdots = \overline{\bar{A} + \bar{B} + \bar{C} + \cdots} \tag{6-24}$$

If we take the complement of both sides of this equation and use Eq. (6-15), we obtain

$$\overline{ABC \cdots} = \bar{A} + \bar{B} + \bar{C} + \cdots \tag{6-25}$$

This equation and its dual,

$$\overline{A + B + C + \cdots} = \bar{A}\bar{B}\bar{C} \cdots \tag{6-26}$$

(which may be proved in a similar manner), are known as De Morgan's laws. These complete the list of basic Boolean identities. For easy future reference, all these relationships are summarized in Table 6-4.

With the aid of Boolean algebra we shall now demonstrate the equivalence of the four EXCLUSIVE OR circuits of the preceding section. Using Eq. (6-25), it is immediately clear that Eq. (6-19) is equivalent to Eq. (6-23). Now the latter equation can be expanded with the aid of Table 6-4 as follows:

$$(A + B)(\bar{A} + \bar{B}) = A\bar{A} + B\bar{A} + A\bar{B} + B\bar{B} = B\bar{A} + A\bar{B} \tag{6-27}$$

TABLE 6-4   Summary of basic Boolean identities

Fundamental laws

OR	AND	NOT
$A + 0 = A$	$A0 = 0$	$A + \bar{A} = 1$
$A + 1 = 1$	$A1 = A$	$A\bar{A} = 0$
$A + A = A$	$AA = A$	$\bar{\bar{A}} = A$
$A + \bar{A} = 1$	$A\bar{A} = 0$	

Associative laws

$$(A + B) + C = A + (B + C) \qquad (AB)C = A(BC)$$

Commutative laws

$$A + B = B + A \qquad AB = BA$$

Distributive law

$$A(B + C) = AB + AC$$

De Morgan's laws

$$\overline{AB \cdots} = \bar{A} + \bar{B} + \cdots$$
$$\overline{A + B + \cdots} = \bar{A}\bar{B} \cdots$$

Auxiliary identities

$$A + AB = A \qquad A + \bar{A}B = A + B$$
$$(A + B)(A + C) = A + BC$$

This result shows that the EXCLUSIVE OR of Eq. (6-20) is equivalent to that of Eq. (6-23). Finally, applying Eq. (6-26) to Eq. (6-21) gives

$$\overline{AB + \bar{A}\bar{B}} = (\overline{AB})(\overline{\bar{A}\bar{B}}) \tag{6-28}$$

From Eq. (6-25), we have

$$(\overline{AB})(\overline{\bar{A}\bar{B}}) = (\bar{A} + \bar{B})(\bar{\bar{A}} + \bar{\bar{B}}) = (\bar{A} + \bar{B})(A + B) \tag{6-29}$$

where use is made of the identity $\bar{\bar{A}} = A$. Comparing Eqs. (6-28) and (6-29) shows that the EXCLUSIVE OR of Eq. (6-21) is equivalent to that of Eq. (6-23).

With the aid of De Morgan's law we can show that *an* AND *circuit for positive logic also functions as an* OR *gate for negative logic.* Let Y be the output and $A, B, \ldots, N$ be the inputs to a positive AND so that

$$Y = AB \cdots N \tag{6-30}$$

Then, by Eq. (6-25),

$$\bar{Y} = \bar{A} + \bar{B} + \cdots + \bar{N} \tag{6-31}$$

If the output and all inputs of a circuit are complemented so that a 1 becomes a 0 and vice versa, then positive logic is changed to negative logic (refer to Fig. 6-1). Since Y and $\bar{Y}$ represent the *same* output terminal, A and $\bar{A}$ the *same* input terminal, etc., the circuit which performs the positive AND logic in Eq. (6-30) also operates as the negative OR gate of Eq. (6-31). Similar reasoning is used to verify that the same circuit is either a negative AND or a positive OR, depending upon how the binary levels are defined. We verified

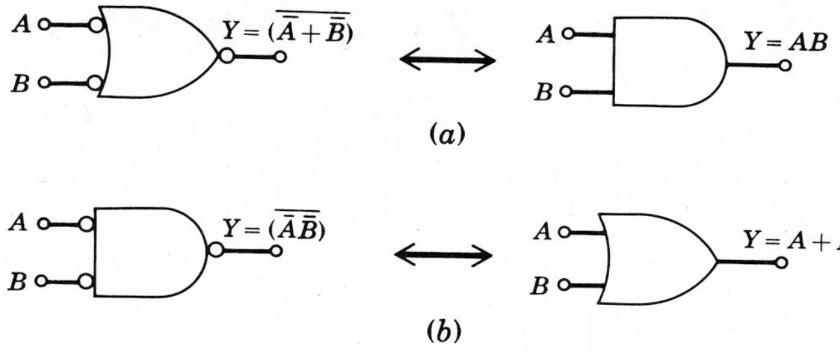

$$Y = (\overline{\overline{A} + \overline{B}}) \qquad \Longleftrightarrow \qquad Y = AB$$

(a)

$$Y = (\overline{\overline{A}\,\overline{B}}) \qquad \Longleftrightarrow \qquad Y = A + B$$

(b)

Fig. 6-13   (a) An OR is converted into an AND by inverting all inputs and also the output.   (b) An AND becomes an OR if all inputs and the output are complemented.

this result for diode logic in Sec. 6-3, but the present proof is independent of how the circuit is implemented.

It should now be clear that it is really not necessary to use all three connectives OR, AND, and NOT. The OR and the NOT are sufficient because, from the De Morgan law of Eq. (6-24), the AND can be obtained from the OR and the NOT, as is indicated in Fig. 6-13a. Similarly, the AND and the NOT may be chosen as the basic logic circuits, and from the De Morgan law of Eq. (6-26), the OR may be constructed as shown in Fig. 6-13b. This figure makes clear once again that an OR (AND) circuit negated at input and output performs the AND (OR) logic.

6-8   THE NAND AND NOR DIODE–TRANSISTOR LOGIC (DTL) GATES

In Fig. 6-11a the negation before the second AND could equally well be put at the output of the first AND without changing the logic. Such an AND-NOT sequence is also present in Fig. 6-13b and in many other logic operations. This negated AND is called a NOT-AND, or a NAND, gate. The logic symbol, Boolean expression, and truth table for the NAND are given in Fig. 6-14. The NAND may be implemented by placing a transistor NOT circuit *after* the diode AND. Such a transposition is shown in Fig. 6-15. Circuits involving diodes and transistors as in Fig. 6-15 are called *diode-transistor logic* (DTL) *gates.*

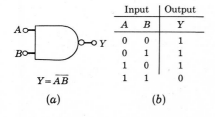

Input		Output
A	B	Y
0	0	1
0	1	1
1	0	1
1	1	0

$$Y = \overline{AB}$$

(a)   (b)

Fig. 6-14   (n) The logic symbol and Boolean expression for a two-input NAND gate; (b) the truth table.

EXAMPLE (*a*) Verify that the circuit of Fig. 6-15 is a positive NAND for the binary levels 0 and 12 V. Neglect source impedance and junction saturation voltages and diode voltages in the forward direction. Find the minimum h_{FE}. (*b*) Will the circuit operate properly if the inputs are obtained from the outputs of similar NAND gates? Assume silicon transistors and diodes and neglect collector saturation resistance. The drop across a conducting diode is 0.7 V.

Solution *a.* If any input is at 0 V, then the junction point P of the diodes is at 0 V because a diode conducts and clamps this point to $V(0) = 0$. The base voltage of the transistor is then

$$V_B = -(12) \left(\frac{15}{115} \right) = -1.56 \text{ V}$$

Hence Q is cut off and Y is at 12 V, or $Y = 1$. This result confirms the first three rows of the truth table of Fig. 6-14*b*.

If all inputs are at $V(1) = 12$ V, assume that all diodes are reverse-biased and that the transistor is in saturation. We shall now verify that these assumptions are indeed correct. If Q is in saturation, then with $V_{BE} = 0$, the voltage at P is $(12)(\frac{15}{30}) = 6$ V. Hence, with 12 V at each input, all diodes are reverse-biased by 6 V. Since the diodes are nonconducting, the two 15-K resistors are in series and the base current of Q is

$$\frac{12}{30} - \frac{12}{100} = 0.40 - 0.12 = 0.28 \text{ mA}$$

Since the collector current is

$$I_C = \frac{12}{2.2} = 5.45 \text{ mA} \quad \text{and} \quad (h_{FE})_{\min} = \frac{5.45}{0.28} = 19$$

then Q will indeed be in saturation if $h_{FE} \geq 19$. Under these circumstances the output is at ground, or $Y = 0$. This result confirms the last row of the truth table.

b. If the inputs are high, the situation is exactly as in part *a*. With respect to keeping the base node at a low voltage when there is no coincidence, the worst situation occurs when all but one input are high. The low input now comes from a transistor in saturation, and $V_{CE,\text{sat}} \approx 0.2$ V. The open-circuit voltage at the

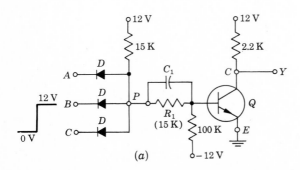

(*a*)

Fig. 6-15 A three-input positive NAND (or negative NOR) gate.

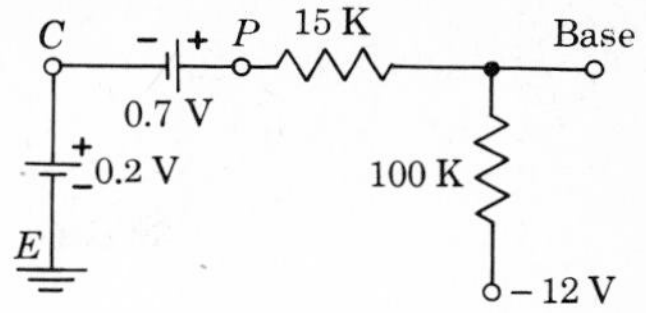

Fig. 6-16 Relating to the calculation of the base
voltage of the transistor in the circuit of Fig. 6-15.

base of Q is, from Fig. 6-16, using superposition

$$V_B = -(12)\left(\frac{15}{115}\right) + (0.9)\left(\frac{100}{115}\right) = -0.78 \text{ V}$$

which cuts off Q and $Y = 1$, as it should.

A NOR Gate A negation following an OR is called a NOT-OR, or a NOR gate. The logic symbol, Boolean expression, and truth table for the NOR are given in Fig. 6-17. A positive NOR circuit is implemented by a cascade of a diode OR and a transistor INVERTER.

The circuit of Fig. 6-15 employs *diode-transistor logic* (DTL). The NAND and NOR may also be implemented in other configurations, as is indicated in Secs. 6-10 and 6-12. With the aid of De Morgan's laws, it can be shown that, regardless of the hardware involved, a positive NAND is also a negative NOR, whereas a negative NAND may equally well be considered a positive NOR.

It is clear that a single input NAND is a NOT. Also, a NAND followed by a NOT is an AND. In Sec. 6-7 it is pointed out that all logic can be performed by using only the two connectives AND and NOT. Therefore we now conclude that, by repeated use of the NAND circuit alone, any logical function can be carried out. A similar argument leads equally well to the result that all logic can be performed by using only the NOR circuit.

EXAMPLE Verify that two-level AND-OR topology is equivalent to a NAND-NAND system.

Solution The AND-OR logic is indicated in Fig. 6-18a. Since $X = \bar{\bar{X}}$, then inverting the output of an AND and simultaneously negating the input to the following OR does not change the logic. These modifications are made in Fig. 6-18b. We have also negated the output of the OR gate and, at the same time,

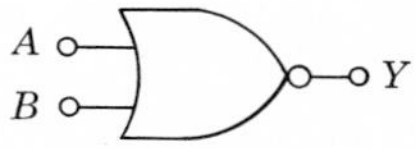

Fig. 6-17 (a) The logic symbol and Boolean expression for a two-input NOR gate; (b) the truth table.

$Y = \overline{A + B}$

(a)

Input		Output
A	B	Y
0	0	1
0	1	0
1	0	0
1	1	0

(b)

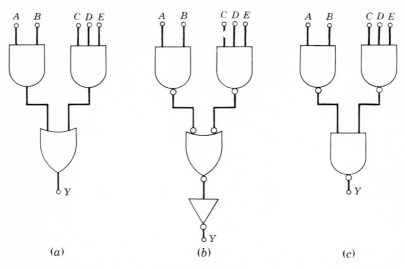

Fig. 6-18 A two-level AND-OR is equivalent to a NAND-NAND configuration.

have added an INVERTER to Fig. 6-18b, so that once again the logic is unaffected. An OR gate negated at each terminal is an AND circuit (Fig. 6-13a). Since an AND followed by an INVERTER is a NAND then Fig. 6-18c is equivalent to Fig. 6-18b. Hence, the NAND-NAND of Fig. 6-18c is equivalent to the AND-OR of Fig. 6-18a.

If any of the inputs in Fig. 6-18 are obtained from the output of another gate then the resultant topology is referred to as *three-level logic*.

6-9   MODIFIED (INTEGRATED–CIRCUIT) DTL GATES

Most logic gates are fabricated as an *integrated circuit* (IC). This process is described in Chap. 5, where it is found that large values of resistance (above 30 K) and of capacitance (above 100 pF) cannot be fabricated economically. On the other hand, transistors and diodes may be constructed very inexpensively. In view of these facts, the NAND gate of Fig. 6-15 is modified for integrated-circuit implementation by eliminating the capacitor C_1, reducing the resistance values drastically, and using diodes or transistors to replace resistors wherever possible. At the same time the power-supply requirements are simplified so that only a single 5-V supply is used. The resulting circuit is indicated in Fig. 6-19.

The operation of this positive NAND gate is easily understood qualitatively. If at least one of the inputs is low (the 0 state), the diode D connected to this input conducts and the voltage V_P at point P is low. Hence diodes $D1$ and $D2$ are nonconducting, $I_B = 0$, and the transistor is OFF.

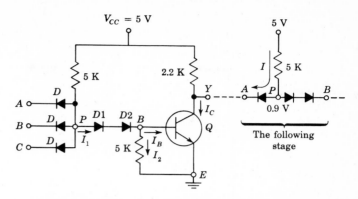

Fig. 6-19   An integrated positive DTL NAND **gate.**

Therefore the output of Q is high and Y is in the 1 state. This logic satisfies the first three rows of the truth table in Fig. 6-14. Consider now the case where all inputs are high (1) so that all input diodes D are cut off. Then V_P tries to rise toward V_{CC}, and a base current I_B results. If I_B is sufficiently large, Q is driven into saturation and the output Y falls to its low (0) state, thus satisfying the fourth row of the truth table.

This NAND gate is considered quantitatively in the following illustrative example. The necessity for using two diodes $D1$ and $D2$ in series is explained. False logic can be caused by switching transients, power-supply noise spikes, coupling between leads, etc. The noise margins that this circuit can tolerate are calculated below.

EXAMPLE   (*a*) For the transistor in Fig. 6-19 assume (Table 4-1) that $V_{BE,\text{sat}} = 0.8$ V, $V_\gamma = 0.5$ V, and $V_{CE,\text{sat}} = 0.2$ V. The drop across a conducting diode is 0.7 V and V_γ (diode) $= 0.6$ V. The inputs of this switch are obtained from the outputs of similar gates. Verify that the circuit functions as a positive NAND and calculate $(h_{FE})_{\text{min}}$. (*b*) Will the circuit operate properly if $D2$ is not used? (*c*) If all inputs are high, what is the magnitude of the noise voltage at the input which will cause the gate to malfunction? (*d*) Repeat part *c* if at least one input is low. Assume, for the moment, that Q is not loaded by a following stage.

Solution   *a.* The logic levels are $V_{CE,\text{sat}} = 0.2$ V for the 0 state and $V_{CC} = 5$ V for the 1 state. If at least one input is in the 0 state, its diode conducts and $V_P = 0.2 + 0.7 = 0.9$ V. Since, in order for $D1$ and $D2$ to be conducting, a voltage of $(2)(0.7) = 1.4$ V is required, these diodes are cut off, and $V_{BE} = 0$. Since the cutin voltage of Q is $V_\gamma = 0.5$ V, then Q is OFF, the output rises to 5 V, and $Y = 1$. This confirms the first three rows of the NAND truth table.

   If all inputs are at $V(1) = 5$ V, then we shall assume that all input diodes are OFF, that $D1$ and $D2$ conduct, and that Q is in saturation. If these conditions are true, the voltage at P is the sum of two diode drops plus $V_{BE,\text{sat}}$, or $V_P = 0.7 + 0.7 + 0.8 = 2.2$ V. The voltage across each input diode is $5 - 2.2 = 2.8$ V in the reverse direction, thus justifying the assumption that D is OFF. We now find $(h_{FE})_{min}$ to put Q into saturation.

$$I_1 = \frac{V_{CC} - V_P}{5} = \frac{5 - 2.2}{5} = 0.56 \text{ mA}$$

$$I_2 = \frac{V_{BE,sat}}{5} = \frac{0.8}{5} = 0.16 \text{ mA}$$

$$I_B = I_1 - I_2 = 0.56 - 0.16 = 0.40 \text{ mA}$$

$$I_C = \frac{V_{CC} - V_{CE,sat}}{2.2} = \frac{5 - 0.2}{2.2} = 2.18 \text{ mA}$$

and

$$(h_{FE})_{min} = \frac{I_C}{I_B} = \frac{2.18}{0.40} = 5.5$$

If $h_{FE} > (h_{FE})_{min}$, then $Y = V(0)$ for all inputs at $V(1)$, thus verifying the last line in the truth table in Fig. 6-14.

b. If at least one input is at $V(0)$, then $V_P = 0.2 + 0.7 = 0.9$ V. Hence, if only one diode $D1$ is used between P and B, then $V_{BE} = 0.9 - 0.6 = 0.3$ V, where 0.6 V represents the diode cutin voltage. Since the cutin base voltage is $V_\gamma = 0.5$ V, then theoretically Q is cut off. However, this is not a very conservative design because a small (>0.2 V) spike of noise will turn Q ON. An even more conservative design uses three diodes in series, instead of the two indicated in Fig. 6-19.

c. If all inputs are high, then from part a, $V_P = 2.2$ V and each input diode is reverse-biased by 2.8 V. A diode starts to conduct when it is forward-biased by 0.6 V. Hence a negative noise spike in excess of $2.8 + 0.6 = 3.4$ V must be present at the input before the circuit malfunctions. Such a large noise voltage is improbable.

d. If at least one input is low, then from part a, $V_P = 0.9$ V and Q is OFF. If a noise spike takes Q just into its active region, $V_{BE} = V_\gamma = 0.5$ and V_P must increase to $0.5 + 0.6 + 0.6 = 1.7$ V. Hence the noise margin is $1.7 - 0.9 = 0.8$ V. If only one diode $D1$ were used, the noise voltage would be reduced by 0.6 V (the drop across $D2$ at cutin) to $0.8 - 0.6 = 0.2$ V. This confirms the value obtained in part b.

Fan-out In the foregoing discussion we have unrealistically assumed that the NAND gate is unloaded. If it drives N similar gates, we say that the *fan-out* is N. The output transistor now acts as a *sink* for the current in the input to the gates it drives. In other words, when Q is in saturation ($Y = 0$), the input current I in Fig. 6-19 of a following stage adds to the collector current of Q. Assume that all the input diodes to a following stage (which is now considered to be a *current source*) are high except the one driven by Q. Then the current in this diode is $I = (5 - 0.9)/5 = 0.82$ mA. This current is called a *standard load*. The total collector current of Q is now $I_C = 0.82N + 2.18$

mA, where 2.18 mA is the unloaded collector current found in part *a* of the preceding example. Since the base current is almost independent of loading, I_B remains at its previous value of 0.4 mA. If we assume a reasonable value for $(h_{FE})_{\min}$ of 30, the fan-out is given by $I_C = h_{FE}I_B$, or

$$I_C = 0.82N + 2.18 = (30)(0.40) = 12.0 \text{ mA} \tag{6-32}$$

and $N = 12$. Of course, the current rating of Q must exceed 12.0 mA if we are to drive 12 gates.

The *fan-in M* of a logic gate gives the number of inputs to the switch. For example, in Fig. 6.19, $M = 3$.

Wired Logic It is possible to connect the outputs of several DTL gates together, as in Fig. 6-20, to perform additional logic without additional hardware. If positive NAND logic is under consideration, this connection is called a *wired*-AND, *phantom*-AND, *dotted*-AND, or *implied*-AND. Thus, if both $Y_1 = 1$ and $Y_2 = 1$, then $Y = 1$, whereas if $Y_1 = 0$ and/or $Y_2 = 0$, then $Y = 0$.

The circuit of Fig. 6-19 also represents a negative NOR gate, and connecting two outputs together as in Fig. 6-20 now represents negative wired-OR logic. If Y_1 and/or Y_2 is in the low state (which is now the 1 state), then Y is also in the low state ($Y = 1$), whereas if Y_1 and Y_2 are both high (the 0 state), then Y is also high ($Y_2 = 0$).

Consider two positive NAND gates wired-AND together as in Fig. 6-20*b*. Then $Y_1 = \overline{AB}$ and $Y_2 = \overline{CDE}$. Hence

$$Y = Y_1Y_2 = (\overline{AB})(\overline{CDE}) = \overline{AB + CDE} \tag{6-33}$$

where use is made of De Morgan's law. Note that the wired-AND has led

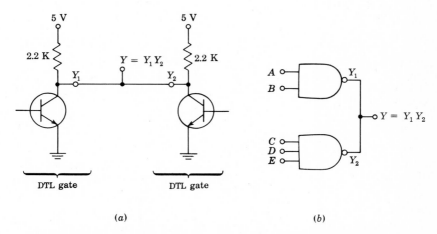

(a)   (b)

Fig. 6-20 (*a*) Wired-AND logic is obtained by connecting the outputs of positive NAND gates together; (*b*) a two-input and a three-input NAND gate wired-AND together to perform the logic in Eq. (6-33).

to an implementation of the AOI two-level logic (Sec. 6-6). Because of the + sign in Eq. (6-33), the connection in Fig. 6-20 is often incorrectly referred to as a positive wired-OR.

Note that the wired-AND connection places the collector resistors in Fig. 6-20a in parallel. This reduction in resistance increases the power dissipation in the ON state. In order to avoid this condition *open-collector* gates are available specifically for wired-AND applications.

6-10   TRANSISTOR–TRANSISTOR–LOGIC (TTL) GATE

The fastest-saturating logic circuit is the transistor-transistor-logic gate (TTL, or T²L), shown in Fig. 6-21. This switch uses a multiple-emitter transistor which is easily and economically fabricated using integrated-circuit techniques (Sec. 5-7). The TTL circuit has the topology of the DTL circuit of Fig. 6-19, with the emitter junctions of $Q1$ acting as the input diodes D of the DTL gate and the collector junction of $Q1$ replacing the diode $D1$ of Fig. 6-19. The base-to-emitter diode of $Q2$ is used in place of the diode $D2$ of the DTL gate, and both circuits have an output transistor ($Q3$ or Q).

The explanation of the operation of the TTL gate parallels that of the DTL switch. Thus, if at least one input is at $V(0) = 0.2$ V, then

$$V_P = 0.2 + 0.7 = 0.9 \text{ V}$$

For the collector junction of $Q1$ to be forward-biased and for $Q2$ and $Q3$ to be ON requires about $0.7 + 0.7 + 0.7 = 2.1$ V. Hence these are OFF; the output rises to $V_{CC} = 5$ V, and $Y = V(1)$. On the other hand, if all inputs are high (at 5 V), the input diodes (the emitter junctions) are reverse-biased and V_P rises toward V_{CC} and drives $Q2$ and $Q3$ into saturation. Then the output is $V_{CE,\text{sat}} = 0.2$ V, and $Y = V(0)$ (and V_P is clamped at about 2.3 V).

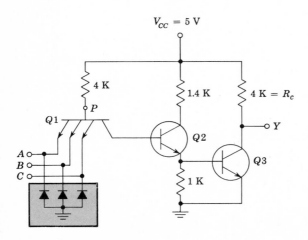

$V_{cc} = 5$ V

Fig. 6-21 An IC positive TTL NAND gate. (Neglect the diodes in the shaded block.)

The above explanation indicates that $Q1$ acts like isolated back-to-back diodes, and not as a transistor. However, we shall now show that, during turnoff, $Q1$ does exhibit transistor action, thereby reducing storage time considerably. Note that the base voltage of $Q2$, which equals the collector voltage of $Q1$, is at $0.8 + 0.8 = 1.6$ V during saturation of $Q2$ and $Q3$. If now any input drops to 0.2 V, then $V_P = 0.9$ V, and hence the base of $Q1$ is at 0.9 V. At this time the collector junction is reverse-biased by $1.6 - 0.9 = 0.7$ V, the emitter junction is forward-biased, and $Q1$ is in its active region. The large collector current of $Q1$ now quickly removes the stored charge in $Q2$ and $Q3$. It is this transistor action which gives TTL the highest speed of any saturated logic.

Clamping diodes (shown in the shaded block in Fig. 6-21) are often included from each input to ground, with the anode grounded. These diodes are effectively out of the circuit for positive input signals, but they limit negative voltage excursions at the input to a safe value. These negative signals may arise from ringing caused by lead inductance resonating with shunt capacitance.

6-11   OUTPUT STAGES

At the output terminal of the DTL or TTL gate there is a capacitive load C_L, consisting of the capacitances of the reverse-biased diodes of the fan-out gates and any stray wiring capacitance. If the collector-circuit resistor of the inverter is R_c (called a *passive pull-up*), then, when the output changes from the low to the high state, the output transistor is cut off and the capacitance charges exponentially from $V_{CE,\text{sat}}$ to V_{CC}. The time constant $R_c C_L$ of this waveform may introduce a prohibitively long delay time into the operation of these gates.

The output delay may be reduced by decreasing R_c, but this will increase the power dissipation when the output is in its low state and the voltage across R_c is $V_{CC} - V_{CE,\text{sat}}$. A better solution to this problem is indicated in Fig. 6-22, where a transistor acts as an *active pull-up* circuit, replacing the passive pull-up resistance R_c. This output configuration is called a *totem-pole* amplifier because the transistor $Q4$ "sits" upon $Q3$. It is also referred to as a power-driver, or power-buffer, output stage.

The transistor $Q2$ acts as a *phase splitter*, since the emitter voltage is out of phase with the collector voltage (for an increase in base current, the emitter voltage increases and the collector voltage decreases). We now explain the operation of this driver circuit in detail, with reference to the TTL gate of Fig. 6-22.

The output is in the low-voltage state when $Q2$ and $Q3$ are driven into saturation. For this state we should like $Q4$ to be OFF. Is it? Note that the collector voltage V_{CN2} of $Q2$ with respect to ground N is given by

$$V_{CN2} = V_{CE2,\text{sat}} + V_{BE3,\text{sat}} = 0.2 + 0.8 = 1.0 \text{ V} \tag{6-34}$$

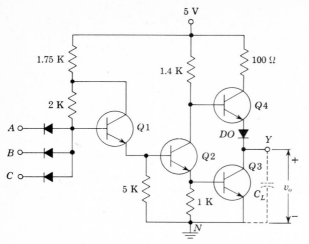

Fig. 6-22　A TTL gate with a totem-pole output driver.

Since the base of $Q4$ is tied to the collector of $Q2$, then $V_{BN4} = V_{CN2} = 1.0\,\text{V}$. *If the output diode DO were missing*, the base-to-emitter voltage of $Q4$ would be

$$V_{BE4} = V_{BN4} - V_{CE3,\text{sat}} = 1.0 - 0.2 = 0.8 \text{ V}$$

which would put $Q4$ into saturation. Under these circumstances the steady current through $Q4$ would be

$$\frac{V_{CC} - V_{CE4,\text{sat}} - V_{CE3,\text{sat}}}{100} = \frac{5 - 0.2 - 0.2}{100} \text{ A} = 46 \text{ mA} \qquad (6\text{-}35)$$

which is excessive and wasted current. The necessity for adding DO is now clear. With it in place, the sum of V_{BE4} and V_{DO} is 0.8 V. Hence both $Q4$ and DO are at cutoff. In summary, if C_L is at the high voltage $V(1)$ and the gate is excited, $Q4$ and DO go off, and $Q3$ conducts. Because of its large active-region current, $Q3$ quickly discharges C_L, and as v_o approaches $V(0)$, $Q3$ enters saturation. The bottom transistor $Q3$ of the totem pole is referred to as a *current sink*, which discharges C_L.

Assume now that with the output at $V(0)$, there is a change of state, because one of the inputs drops to its low state. Then $Q2$ is turned off, which causes $Q3$ to go to cutoff because V_{BE3} drops to zero. The output v_o remains momentarily at 0.2 V because the voltage across C_L cannot change instantaneously. Now $Q4$ goes into saturation and DO conducts, as we can verify:

$$V_{BN4} = V_{BE4,\text{sat}} + V_{DO} + v_o = 0.8 + 0.7 + 0.2 = 1.7 \text{ V}$$

and the base and collector currents of $Q4$ are

$$I_{B4} = \frac{V_{CC} - V_{BN4}}{1.4} = \frac{5 - 1.7}{1.4} = 2.36 \text{ mA}$$

$$I_{C4} = \frac{V_{CC} - V_{CE4,\text{sat}} - V_{DO} - v_o}{0.1} = \frac{5 - 0.2 - 0.7 - 0.2}{0.1} = 39.0 \text{ mA}$$

Hence, if h_{FE} exceeds $(h_{FE})_{min} = I_{C4}/I_{B4} = 39.0/2.36 = 16.5$, then $Q4$ is in saturation. The transistor $Q4$ is referred to as a *source*, supplying current to C_L. As long as $Q4$ remains in saturation, the output voltage rises exponentially toward V_{CC} with the very small time constant $(100 + R_{CS4} + R_f)C_L$, where R_{CS4} is the saturation resistance (Sec. 4-8) of $Q4$, and where R_f (a few ohms) is the diode forward resistance. As v_o increases, the currents in $Q4$ decrease, and $Q4$ comes out of saturation and finally v_o reaches a steady state when $Q4$ is at the cutin condition. Hence the final value of the output voltage is

$$v_o = V_{CC} - V_{BE4,\text{cutin}} - V_{DO,\text{cutin}} \approx 5 - 0.5 - 0.6 = 3.9 \text{ V} = V(1)$$
$$(6\text{-}36)$$

If the 100-Ω resistor were omitted, there would result a faster change in output from $V(0)$ to $V(1)$. However, the 100-Ω resistor is needed to limit the current spikes during the turn-on and turnoff transients. In particular, $Q3$ does not turn off (because of storage time) as quickly as $Q4$ turns on. With both totem-pole transistors conducting at the same time, the supply voltage would be short-circuited if the 100-Ω resistor were missing. The peak current drawn from the supply during the transient is limited to $I_{C4} + I_{B4} = 39 + 2.4 \approx 41$ mA if the 100-Ω resistor is used. These current spikes generate noise in the power-supply distribution system, and also result in increased power consumption at high frequencies.

Wired Logic It should be emphasized that the wired-AND connection must *not* be used with the totem-pole driver circuit. If the output from one gate is high while that from a second gate is low, and if these two outputs are tied together, we have exactly the situation just discussed in connection with transient current spikes. Hence, if the wired-AND were used, the power supply would deliver a *steady* current of 41 mA under these circumstances.

6-12   RESISTOR–TRANSISTOR LOGIC (RTL)

In addition to TTL (the most popular logic family) and DTL there are available several additional families, called *resistor-transistor logic* (RTL), *emitter-coupled logic* (ECL), and *metal-oxide-semiconductor* (MOS) *logic*. Since MOS logic uses the field-effect transistor (FET), the discussion of these gates is postponed until Chap. 9, where the MOSFET is introduced. Since ECL requires an understanding of the *differential amplifier*, this type of (nonsaturating) logic is considered in Sec. 14-6, after the differential amplifier is studied. A discussion of RTL now follows.

Resistor-Transistor Logic (RTL) This configuration is indicated in Fig. 6-23a, which represents a three-input positive NOR gate with a fan-out of 5. If any input is high, the corresponding transistor is driven into saturation and the output is low, $v_o = V_{CE,\text{sat}} \approx 0.2$ V. However, if all inputs are low, then

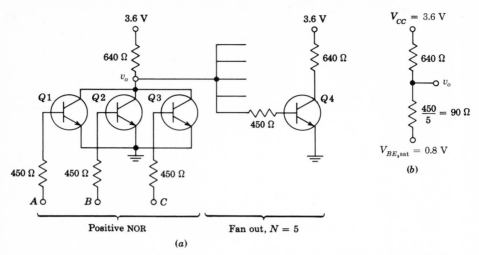

Fig. 6-23 (*a*) An RTL positive NOR gate with a fan-in of 3 and a fan-out of 5. (*b*) The equivalent circuit from which to calculate v_o in the high state.

all input transistors are cut off by $V_\gamma - V(0) = 0.5 - 0.2 = 0.3$ V and the output v_o is high. (Note the low noise margin.) The preceding two statements confirm that the gate performs positive NOR (or negative NAND) logic.

The value of $V(1)$ depends upon the fan-out. For example, if $N = 5$, then the output of the NOR gate is loaded by five 450-Ω resistors in parallel (or 90 Ω), which is tied to $V_{BE,\text{sat}} \approx 0.8$ V, as shown in Fig. 6-23*b*. Under these circumstances (using superposition),

$$v_o = \frac{3.6 \times 90}{90 + 640} + \frac{0.8 \times 640}{90 + 640} = 1.14 \text{ V} = V(1) \tag{6-37}$$

This voltage must be large enough so that the base current can drive each of the five transistors into saturation. Since

$$I_B = \frac{1.14 - 0.8}{0.45} = 0.755 \text{ mA} \qquad I_C = \frac{3.6 - 0.2}{0.64} = 5.31 \text{ mA}$$

then the circuit will operate properly if $h_{FE} > (h_{FE})_{\min} = 5.31/0.76 = 7.0$.

Resistor-transistor logic uses the minimum space (for a standard digital function) on a silicon wafer, and hence is very economical.

6-13   PROPAGATION DELAY TIMES

In the discussion of each logic configuration some of its advantages and disadvantages have been listed. An exhaustive comparison is extremely difficult because we must take into account all the following characteristics: (1) speed (propagation time delay), (2) noise immunity, (3) fan-in and fan-out capabilities, (4) power-supply requirements, (5) power dissipation per gate, (6) suitability for integrated fabrication, (7) operating temperature range, (8) number of functions available, and (9) cost. Also to be considered is the personal prejudice of the engineer, who is always strongly influenced by past experience. Items 1 and 8 require some explanation; all others have already been defined or are self-evident.

Propagation Delay As the input voltage to a positive NAND gate rises from $V(0)$ toward $V(1)$, then at some *switching threshold voltage* $V'(0)$ (Fig. 6-24), conditions within the gate are modified, so that a change of state of the output from $V(1)$ to $V(0)$ is initiated. Similarly, as the input falls from $V(1)$ toward $V(0)$, then at some other *switching threshold voltage* $V'(1)$, the initiation of the change of state from $V(0)$ to $V(1)$ takes place. We now define (as in Fig. 6-24) the propagation delay ON time T_{ON} as the interval between the time when the input v_i reaches $V'(0)$ and the output falls to $V'(1)$. Also, the propagation delay OFF time T_{OFF} is defined as the interval between the time when the input equals $V'(1)$ and the output rises to $V'(0)$. Because of minority-carrier storage time, $T_{ON} \neq T_{OFF}$. Hence the *propagation delay time T_{PD}* is usually defined as the average of these two times, or

$$T_{PD} \equiv \tfrac{1}{2}(T_{ON} + T_{OFF}) \tag{6-38}$$

In passing, we note that some authors arbitrarily assume the two threshold voltages to be equal: $V'(0) = V'(1) = \tfrac{1}{2}[V(0) + V(1)]$. Logic gates have values of T_{PD} which range from one to several hundred nanoseconds.

Functions The basic AND, OR, NAND, and NOR gates are combined in one integrated chip in various combinations to perform specific circuit functions. These *system building blocks* are discussed in Chaps. 7, 8, and 9 and include flip-flops, counters, arithmetic functions, decoders, shift registers, etc.

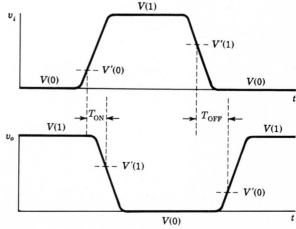

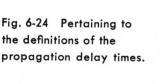

Fig. 6-24   Pertaining to the definitions of the propagation delay times.

REVIEW QUESTIONS

6-1 Express the following decimal numbers in binary form: (*a*) 28; (*b*) 100; (*c*) 5,127.

6-2 Define (*a*) *positive logic;* (*b*) *negative logic.*

6-3 Define an OR gate and give its truth table.

6-4 Draw a positive diode OR gate and explain its operation.

6-5 Evaluate the following expressions: (*a*) $A + 1$; (*b*) $A + A$; (*c*) $A + 0$.

6-6 Define an AND gate and give its truth table.

6-7 Draw a positive-diode AND gate and explain its operation.

6-8 Evaluate the following expressions: (*a*) $A1$; (*b*) AA; (*c*) $A0$; (*d*) $A + AB$.

6-9 Define a NOT gate and give its truth table.

6-10 Draw a positive-logic NOT gate and explain its operation.

6-11 Evaluate the following expressions: (*a*) $\bar{\bar{A}}$; (*b*) $\bar{A}A$; (*c*) $\bar{A} + A$.

6-12 Define an INHIBITOR and give the truth table for $AB\bar{S}$.

6-13 Define an EXCLUSIVE OR and give its truth table.

6-14 Show two logic block diagrams for an EXCLUSIVE OR.

6-15 Verify that the following Boolean expressions represent an EXCLUSIVE OR: (*a*) $\overline{AB + \bar{A}\bar{B}}$; (*b*) $(A + B)(\bar{A} + \bar{B})$.

6-16 State the two forms of De Morgan's laws.

6-17 Show how to implement an AND with OR and NOT gates.

6-18 Show how to implement an OR with AND and NOT gates.

6-19 Define a NAND gate and give its truth table.

6-20 Draw a positive NAND gate with diodes and a transistor (DTL) and explain its operation.

6-21 Define a NOR gate and give its truth table.

6-22 Repeat Rev. 6-20 for a NOR gate.

6-23 Draw the circuit of an IC DTL gate and explain its operation.

6-24 Define (*a*) *fan-out;* (*b*) *fan-in;* (*c*) *standard load;* (*d*) *current sink;* (*e*) *current source.*

6-25 What logic is performed if the outputs of two DTL gates are connected together? Explain.

6-26 Draw the circuit of a TTL gate and explain its operation.

6-27 Draw a totem-pole output buffer with a TTL gate. Explain its operation.

6-28 Draw the circuit of an RTL gate and explain its operation for positive logic.

6-29 Repeat Rev. 6-28 for negative logic.

7 / COMBINATIONAL DIGITAL SYSTEMS

A digital system is constructed from very few types of basic network configurations, these elementary types being used over and over again in various topological combinations. As emphasized in Sec. 6-7, it is possible to perform all logic operations with a single type of circuit (for example, a NAND gate). A digital system must store binary numbers in addition to performing logic. To take care of this requirement, a memory cell, called a FLIP-FLOP, is introduced in the next chapter.

Theoretically, any digital system can be constructed entirely from NAND gates and FLIP-FLOPS. Some functions (such as binary addition) are present in many systems, and hence the combination of gates and/or FLIP-FLOPS required to perform this function is available on a single chip. These integrated circuits form the practical (commercially available) basic building blocks for a digital system. The number of such different ICs is not large, and these chips perform the following functions: binary addition, decoding (demultiplexing), data selection (multiplexing), counting, storage of binary information (memories and registers), digital-to-analog (D/A) and analog-to-digital (A/D) conversion, and a few other related operations. Those building blocks which depend upon combinational logic are described in this chapter.

7-1 STANDARD GATE ASSEMBLIES

Since the fundamental gates are used in large numbers even in a relatively simple digital system, they are not packaged individually; rather, several gates are constructed within a single chip. The following list of standard digital IC gates is typical, but far from exhaustive:

Quad two-input NAND	Quad two-input NOR
Triple three-input NAND	Quad two-input AND
Dual four-input NAND	Hex inverter buffer
Single eight-input NAND	Dual two-wide, two-input AOI

These combinations are available in most logic families (TTL, DTL, etc.) listed in Sec. 6-12. The limitation on the number of gates per chip is usually set by the number of pins available. The most common packages are the *flat pack* and the *dual-in-line* (plastic package, type N, or ceramic package, type J), each of which has fourteen leads, seven brought out to each side of the IC. The dimensions of the assembly, which is much larger than the chip size, are approximately 0.8 by 0.3 by 0.2 in. A schematic of the triple three-3 output leads, a power-supply lead, and a ground lead; a total of 14 leads input NAND is shown in Fig. 7-1a. Note that there are $3 \times 3 = 9$ input leads, are used.

In Fig. 7-1b is indicated the dual two-wide, two-input AOI (Sec. 6-6). This combination needs 4 input leads and 1 output lead per AOI, or 10 for the dual array. If 1 power-supply lead and 1 ground lead are added, we see that 12 of the 14 available pins are used.

The circuit diagram for this AOI gate is given in Fig. 7-2, implemented in TTL logic. The operation of this network should be clear from the discussion in Chap. 6. Thus $Q1$ and the input to $Q2$ (corresponding to the similarly numbered transistors of Fig. 6-21) constitute an AND gate. The identical arrangement of $Q5$ and $Q6$ constitutes a second AND gate. Since the collectors of $Q2$ and $Q6$ are tied together at P, the output at this node corresponds to either the inputs ② AND ③ OR ④ AND ⑤. Also, because of the inversion through a transistor, the NOT operation is performed at P. The result is AND-OR-INVERT (AOI) logic ($\overline{AB + CD}$). Finally, note that $Q3$, DO, and $Q4$ are the totem-pole output stage of Fig. 6-22.

An alternative way of analyzing the circuit of Fig. 7-2 is to consider $Q1$ and $Q2$ (with the output at P) to constitute a NAND circuit. Similarly, $Q5$

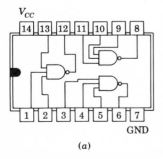 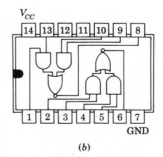

(a)   (b)

Fig. 7-1 The lead connections (top view) of (a) the TI 7410 triple three-input NAND and (b) the TI 7451 dual two-wide, two-input AOI gate.

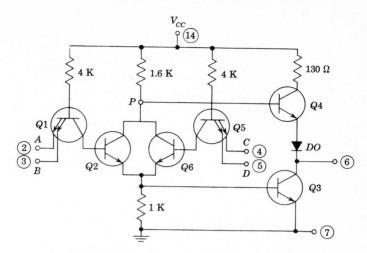

Fig. 7-2   The circuit configuration for a TTL AND-OR-INVERT **gate.**

and $Q6$ form a second NAND gate. The outputs of these two NAND configurations are shorted together by the lead connecting the collectors of $Q2$ and $Q6$ to form a wired-AND (Sec. 6-9). Hence the output at P is, using De Morgan's law Eq. (6-26)

$$(\overline{AB})(\overline{CD}) = \overline{AB + CD}$$

which confirms that AOI logic is performed.

Some of the more complicated functions to be described in this chapter require in excess of 14 pins, and these ICs are packaged with either 16 or 24 leads. The latter has the dimensions 1.3 by 0.6 by 0.2 in.

The standard combinations considered in this section are examples of *small-scale integration* (SSI). Less than 100 individual circuit components (about 12 gates) on a chip is considered SSI. The FLIP-FLOPS discussed in Sec. 8-1 are also SSI packages. Most other functions (using BJTs) discussed in this chapter are examples of *medium-scale integration* (MSI), defined to have more than 100, but less than 1,000 components (about 100 gates) per chip. The BJT memories of Sec. 9-7 and the MOSFET arrays of Sec. 9-6 may contain in excess of 1,000 components and are defined as *large-scale integration* (LSI).

7-2   BINARY ADDERS

A digital computer must obviously contain circuits which will perform arithmetic operations, i.e., addition, subtraction, multiplication, and division. The basic operations are addition and subtraction, since multiplication is essentially repeated addition, and division is essentially repeated subtraction. It is entirely possible to build a computer in which an *adder* is the only arithmetic unit present. Multiplication, for example, may then be performed by *programming;* i.e., the computer may be given instructions telling it how to use the adder repeatedly to find the product of two numbers.

Suppose we wish to sum two numbers in decimal arithmetic and obtain, say, the hundreds digit. We must add together not only the hundreds digit of each number but also a carry from the tens digit (if one exists). Similarly, in binary arithmetic we must add not only the digit of like significance of the two numbers to be summed, but also the carry bit (should one be present) of the next lower significant digit. This operation may be carried out in two steps: first, add the two bits corresponding to the 2^n digit, and then add the resultant to the carry from the 2^{n-1} digit. A two-input adder is called a *half adder*, because to complete an addition requires two such half adders.

We shall first show how a *half adder* is constructed from the basic logic gates and then indicate how the *full*, or *complete*, *adder* is assembled. A half adder has two inputs—A and B—representing the bits to be added, and two outputs—D (for the digit of the same significance as A and B represent) and C (for the carry bit).

Half Adder The symbol for a half adder is given in Fig. 7-3a, and the truth table in Fig. 7-3b. Note that the D column gives the sum of A and B as long as the sum can be represented by a single digit. When, however, the sum is larger than can be represented by a single digit, then D gives the digit in the result which is of the same significance as the individual digits being added. Thus, in the first three rows of the truth table, D gives the sum of A and B directly. Since the decimal equation "1 plus 1 equals 2" is written in binary form as "01 plus 01 equals 10," then in the last row $D = 0$. Because a 1 must now be carried to the place of next higher significance, $C = 1$.

From Fig. 7-3b we see that D obeys the EXCLUSIVE-OR function and C follows the logic of an AND gate. These functions are indicated in Fig. 7-3c, and may be implemented in many different ways with the circuitry discussed

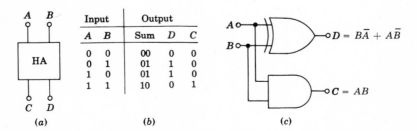

Input		Output		
A	B	Sum	D	C
0	0	00	0	0
0	1	01	1	0
1	0	01	1	0
1	1	10	0	1

$D = B\bar{A} + A\bar{B}$

$C = AB$

(a)   (b)   (c)

Fig. 7-3 (a) The symbol for a half adder; (b) the truth table for the digit D and the carry C; (c) the implementation for D with an EXCLUSIVE-OR gate and C with an AND gate.

in Chap. 6. For example, the EXCLUSIVE-OR gate can be constructed with any of the four topologies of Sec. 6-6. The configuration in Fig. 6-11b ($Y = A\bar{B} + B\bar{A}$) is implemented in TTL logic with the AOI circuit of Fig. 7-2. The inverter for B (or A) is a single-input NAND gate. Since Y has an AND-OR (rather than an AND-OR-INVERT) topology, a transistor inverter is placed between node P and the base of $Q4$ of Fig. 7-2.

Parallel Operation Two multidigit numbers may be added serially (one column at a time) or in parallel (all columns simultaneously). Consider parallel operation first. For an N-digit binary number there are (in addition to a common ground) N signal leads in the computer for each number. The nth line for number A (or B) is excited by A_n (or B_n), the bit for the 2^n digit ($n = 0, 1, \ldots, N - 1$). A parallel binary adder is indicated in Fig. 7-4. Each digit except the least-significant one (2^0) requires a complete adder consisting of two half adders in cascade. The sum digit for the 2^0 bit is $S_0 = D_0$ of a half adder because there is no carry to be added to A_0 plus B_0. The sum S_n ($n \neq 0$) of A_n plus B_n is made in two steps. First the digit D_n is obtained from one half adder, and then D_n is summed with the carry C_{n-1} which may have resulted from the next lower place. As an example, consider $n = 2$ in Fig. 7-4. There the carry bit C_1 may be the result of the direct sum of A_1 plus B_1 if each of these is 1. This first carry is called C_{11} in Fig. 7-4. A second possibility is that $A_1 = 1$ and $B_1 = 0$ (or vice versa), so that $D_1 = 1$, but that there is a carry C_0 from the next lower significant bit. The sum of $D_1 = 1$ and $C_0 = 1$ gives rise to the carry bit designated C_{12}. It should be clear that C_{11} and C_{12} cannot both be 1, although they will both be 0 if $A_1 = 0$ and $B_1 = 0$. Since either C_{11} or C_{12} must be transmitted to the next stage, an OR gate must be interposed between stages, as indicated in Fig. 7-4.

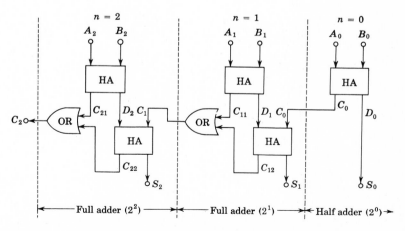

Fig. 7-4 A parallel binary adder consisting of half adders.

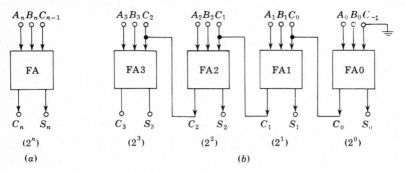

Fig. 7-5 (a) The symbol for a full adder. (b) A 4-bit parallel binary adder constructed from cascaded full adders.

Full Adder In integrated-circuit implementation, addition is performed using a complete adder, which (for reasons of economy of components) is not constructed from two half adders. The symbol for the nth full adder (FA) is indicated in Fig. 7-5a. The circuit has three inputs: the addend A_n, the augend B_n, and the input carry C_{n-1} (from the next lower bit). The outputs are the sum S_n (sometimes designated Σ_n) and the output carry C_n. A parallel 4-bit adder is indicated in Fig. 7-5b. Since FA0 represents the least significant bit (LSB), it has no input carry; hence $C_{-1} = 0$.

The circuitry within the block FA may be determined from Fig. 7-6, which is the truth table for adding 3 binary bits. From this table we can verify that the Boolean expressions for S_n and C_n are given by

$$S_n = \bar{A}_n\bar{B}_nC_{n-1} + \bar{A}_nB_n\bar{C}_{n-1} + A_n\bar{B}_n\bar{C}_{n-1} + A_nB_nC_{n-1} \tag{7-1}$$

$$C_n = \bar{A}_nB_nC_{n-1} + A_n\bar{B}_nC_{n-1} + A_nB_n\bar{C}_{n-1} + A_nB_nC_{n-1} \tag{7-2}$$

Note that the first term of S_n corresponds to line 1 of the table, the second term to line 2, the third term to line 4, and the last term to line 7. (These are the only rows where $S_n = 1$.) Similarly, the first term of C_n corresponds to line 3 (where $C_n = 1$), the second term to line 5, etc.

The AND operation ABC is sometimes called the *product* of A and B and C. Also, the OR operation $+$ is referred to as *summation*. Hence expressions such

Line	Inputs			Outputs	
	A_n	B_n	C_{n-1}	S_n	C_n
0	0	0	0	0	0
1	0	0	1	1	0
2	0	1	0	1	0
3	0	1	1	0	1
4	1	0	0	1	0
5	1	0	1	0	1
6	1	1	0	0	1
7	1	1	1	1	1

Fig. 7-6 Truth table for a three-input adder.

as those in Eqs. (7-1) and (7-2) represent a *Boolean sum of products*. Such an equation is said to be in a *standard*, or *canonical*, *form*, and each term in the equation is called a *minterm*. A minterm contains the product of all Boolean variables, or their complements.

The expression for C_n can be simplified considerably as follows: Since $Y + Y + Y = Y$, then Eq. (7-2), with $Y = A_n B_n C_{n-1}$, becomes

$$C_n = (\bar{A}_n B_n C_{n-1} + A_n B_n C_{n-1}) + (A_n \bar{B}_n C_{n-1} + A_n B_n C_{n-1})$$
$$+ (A_n B_n \bar{C}_{n-1} + A_n B_n C_{n-1}) \quad (7\text{-}3)$$

Since $\bar{X} + X = 1$ where $X = A_n$ for the first parentheses, $X = B_n$ for the second parentheses, and $X = C_{n-1}$ for the third parentheses, then Eq. (7-3) reduces to

$$C_n = B_n C_{n-1} + C_{n-1} A_n + A_n B_n \quad (7\text{-}4)$$

This expression could have been written down directly from the truth table of Fig. 7-6 by noting that $C_n = 1$ if and only if at least two out of the three inputs is 1.

It is interesting to note that if all 1s are changed to 0s and all 0s to 1s, then lines 0 and 7 are interchanged, as are 1 and 6, 2 and 5, and also 3 and 4. Because this switching of 1s and 0s leaves the truth table unchanged, whatever logic is represented by Fig. 7-6 is equally valid if all inputs and outputs are complemented. Therefore Eq. (7-3) is true if all variables are negated, or

$$\bar{C}_n = \bar{B}_n \bar{C}_{n-1} + \bar{C}_{n-1} \bar{A}_n + \bar{A}_n \bar{B}_n \quad (7\text{-}5)$$

This same result is obtained (Prob. 7-3) by Boolean manipulation of Eq. (7-4).

By evaluating $D_n \equiv (A_n + B_n + C_{n-1})\bar{C}_n$ and comparing the result with Eq. (7-1), we find that $S_n \equiv D_n + A_n B_n C_{n-1}$, or

$$S_n = A_n \bar{C}_n + B_n \bar{C}_n + C_{n-1} \bar{C}_n + A_n B_n C_{n-1} \quad (7\text{-}6)$$

Equations (7-4) and (7-6) are implemented in Fig. 7-7 using AOI gates of the type shown in Fig. 7-2.

MSI Adders   There are commercially available 1-bit, 2-bit, and 4-bit full adders, each in one package. In Fig. 7-8 is indicated the logic topology for 2-bit addition. The inputs to the first stage are A_0 and A_1; the input marked C_{-1} is grounded. The output is the sum S_0. The carry C_0 is connected internally and is not brought to an output pin. This 2^0 stage (LSB) is identical with that in Fig. 7-7 with $n = 0$. The abbreviation LSB means least significant bit.

Since the carry from the first stage is $\bar{C}_0$, it should be negated before it is fed to the 2^1 stage. However, the delay introduced by this inversion is undesirable, because the limitation upon the maximum speed of operation is the propagation delay (Sec. 6-13) of the carry through all the bits in the adder.

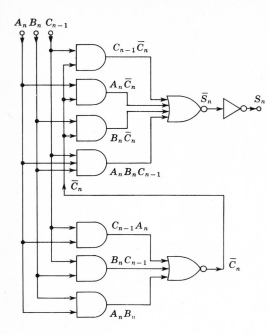

Fig. 7-7 Block-diagram implementation of the nth stage of a full adder.

The NOT-gate delay is eliminated completely in the carry by connecting $\bar{C}_0$ directly to the following stage and by complementing the inputs A_1 and B_1 before feeding these to this stage. This latter method is used in Fig. 7-8. Note that now the outputs S_1 and C_1 are obtained directly without requiring inverters. The logic followed by this second stage for the carry is given by Eq. (7-5), and for the sum by the modified form of Eq. (7-6), where each symbol is replaced by its complement.

In a 4-bit adder C_1 is not brought out but is internally connected to the third stage, which is identical with the first stage. Similarly, the fourth and second stages have identical logic topologies. A 4-bit adder requires a 16-pin package: 8 inputs, 4 sum outputs, a carry output, a carry input, the power-supply input, and ground. The carry input is needed only if two arithmetic units are cascaded; for example, cascading a 2-bit with a 4-bit adder gives the sum of two 6-bit numbers. If the 2-bit unit is used for the 2^4 and 2^5 digits, then 4 must be added to all the subscripts in Fig. 7-8. For example, C_{-1} is now called C_3 and is obtained from the output carry of the 4-bit adder.

The MSI chip (TI 74LS83†) for a 4-bit binary full adder contains somewhat over 200 components (resistors, diodes, or transistors). For high-speed, low-power operation, Schottky transistors and diodes (Sec. 5-10) are used in each AOI block of Fig. 7-2. The NOT circuit for S_0 is simply a transistor

---

† The specific designations given in this chapter refer to Texas Instrument units. However, equivalent units are available from other vendors, such as Fairchild Semiconductor, Motorola, Inc., National Semiconductor, RCA, Signetics, etc.

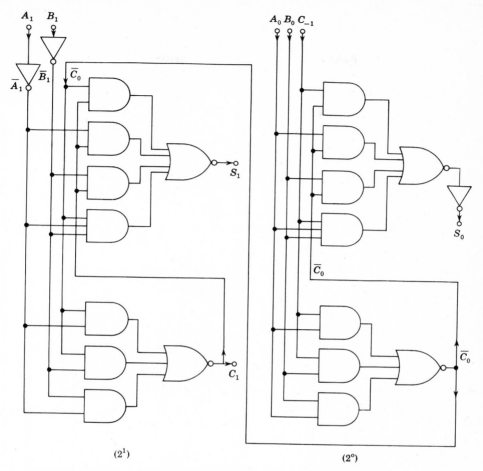

Fig. 7-8 Logic diagram of an integrated 2-bit full adder (TI 7482).

inverter placed between node P and the base of $Q4$ of Fig. 7-2. The NOT circuit for inverting A_1 (or B_1) is a single-input NAND gate. The propagation delay time of the carry is typically 50 ns, and the power dissipation is 75 mW.

Serial Operation In a serial adder the inputs A and B are synchronous pulse trains on two lines in the computer. Figure 7-9a and b shows typical pulse trains representing, respectively, the decimal numbers 13 and 11. Pulse trains representing the sum (24) and difference (2) are shown in Fig. 7-9c and d, respectively. A serial *adder* is a device which will take as inputs the two waveforms of Fig. 7-9a and b and deliver the output waveform in Fig. 7-9c. Similarly, a *subtractor* will yield the output shown in Fig. 7-9d.

We have already emphasized that the sum of two multidigit numbers may

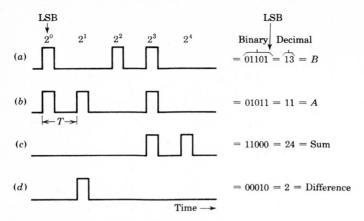

Fig. 7-9 (a,b) Pulse waveforms representing numbers B and A; (c,d) waveforms representing sum and difference. (LSB = least significant bit.)

be formed by adding to the sum of the digits of like significance the carry (if any) which may have resulted from the next lower place. With respect to the pulse trains of Fig. 7-9, the above statement is equivalent to saying that, at any instant of time, we must add (in binary form) to the pulses A and B the carry pulse (if any) which comes from the resultant formed one period T earlier. The logic outlined above is performed by the full-adder circuit of Fig. 7-10. The circuit differs from the configuration in the parallel adder of Fig. 7-5 by the inclusion of a time delay TD which is equal to the time T between pulses. Hence the carry pulse is delayed a time T and added to the digit pulses in A and B, exactly as it should be.

A comparison of Figs. 7-5 and 7-10 indicates that parallel addition is faster than serial because all digits are added simultaneously in the former, but in sequence in the latter. However, whereas only one full adder is needed for serial arithmetic, we must use a full adder for each bit in parallel addition. Hence parallel addition is much more expensive than serial operation.

The time delay unit TD is a type D FLIP-FLOP, and the serial numbers A_n, B_n, and S_n are stored in *shift registers* (Secs. 8-2 and 8-3).

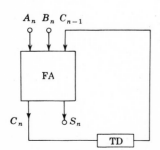

Fig. 7-10 A serial binary full adder.

TABLE 7-1   BCD representation for the decimal number 264

Weighting factor	800	400	200	100	80	40	20	10	8	4	2	1
BCD code	0	0	1	0	0	1	1	0	0	1	0	0
Decimal digits			2				6				4	

7-3   DECODER/DEMULTIPLEXER

In a digital system, instructions as well as numbers are conveyed by means of binary levels or pulse trains. If, say, 4 bits of a character are set aside to convey instructions, then 16 different instructions are possible. This information is *coded* in binary form. Frequently a need arises for a multiposition switch which may be operated in accordance with this code. In other words, for each of the 16 codes, one and only one line is to be excited. This process of identifying a particular code is called *decoding*.

Binary-Coded-Decimal (BCD) System   This code translates decimal numbers by replacing each decimal digit with a combination of 4 binary digits. Since there are 16 distinct ways in which the 4 binary digits can be arranged in a row, any 10 combinations can be used to represent the decimal digits from 0 to 9. Thus we have a wide choice of BCD codes. One of these, called the "natural binary-coded-decimal," is the 8421 code illustrated by the first *Par* 10 entries in Table 6-2. This is a weighted code because the decimal digit in the 8421 code is equal to the sum of the products of the bits in the coded words times the successive powers of two starting from the right (LSB). We need N 4-bit sets to represent in BCD notation an N-digit decimal number. The first 4-bit set on the right represents units, the second represents tens, the third hundreds, and so on. For example, the decimal number 264 requires three 4-bit sets, as shown in Table 7-1. Note that this three-decade BCD code can represent any number between 0 and 999; hence it has a resolution of 1 part in 1,000, or 0.1 percent. It requires 12 bits, which in a straight binary code can resolve one part in $2^{12} = 4,096$, or 0.025 percent.

BCD-to-Decimal Decoder   Suppose we wish to decode a BCD instruction representing one decimal digit, say 5. This operation may be carried out with a four-input AND gate excited by the 4 BCD bits. For example, the output of the AND gate in Fig. 7-11 is 1 if and only if the BCD inputs are $A = 1$ (LSB), $B = 0$, $C = 1$, and $D = 0$. Since this code represents the decimal number 5, the output is labeled "line 5."

Fig. 7-11   The output is 1 if the BCD input is 0101 and is 0 for any other input instruction.

A BCD-to-decimal decoder is indicated in Fig. 7-12. This MSI unit has four inputs, A, B, C, and D, and 10 output lines. (Ignore the dashed lines, for the moment.) In addition, there must be a ground and a power-supply connection, and hence a 16-pin package is required. The complementary inputs $\bar{A}$, $\bar{B}$, $\bar{C}$, and $\bar{D}$ are obtained from inverters on the chip. Since NAND gates are used, an output is 0 (low) for the correct BCD code and is 1 (high) for any other (invalid) code. The system in Fig. 7-12 is also referred to as a "4-to-10-line decoder" designating that a 4-bit input code selects 1 of 10 output lines. In other words, the decoder acts as a 10-position switch which responds to a BCD input instruction.

It is sometimes desired to decode only during certain intervals of time. In such applications an additional input, called a *strobe*, is added to each NAND gate. All strobe inputs are tied together and are excited by a binary signal S, as indicated by the dashed lines in Fig. 7-12. If $S = 1$, a gate is *enabled* and decoding takes place, whereas if $S = 0$, no coincidence is possible and decoding is inhibited. The strobe input can be used with a decoder having any number of inputs or outputs.

Demultiplexer A *demultiplexer* is a system for transmitting a binary signal (serial data) on one of N lines, the particular line being selected by

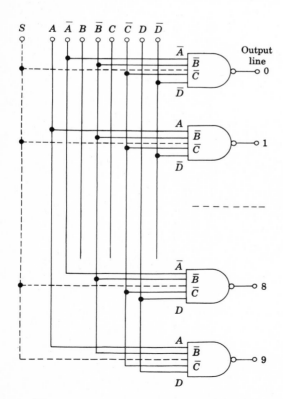

Fig. 7-12 A BCD-to-decimal decoder; assume that $S = 1$. (Lines 2 to 7 are not indicated.) The dashed lines convert the system into a demultiplexer if S represents the input signal.

Fig. 7-13   A decoder is converted into a demultiplexer (with an enabling input) if the S terminal
in Fig. 7-12 is obtained from the above AND
gate output.

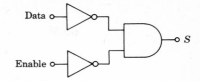

means of an address.   A decoder is converted into a demultiplexer by means
of the dashed connections in Fig. 7-12.   If the data signal is applied at S,
then the output will be the complement of this signal (because the output is
0 if all inputs are 1) and will appear only on the addressed line.

An enabling signal may be applied to a demultiplexer by cascading the
system of Fig. 7-12 with that indicated in Fig. 7-13.   If the *enable* input is
0, then S is the complement of the data.   Hence, the data will appear (without
inversion) on the line with the desired code.   If the enable input is 1, $S = 0$,
the data are inhibited from appearing on any line and all inputs remain at 1.

4-to-16-Line Decoder/Demultiplexer   If an address corresponding to a
decimal number in excess of 9 is applied to the inputs in Fig. 7-12, this instruction is rejected; that is, all 10 outputs remain at 1.   If it is desired to select
1 of 16 output lines, the system is expanded by adding 6 more NAND gates
and using all 16 codes possible with 4 binary bits.

The TI 74154 is a 4-to-16-line decoder/demultiplexer.   It has 4 address
lines, 16 output lines, an enable input, a data input, a ground pin, and a
power-supply lead, so that a 24-pin package is required.

A 2-to-4-line and a 3-to-8-line decoder/demultiplexer are also available
in individual IC packages.

Decoder/Lamp Driver   Some decoders are equipped with special output
stages so that they can drive lamps such as the Burroughs Nixie tube.   A
Nixie indicator is a cold-cathode gas-discharge tube with a single anode and
10 cathodes, which are wires shaped in the form of numerals 0 to 9.   These
cathodes are connected to output lines 0 to 9, respectively, and the anode is
tied to a fixed supply voltage.   The decoder/lamp driver/Nixie indicator combination makes visible the decimal number corresponding to the BCD number
applied.   Thus, if the input is 0101, the numeral 5 will glow in the lamp.

A decoder for seven-segment numerals made visible by using light-emitting diodes is discussed in Sec. 7-7.

7-4   DATA SELECTOR/MULTIPLEXER

The function performed by a *multiplexer* is to select 1 out of N input data
sources and to transmit the selected data to a single information channel.
Since in a demultiplexer there is only one input line and these data are caused

to appear on 1 out of N output lines, a multiplexer performs the inverse process of a demultiplexer.

The demultiplexer of Fig. 7-12 is converted into a multiplexer by making the following two changes: (1) Add a NAND gate whose inputs include all N outputs of Fig. 7-12 and (2) augment each NAND gate with an individual data input $X_0, X_1, \ldots, X_N$. The logic system for a 4-to-1-line data-selector multiplexer is drawn in Fig. 7-14. This AND-OR logic is equivalent to the NAND-NAND logic as described in the above steps 1 and 2. (See Fig. 6-18.) Note that the same decoding configuration is used in both the multiplexer and demultiplexer. If the select code is 01, then X_1 appears at the output Y, if the address is 11, then $Y = X_3$, etc., provided that the system is enabled ($S = 0$). Multiplexers are also available for selecting 1 of 8 or 1 of 16 data sources. The latter (TI 74150) is a 24-pin IC with 16 data inputs, a 4-bit select code, a strobe input, one output, a power-supply lead, and a ground terminal. For a 16-to-1 line multiplexer, Fig. 7-14 is extended from four 4-input AND gates to sixteen 6-input AND circuits. Two 16-data-input multiplexers may be interconnected to select 1 out of 32 information sources (Prob. 7-11).

Parallel-to-Serial Conversion Consider a 16-bit word available in parallel form so that X_0 represents the 2^0 bit, X_1 the 2^1 bit, etc. By means of a counter (Sec. 8-4), it is possible to change the select code so that it is 0000 for the first T s, 0001 for the second T s, 0010 for the third interval T, etc. With such excitation of the address, the output of the multiplexer will be X_0 for the first T s, X_1 for the next interval T, X_2 for the third period, etc. The output Y is a waveform which represents serially the binary data applied in parallel at the input. In other words, a parallel-to-serial conversion is accomplished of one 16-bit word. This process takes 16 T s.

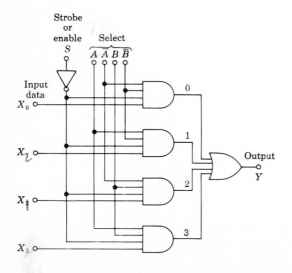

Fig. 7-14 A 4-to-1-line multiplexer.

Sequential Data Selection   By changing the address with a counter in the manner indicated in the preceding paragraph, the operation of an electro-mechanical stepping switch is simulated.   If the data inputs are pulse trains, this information will appear sequentially on the output channel: in other words, pulse train X_0 will appear for T s, followed by X_1 for the next T s, etc.   If the number of data sources is M, then X_0 is again selected during the interval $MT < t < (M + 1)T$.

7-5   ENCODER

A decoder is a system which accepts an M-bit word and establishes the state 1 on one (and only one) of 2^M output lines (Sec. 7-3).   In other words, a decoder identifies (recognizes) a particular code.   The inverse process is called *encoding*.   An encoder has a number of inputs, only one of which is in the 1 state, and an N-bit code is *generated*, depending upon which of the inputs is excited.

Consider, for example, that it is required that a binary code be transmitted with every stroke of an alphanumeric keyboard (a typewriter or teletype). There are 26 lowercase and 26 capital letters, 10 numerals, and about 22 special characters on such a keyboard so that the total number of codes necessary is approximately 84.   This condition can be satisfied with a minimum of 7 bits ($2^7 = 128$, but $2^6 = 64$).   Let us modify the keyboard so that, if a key is depressed, a switch is closed, thereby connecting a 5-V supply (corresponding to the 1 state) to an input line.   A block diagram of such an encoder is indicated in Fig. 7-15.   Inside the shaded block there is a rectangular array (or matrix) of wires, and we must determine how to interconnect these wires so as to generate the desired codes.

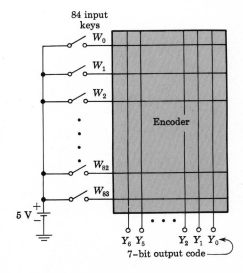

Fig. 7-15   A block diagram of an encoder for generating an output code (word) for every character on a keyboard.

To illustrate the design procedure for constructing an encoder, let us simplify the above example by limiting the keyboard to only 10 keys, the numerals 0, 1, . . . , 9. A 4-bit output code is sufficient in this case, and let us choose BCD words for the output codes. The truth table defining this encoding is given in Table 7-2. Input W_n $(n = 0, 1, . . . , 9)$ represents the nth key. When $W_n = 1$, key n is depressed. Since it is assumed that no more than one key is activated simultaneously, then in any row every input except one is a 0. From this truth table we conclude that $Y_0 = 1$ if $W_1 = 1$ or if $W_3 = 1$ or if $W_5 = 1$ or if $W_7 = 1$ or if $W_9 = 1$. Hence, in Boolean notation,

$$Y_0 = W_1 + W_3 + W_5 + W_7 + W_9 \qquad (7\text{-}7)$$

Similarly,

$$\begin{aligned}
Y_1 &= W_2 + W_3 + W_6 + W_7 \\
Y_2 &= W_4 + W_5 + W_6 + W_7 \\
Y_3 &= W_8 + W_9
\end{aligned} \qquad (7\text{-}8)$$

The OR gates in Eqs. (7-7) and (7-8) are implemented with diodes in Fig. 7-16. (Compare with Fig. 6-3, but with the diodes reversed, because we are now considering positive logic.) An encoder array such as that in Fig. 7-16 is called a *rectangular diode matrix*.

Incidentally, a decoder can also be constructed as a rectangular diode matrix (Prob. 7-14). This statement follows from the fact that a decoder consists of AND gates (Fig. 7-11), and it is possible to implement AND gates with diodes (Fig. 6-5b).

Each diode of the encoder of Fig. 7-16 may be replaced by the base-emitter diode of a transistor. If the collector is tied to the supply voltage V_{CC},

TABLE 7-2   The truth table for encoding the decimal numbers 0 to 9

Inputs										Outputs			
W_9	W_8	W_7	W_6	W_5	W_4	W_3	W_2	W_1	W_0	Y_3	Y_2	Y_1	Y_0
0	0	0	0	0	0	0	0	0	1	0	0	0	0
0	0	0	0	0	0	0	0	1	0	0	0	0	1
0	0	0	0	0	0	0	1	0	0	0	0	1	0
0	0	0	0	0	0	1	0	0	0	0	0	1	1
0	0	0	0	0	1	0	0	0	0	0	1	0	0
0	0	0	0	1	0	0	0	0	0	0	1	0	1
0	0	0	1	0	0	0	0	0	0	0	1	1	0
0	0	1	0	0	0	0	0	0	0	0	1	1	1
0	1	0	0	0	0	0	0	0	0	1	0	0	0
1	0	0	0	0	0	0	0	0	0	1	0	0	1

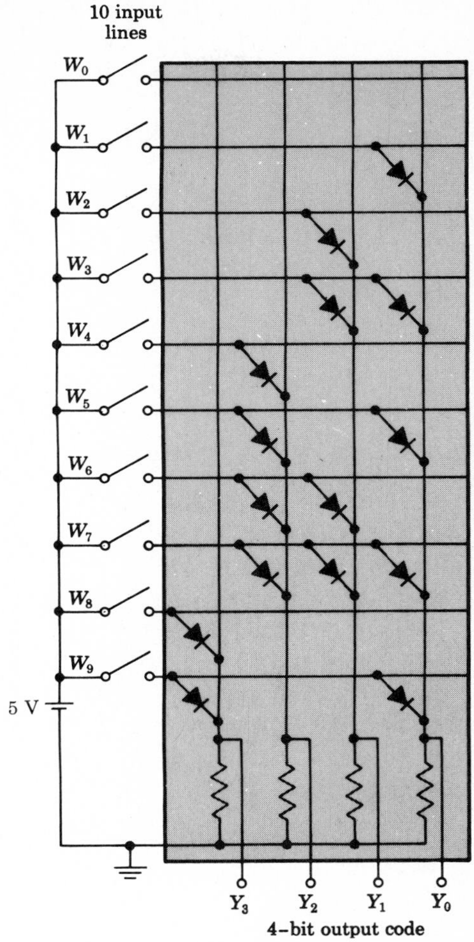

10 input
lines

Fig. 7-16 An encoding matrix to transform a decimal number into a binary code (BCD).

Y_3 Y_2 Y_1 Y_0
4–bit output code

then the circuit of Fig. 7-17a, called an *emitter-follower* OR gate results. Such a configuration is indicated for the output Y_2 of Eq. (7-8). Note that if either W_4 or W_5 or W_6 or W_7 is high, then the emitter follows the high-input base voltage and the output is high, thus verifying that $Y_2 = W_4 + W_5 + W_6 + W_7$, as required by Eq. (7-8).

Only one transistor (with multiple emitters) is required for each encoder input. The base is tied to the input line, and each emitter is connected to a different output line, as dictated by the encoder logic. For example, since in Fig. 7-16 line W_7 is tied to three diodes whose cathodes go to Y_0, Y_1, and Y_2, then this combination may be replaced by the three-emitter transistor $Q7$ connected as in Fig. 7-17b. The maximum number of emitters that may be required equals the number of bits in the output code. For the particular encoder sketched in Fig. 7-16, $Q1$, $Q2$, $Q4$, and $Q8$ each have one emitter, $Q3$, $Q5$, $Q6$, and $Q9$ have two emitters each, and $Q7$ has three emitters.

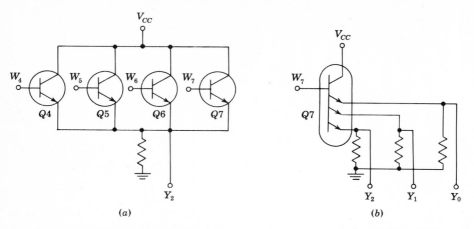

Fig. 7-17 (a) An emitter-follower OR gate. (b) Line W_7 in the encoder of Fig. 7-16 is connected to the base of a three-emitter transistor.

7-6   READ–ONLY MEMORY (ROM)

Consider the problem of converting one binary code into another. Such a code-conversion system (designated ROM and sketched in Fig. 7-18a) has M inputs $(X_0, X_1, \ldots, X_{M-1})$ and N outputs $(Y_0, Y_1, \ldots, Y_{N-1})$, where N may be greater than, equal to, or less than M. A definite M-bit code is to result in a specific output code of N bits. This code translation is achieved, as indicated in Fig. 7-18b, by first decoding the M inputs onto $2^M \equiv \mu$ word

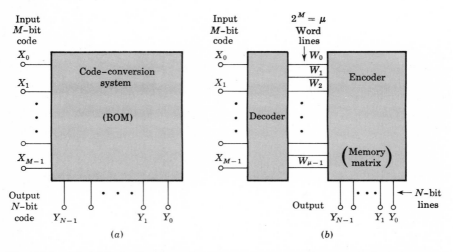

Fig. 7-18 (a) A block diagram of a system for converting one code into another; a read-only memory (ROM). (b) An ROM may be considered to be a decoder for the input code followed by an encoder for the output code.

lines $(W_0, W_1, \ldots, W_{\mu-1})$ and then encoding each line into the desired output word. If the inputs assume all possible combinations of 1s and 0s, then μ N-bit words are "read" at the output (not all these 2^M words need be unique, since it may be desirable to have the same output code for several different input words).

The functional relationship between output and input words is built into hardware in the encoder block of Fig. 7-18. Since this information is thus stored permanently, we say that the system has "memory." The *memory elements* are the diodes in Fig. 7-16 or the emitters of transistors in Fig. 7-17. The output word for any input code may be read as often as desired. However, since the stored relationship between output and input codes cannot be modified without adding or subtracting memory elements (hardware), this system is called a *read-only memory*, abbreviated ROM.

A typical bipolar ROM (MM 6280, available from Monolithic Memories) has $M = 10$ and $N = 8$, resulting in $2^M = 2^{10} = 1,024$ words of 8 bits each. This size is referred to as a $8 \times 1,024 = 8,192$-bit memory, and is an example of large-scale integration (LSI). Read-only memories using MOSFETs as memory elements are discussed in Sec. 9-6.

Code Converters The truth table for translating from a binary to a Gray code is given in Table 7-3. The input bits are decoded in an ROM into the word lines $W_0, W_1, \ldots, W_{15}$, as indicated in Fig. 7-18b, and then

TABLE 7-3   Conversion from a binary to a Gray code

Binary code inputs				Decoded word	Gray code outputs			
X_3	X_2	X_1	X_0	W_n	Y_3	Y_2	Y_1	Y_0
0	0	0	0	W_0	0	0	0	0
0	0	0	1	W_1	0	0	0	1
0	0	1	0	W_2	0	0	1	1
0	0	1	1	W_3	0	0	1	0
0	1	0	0	W_4	0	1	1	0
0	1	0	1	W_5	0	1	1	1
0	1	1	0	W_6	0	1	0	1
0	1	1	1	W_7	0	1	0	0
1	0	0	0	W_8	1	1	0	0
1	0	0	1	W_9	1	1	0	1
1	0	1	0	W_{10}	1	1	1	1
1	0	1	1	W_{11}	1	1	1	0
1	1	0	0	W_{12}	1	0	1	0
1	1	0	1	W_{13}	1	0	1	1
1	1	1	0	W_{14}	1	0	0	1
1	1	1	1	W_{15}	1	0	0	0

are encoded into the desired Gray code $Y_3Y_2Y_1Y_0$. The W's are the minterm outputs of the decoder. For example,

$$W_0 = \bar{X}_3\bar{X}_2\bar{X}_1\bar{X}_0 \qquad W_5 = \bar{X}_3X_2\bar{X}_1X_0 \qquad W_9 = X_3\bar{X}_2\bar{X}_1X_0 \qquad (7\text{-}9)$$

From the truth table 7-3, we conclude that

$$Y_0 = W_1 + W_2 + W_5 + W_6 + W_9 + W_{10} + W_{13} + W_{14} \qquad (7\text{-}10)$$

This equation is implemented by connecting eight diodes with their cathodes all tied to Y_0 and their anodes connected to the decoder lines W_1, W_2, W_5, W_6, W_9, W_{10}, W_{13}, and W_{14}, respectively (or the base-emitter diodes of transistors may be used in an analogous manner to form an emitter-follower OR gate, as in Fig. 7-17a). Similarly, from Table 7-3, we may write the Boolean expressions for the other output bits. For example,

$$Y_3 = W_8 + W_9 + W_{10} + W_{11} + W_{12} + W_{13} + W_{14} + W_{15} \qquad (7\text{-}11)$$

Consider the inverse code translation, from Gray to binary. The Gray code inputs are arranged in the order W_0, W_1, . . . , W_{15} (corresponding to decimal numbers 0 to 15). The binary code corresponding to a given input word W_n is listed as the output code for that line. For example, from Table 7-3, we find that the Gray code 1001 corresponds to the binary code 1110, and this relationship is maintained in Table 7-4 on line W_9. From this table

TABLE 7-4   Conversion from a Gray to a binary code

X_3	X_2	X_1	X_0	W_n	Y_3	Y_2	Y_1	Y_0
\multicolumn{4}{c}{Gray code inputs}	Decoded word	\multicolumn{4}{c}{Binary code outputs}			
0	0	0	0	W_0	0	0	0	0
0	0	0	1	W_1	0	0	0	1
0	0	1	0	W_2	0	0	1	1
0	0	1	1	W_3	0	0	1	0
0	1	0	0	W_4	0	1	1	1
0	1	0	1	W_5	0	1	1	0
0	1	1	0	W_6	0	1	0	0
0	1	1	1	W_7	0	1	0	1
1	0	0	0	W_8	1	1	1	1
1	0	0	1	W_9	1	1	1	0
1	0	1	0	W_{10}	1	1	0	0
1	0	1	1	W_{11}	1	1	0	1
1	1	0	0	W_{12}	1	0	0	0
1	1	0	1	W_{13}	1	0	0	1
1	1	1	0	W_{14}	1	0	1	1
1	1	1	1	W_{15}	1	0	1	0

we obtain the relationship between output and input bits. For example,

$$Y_0 = W_1 + W_2 + W_4 + W_7 + W_8 + W_{11} + W_{13} + W_{14} \qquad (7\text{-}12)$$

This equation defines how the memory elements are to be arranged in the encoder. Note that the ROM for Table 7-4 uses the same decoding arrangement as that for Table 7-3, but the encoders are completely different. In other words, the IC chips for these two ROMs are quite distinct since individual masks must be used for the encoder matrix of memory elements.

Programming the ROM Consider a 256-bit read-only memory (TI 7488A) arranged in 32 words of 8 bits each. The decoder input is a 5-bit binary select code, and its outputs are the 32 word lines. The encoder consists of 32 transistors (each base is tied to a different line) and with 8 emitters in each transistor. The customer fills out the truth table he wishes the ROM to satisfy, and then the vendor makes a mask for the metalization so as to connect one emitter of each transistor to the proper output line, or alternatively, to leave it floating. For example, for the Gray-to-binary-code conversion, Eq. (7-12) indicates that one emitter from each of transistors $Q1$, $Q2$, $Q4$, $Q7$, $Q8$, $Q11$, $Q13$, and $Q14$ is connected to line Y_0, whereas the corresponding emitter on each of the other transistors $Q0$, $Q3$, $Q5$, $Q6$, . . . is left unconnected. The process just described is called *custom programming*, or *mask programming*, of an ROM. Note that *hardware* (not *software*) programming is under consideration. If the sales demand for a particular code is sufficient, this ROM becomes available as an "off-the-shelf" item.

For small quantities of an ROM, the mask cost may be prohibitive, and also the delivery time may be too long. Hence some manufacturers† supply *field-programmable* ROMs, abbreviated pROM, or ROMP (read-only memory, programmable). Such an IC chip has an encoder matrix made with all connections which may possibly be required. For example, the 256-bit memory discussed above is constructed as a pROM with 32 transistors, each having eight emitters (labeled E_0, E_1, . . . , E_7) and with E_0 from each transistor tied to output Y_0, E_1 to Y_1, etc. In series with each emitter there is incorporated a narrow aluminum or nichrome strip which acts as fuse and opens up when a current in excess of a maximum value is passed through this memory element. The user can easily fuse, or "zap," in the field those memory-element links which must be opened in order that the ROM perform the desired functional relationship between output and input.

Diode matrices are also available with fusible links, and these can be used for the encoder portion of a pROM, or also as a decoder (Prob. 7-15).

7-7   ROM APPLICATIONS

As emphasized in the preceding section, an ROM is a code-conversion unit. However, many different practical systems represent a translation from one

---

† For example, Harris Semiconductor, Intel, Intersil, Monolithic Memories, Motorola, Signetics, and Texas Instruments.

code to another. The most important of these ROM applications are discussed below.

Look-up Tables Routine calculations such as trigonometric functions, logarithms, exponentials, square roots, etc., are sometimes required of a computer. If these are repeated often enough, it is more economical to include an ROM as a *look-up table*, rather than to use a subroutine or a software program to perform the calculation. A look-up table for $Y = \sin X$ is a code-conversion system between the input code representing the argument X in binary notation (to whatever accuracy is desired) and the output code giving the corresponding values of the sine function. Clearly, any calculation for which a truth table can be written may be implemented with an ROM—a different ROM for each truth table.

Sequence Generators In a digital system (such as a computer, a data communications system, etc.) several pulse trains are often required for control (gating) purposes. The ROM may be used to supply these binary sequences if the address is changed by means of a counter. As mentioned in Sec. 7-4, the input to the encoder changes from W_0 to W_1 to W_2, etc. every T s. Under this excitation the output Y_1 of the ROM represented by Table 7-4 is

$$Y_1 = 1100001100111100 \qquad \text{(LSB is at the right)} \qquad (7\text{-}13)$$

This equation is obtained by reading the digits in the Y_1 column from top to bottom. It indicates that for the first $2T$ s, Y_1 remains low; for the following $4T$ s, Y_1 is high; for the next $2T$ s, Y_1 is low, for the next $2T$ s, Y_1 is high; for the following $4T$ s, Y_1 is low; for the last $2T$ s, Y_1 is high; and after these $16T$ s, this sequence is repeated (as long as pulses are fed to the counter).

Simultaneously with Y_1, three other synchronous pulse trains, Y_0, Y_2, and Y_3, are created. In general, the number of sequences obtained equals the number of outputs from the ROM. Any desired serial binary waveforms are generated if the truth table is properly specified, i.e., if the ROM is correctly programmed.

Seven-Segment Visible Display It is common practice to make visible the reading of a digital instrument (a frequency meter, digital voltmeter, etc.) by means of the seven-segment numeric indicator sketched in Fig. 7-19a. A wide variety of readouts are commercially available. A solid-state indicator in which the segments obtain their luminosity from light-emitting gallium arsenide or phosphide diodes (Sec. 2-10) is operated at low voltage and low power and hence may be driven directly from IC logic gates.

The first 10 displays in Fig. 7-19b are the numerals 0 to 9, which, in the digital instrument, are represented in BCD form. Such a 4-bit code has 16 possible states, and the displays 10 to 15 of Fig. 7-19b are unique symbols used to identify a nonvalid BCD condition.

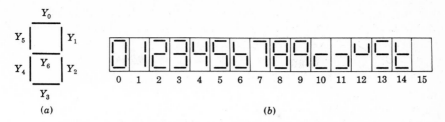

Fig. 7-19 (a) Identification of the segments in a seven-segment LED visible indicator. (b) The display which results from each of the sixteen 4-bit input codes.

The problem of converting from a BCD input to the seven-segment outputs of Fig. 7-19 is easily solved using an ROM. If an excited (luminous) segment is identified as state 0 and a dark segment as the 1 state, then truth table 7-5 is obtained. This table is verified as follows: For word W_0 (corresponding to numeral 0) we see from Fig. 7-19 that $Y_6 = 1$ and all other Y's are 0. For word W_4 (corresponding to the numeral 4) $Y_0 = Y_3 = Y_4 = 1$ and $Y_1 = Y_2 = Y_5 = Y_6 = 0$, and so forth. The ROM is programmed as explained in Sec. 7-6 to satisfy this truth table. For example,

$$Y_0 = W_1 + W_4 + W_6 + W_{10} + W_{11} + W_{12} + W_{14} + W_{15} \qquad (7\text{-}14)$$

TABLE 7-5　　Conversion from a BCD to a seven-segment-indicator code

$X_3 = D$	$X_2 = C$	$X_1 = B$	$X_0 = A$	W_n	Y_6	Y_5	Y_4	Y_3	Y_2	Y_1	Y_0
				Decoded word			Seven-segment-indicator code outputs				
0	0	0	0	W_0	1	0	0	0	0	0	0
0	0	0	1	W_1	1	1	1	1	0	0	1
0	0	1	0	W_2	0	1	0	0	1	0	0
0	0	1	1	W_3	0	1	1	0	0	0	0
0	1	0	0	W_4	0	0	1	1	0	0	1
0	1	0	1	W_5	0	0	1	0	0	1	0
0	1	1	0	W_6	0	0	0	0	0	1	1
0	1	1	1	W_7	1	1	1	1	0	0	0
1	0	0	0	W_8	0	0	0	0	0	0	0
1	0	0	1	W_9	0	0	1	1	0	0	0
1	0	1	0	W_{10}	0	1	0	0	1	1	1
1	0	1	1	W_{11}	0	1	1	0	0	1	1
1	1	0	0	W_{12}	0	0	1	1	1	0	1
1	1	0	1	W_{13}	0	0	1	0	1	1	0
1	1	1	0	W_{14}	0	0	0	0	1	1	1
1	1	1	1	W_{15}	1	1	1	1	1	1	1

Combinational Logic If N logic equations of M variables are given in the sum-of-products canonical form, these equations may be implemented with an M-input, N-output ROM.

Character Generator Alphanumeric characters may be "written" on the face of a cathode-ray tube (a television-type display) with the aid of an ROM.

7-8   DIGITAL COMPARATOR

It is sometimes necessary to know whether a binary number A is greater than, equal to, or less than another number B. The system for making this determination is called a *magnitude digital* (or *binary*) *comparator*. Consider single bit numbers first. As mentioned in Sec. 6-6, the EXCLUSIVE-NOR gate is an *equality detector* because

$$E = \overline{A\bar{B} + \bar{A}B} = \begin{cases} 1 & A = B \\ 0 & A \neq B \end{cases} \tag{7-15}$$

The condition $A > B$ is given by

$$C = A\bar{B} = 1 \tag{7-16}$$

because if $A > B$, then $A = 1$ and $B = 0$, so that $C = 1$. On the other hand, if $A = B$ or $A < B$ ($A = 0$, $B = 1$), then $C = 0$.

Similarly, the restriction $A < B$ is determined from

$$D = \bar{A}B = 1 \tag{7-17}$$

The logic block diagram for the nth bit drawn in Fig. 7-20 has all three desired outputs C_n, D_n, and E_n. It consists of two inverters, two AND gates, and the AOI circuit of Fig. 7-2. Alternatively, Fig. 7-20 may be considered to consist of an EXCLUSIVE-NOR and two AND gates. (Note that the outputs of the AND gates in the AOI block of Fig. 7-2 are not available, and hence additional AND gates must be fabricated to give C_n and D_n.)

Consider now a 4-bit comparator. $A = B$ requires that

$$A_3 = B_3 \quad \text{and} \quad A_2 = B_2 \quad \text{and} \quad A_1 = B_1 \quad \text{and} \quad A_0 = B_0$$

Hence the AND gate E in Fig. 7-21 described by

$$E = E_3 E_2 E_1 E_0 \tag{7-18}$$

implies $A = B$ if $E = 1$ and $A \neq B$ if $E = 0$. (Assume that the input E' is held high; $E' = 1$.)

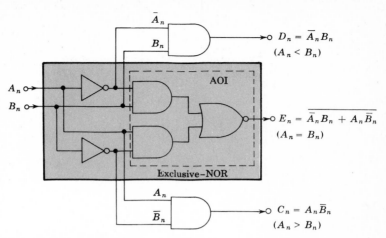

Fig. 7-20   A 1-bit digital comparator.

The inequality $A > B$ requires that

$$A_3 > B_3 \qquad \text{(MSB)}$$

or

$$A_3 = B_3 \qquad \text{and} \qquad A_2 > B_2$$

or

$$A_3 = B_3 \qquad \text{and} \qquad A_2 = B_2 \qquad \text{and} \qquad A_1 > B_1$$

or

$$A_3 = B_3 \qquad \text{and} \qquad A_2 = B_2 \qquad \text{and} \qquad A_1 = B_1 \qquad \text{and} \qquad A_0 > B_0$$

The above conditions are satisfied by the Boolean expression

$$C = A_3\bar{B}_3 + E_3A_2\bar{B}_2 + E_3E_2A_1\bar{B}_1 + E_3E_2E_1A_0\bar{B}_0 \qquad (7\text{-}19)$$

if and only if $C = 1$. The AND-OR gate for C is indicated in Fig. 7-21. (Assume that $C' = 0$.)

The condition that $A < B$ is obtained from Eq. (7-19) by interchanging A and B. Thus

$$D = \bar{A}_3B_3 + E_3\bar{A}_2B_2 + E_3E_2\bar{A}_1B_1 + E_3E_2E_1\bar{A}_0B_0 \qquad (7\text{-}20)$$

implies that $A < B$ if and only if $D = 1$. This portion of the system is obtained from Fig. 7-21 by changing A to B, B to A, and C to D. Alternatively, D may be obtained from $D = \bar{E}\bar{C}$ because, if $A \neq B$ ($E = 0$) and if $A > B$ ($C = 0$), then $A < B$ ($D = 1$). However, this implementation for D introduces the additional propagation delay of an inverter and an AND gate. Hence the logic indicated in Eq. (7-20) for D is fabricated on the same chip as that for C in Eq. (7-19) and E in Eq. (7-18).

The TI 54L85 is an MSI package which performs 4-bit-magnitude comparison. If numbers of greater length are to be compared, several such units can be cascaded. Consider an 8-bit comparator. Designate the $A = B$ out-

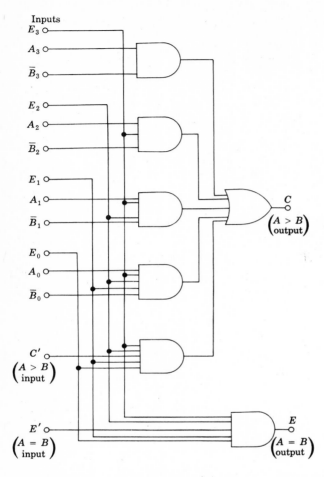

Inputs
E_3
A_3
$\bar{B}_3$

E_2
A_2
$\bar{B}_2$

E_1
A_1
$\bar{B}_1$

E_0
A_0
$\bar{B}_0$

C'
$\begin{pmatrix} A > B \\ \text{input} \end{pmatrix}$

E'
$\begin{pmatrix} A = B \\ \text{input} \end{pmatrix}$

C
$\begin{pmatrix} A > B \\ \text{output} \end{pmatrix}$

E
$\begin{pmatrix} A = B \\ \text{output} \end{pmatrix}$

Fig. 7-21 A 4-bit magnitude comparator. (Assume that $C' = 0$ and $E' = 1$.) If $E = 1$, then $A = B$, and if $C = 1$, then $A > B$. If $D = 1$, then $A < B$, where D has the same logic topology as C but with A and B interchanged. The inputs $\bar{A}_n$, B_n, and D' $(A < B)$ are not indicated. The inputs E_n are obtained from Fig. 7-20.

put terminal of the stage handling the less significant bits by E_L, the $A > B$ output terminal of this stage by C_L, and the $A < B$ output by D_L. Then the connections $E' = E_L$, $C' = C_L$, and $D' = D_L$ (Fig. 7-21) must be made to the stage with the more significant bits (Prob. 7-20). For the stage handling the less significant bits, the outputs C' and D' are grounded ($C' = 0$ and $D' = 0$) and the input E' is tied to the supply voltage ($E' = 1$). Why?

7-9   PARITY CHECKER/GENERATOR

Another arithmetic operation that is often invoked in a digital system is that of determining whether the sum of the binary bits in a word is odd (called *odd parity*) or even (designated *even parity*). The output of an EXCLUSIVE-OR gate is 1 if and only if one input is 1 and the other is 0. Alternatively stated, the output is 1 if the sum of the digits is 1. An ex-

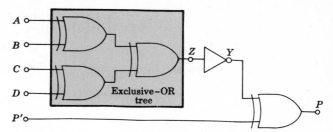

Fig. 7-22   An odd-parity checker, or parity-bit generator system, for a 4-bit input word.   Assume $P' = 0$ and then $P = 0(1)$ represents odd (even) parity.

tension of this concept to the EXCLUSIVE-OR tree of Fig. 7-22 leads to the conclusion that $Z = 1$ (or $Y = 0$) if the sum of the input bits A, B, C, and D is odd.   Hence, if the input P' is grounded ($P' = 0$), then $P = 0$ for odd parity and $P = 1$ for even parity.

The system of Fig. 7-22 is not only a parity checker, but it may also be used to generate a parity bit P.   Independently of the parity of the 4-bit input word, the parity of the 5-bit code A, B, C, D, and P is odd.   This statement follows from the fact that if the sum of A, B, C, and D is odd (even), then P is 0 (1), and therefore the sum of A, B, C, D, and P is always odd.

The use of a parity code is an effective way of increasing the reliability of transmission of binary information.   As indicated in Fig. 7-23, a parity bit P_1 is generated and transmitted along with the N-bit input word.   At the receiver the parity of the augmented $(N + 1)$-bit signal is tested, and if the output P_2 of the checker is 0, it is assumed that no error has been made in transmitting the message, whereas $P_2 = 1$ is an indication that (say, due to

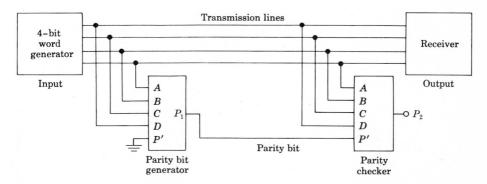

Fig. 7-23   Binary data transmission is tested by generating a parity bit at the input to a line and checking the parity of the transmitted bits plus the generated bit at the receiving end of the system.

noise) the received word is in error.   Note that only errors in an odd number of digits can be detected with a single parity check.

An MSI 8-bit parity generator/checker is available (TI 74180) with control inputs so that it may be used in either odd- or even-parity applications (Prob. 7-23).   For words of length greater than 8 bits, several such units may be cascaded (Prob. 7-25).

The MSI unit TI SN7486 contains four two-input EXCLUSIVE-OR gates.

REVIEW QUESTIONS

7-1 (*a*) How many input leads are needed for a chip containing quad two-input NOR gates?   Explain.   (*b*) Repeat part *a* for dual two-wide, two-input AOI gates.

7-2 Define SSI, MSI, and LSI.

7-3 Draw the circuit configuration for an IC TTL AOI gate.   Explain its operation.

7-4 (*a*) Find the truth table for the *half adder*.   (*b*) Show the implementation for the digit D and the carry C.

7-5 (*a*) Show the system of a three-bit *parallel binary full adder* consisting of half adders.   (*b*) Explain its operation.

7-6 Show the system of a 4-bit parallel binary adder, constructed from single-bit full adders.

7-7 (*a*) Draw the truth table for a three-input adder.   Explain clearly the meaning of the input and output symbols in the table.   (*b*) Write the Boolean expressions for the sum and the carry.   (Do not simplify these.)

7-8 (*a*) Show the system for a *serial* binary full adder.   (*b*) Explain the operation.

7-9 Write the decimal number 749 in the BCD system.

7-10 (*a*) Define a *decoder*.   (*b*) Show how to decode the 4-bit code 1011 (LSB).

7-11 (*a*) Define a *demultiplexer*.   (*b*) Show how to convert a decoder into a demultiplexer.   (*c*) Indicate how to add a strobe to this system.

7-12 (*a*) Define a *multiplexer*.   (*b*) Draw a logic block diagram of a 4-to-1-line multiplexer.

7-13 Show how a multiplexer may be used as (*a*) a *parallel-to-serial converter* and (*b*) a *sequential data selector*.

7-14 (*a*) Define an *encoder*.   (*b*) Indicate a diode matrix encoder to transform a decimal number into a binary code.

7-15 (*a*) Indicate an encoder matrix using emitter followers.   In particular, for an encoder to transform a decimal number into a binary code, show the connections (*b*) to the output Y_1 and (*c*) to the line W_5.

7-16 (*a*) Define a *read-only memory*.   (*b*) Show a block diagram of an ROM. (*c*) What is stored in the memory?   (*d*) What hardware constitutes the memory elements?

7-17 (*a*) Write the truth table for converting from a binary to a Gray code. (*b*) Write the first six lines of the truth table for converting a Gray into a binary code.

7-18 Explain what is meant by *mask-programming* an ROM.

7-19 (*a*) Explain what is meant by a pROM (or ROMP).   (*b*) How is the programming done in the field?

7-20 List three ROM applications and explain these very briefly.

7-21 (*a*) What is a *seven-segment visible display?* (*b*) Show the following two lines in the conversion table from BCD to seven-segment-indicator code: 0011 and 1001.

7-22 Consider two 1-bit numbers A and B. What are the logic gates required to test for (*a*) $A = B$, (*b*) $A > B$, and (*c*) $A < B$?

7-23 (*a*) Consider two 4-bit numbers A and B. If $E = 1$ represents the equality $A = B$, write the Boolean expression for E. Explain. (*b*) If $C = 1$ represents the inequality $A > B$, write the Boolean expression for C. Explain.

7-24 Show the system for a 4-bit odd-parity checker.

7-25 (*a*) Show a system for increasing the reliability of transmission of binary information, using a parity check and generator. (*b*) Explain the operation of the system.

8 / SEQUENTIAL DIGITAL SYSTEMS

Systems which operate in synchronism with a train of pulses and which possess memory are discussed in this chapter. These include FLIP-FLOPS, shift registers, and counters.

8-1   A 1-BIT MEMORY

All the systems discussed in Chap. 7 were based upon combinational logic; the outputs at a given instant of time depend only upon the values of the inputs at the same moment. Such a system is said to have no memory. Note that an ROM is a combinational circuit and, according to the above definition, it has no memory. *The memory of an ROM refers to the fact that it "memorizes" the functional relationship between the output variables and the input variables.* It does *not* store bits of information.

A Sequential System   Many digital systems are pulsed or clocked; i.e., they operate in synchronism with a pulse train of period T, called the system *clock* (abbreviated Ck), such as that indicated in Fig. 8-1. The pulse width t_p is assumed small compared with T. The binary values at each node in the system are assumed to remain constant in each interval between pulses. A transition from one state of the system to another may take place only with the application of a clock pulse. Let Q_n be the output (0 or 1) at a given node in the nth interval (bit time n) preceding the nth clock pulse (Fig. 8-1). Then Q_{n+1} is the corresponding output in the interval immediately after the nth pulse. Such a system where the values Q_1, Q_2, Q_3, . . . , of Q_n are obtained in time sequence at intervals T is called a *sequential* (to distinguish it from a *combinational*) logic system. The value of Q_{n+1} may depend upon the nodal values during the previous (nth) bit time. Under these circumstances a sequential circuit possesses memory.

168

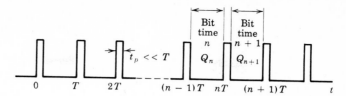

Fig. 8-1 The output of a master oscillator used as a clock pulse train to synchronize a digital sequential system.

A 1-bit Storage Cell The basic digital memory circuit is obtained by cross-coupling two NOT circuits $N1$ and $N2$ (single-input NAND gates) in the manner shown in Fig. 8-2a. The output of each gate is connected to the input of the other, and this feedback combination is called a FLIP-FLOP. The most important property of the FLIP-FLOP is that it can exist in one of two stable states, either $Q = 1$ $(\bar{Q} = 0)$, called the 1 state, or $Q = 0$ $(\bar{Q} = 1)$, referred to as the 0 state. The existence of these stable states is consistent with the interconnections shown in Fig. 8-2a. For example, if the output of $N1$ is $Q = 1$, so also is A_2, the input to $N2$. This inverter then has the state 0 at its output $\bar{Q}$. Since $\bar{Q}$ is tied to A_1, then the input to $N1$ is 0, and the corresponding output is $Q = 1$. This result confirms our original assumption that $Q = 1$. A similar argument leads to the conclusion that $Q = 0$; $\bar{Q} = 1$ is also a possible state. It is readily verified that the situation in which both outputs are in the same state (both 1 or both 0) is not consistent with the interconnection.

Since the FLIP-FLOP has two stable states, it is also called a *binary*, or *bistable* MULTI. Since it may store one bit of information (either $Q = 1$ or $Q = 0$), it is a *1-bit memory unit*, or a *1-bit storage cell*. Since this information is locked, or latched, in place, this FLIP-FLOP is also known as a *latch*.

Suppose it is desired to store a specific state, say $Q = 1$, in the latch. Or conversely, we may wish to remember the state $Q = 0$. We may "write" a

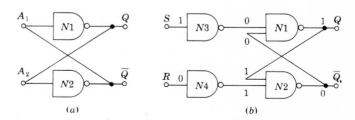

(a) (b)

Fig. 8-2 (a) A 1-bit memory or latch. (b) The FLIP-FLOP provided with means for presetting or for clearing the state of the cell.

1 or 0 into the memory cell by changing the NOT gates of Fig. 8-2a to two-input NAND gates, $N1$ and $N2$, and by feeding this latch through two NOT gates, $N3$ and $N4$, whose inputs are S and R, as in Fig. 8-2b. If we assume that $S = 1$ and $R = 0$, then the state of each gate input and output is indicated on the logic diagram. Since $Q = 1$, we have thus verified that to set the FLIP-FLOP to the 1 state requires inputs $S = 1$ and $R = 0$. The input S is called the *set*, or *preset*, input. In a similar manner it can be demonstrated that to enter a 0 into the memory, it is necessary to choose $S = 0$ and $R = 1$. Hence R is referred to as the *reset*, or *clear*, input. The input combination $S = R = 0$ leads to an undetermined state (Q could be either 1 or 0). Also $S = R = 1$ is not allowed (Prob. 8-1).

The Clocked S-R FLIP-FLOP In a sequential system it is required to set or reset a FLIP-FLOP in synchronism with clock pulses. This is accomplished by changing $N3$ and $N4$ in Fig. 8-2b to two-input NAND gates and by applying the clock pulse train Ck simultaneously to $N3$ and $N4$. Such a triggered set-reset (abbreviated S-R, or R-S) FLIP FLOP is indicated in Fig. 8-3a. The gates $N1$ and $N2$ form a latch, whereas $N3$ and $N4$ are the *control,* or *steering,* gates which program the state of the FLIP-FLOP after the pulse appears.

Note that between clock pulses ($Ck = 0$), the outputs of $N3$ and $N4$ are 1 independently of the values of R or S. Hence the circuit is equivalent to the latch of Fig. 8-2a. If $Q = 1$, it remains 1, whereas if $Q = 0$, it remains 0. In other words, *the* FLIP-FLOP *does not change state between clock pulses;* it is invariant within a bit time.

Now consider the time $t = nT(+)$ when a clock pulse is present ($Ck = 1$). If $S = 0$ and $R = 0$, then the outputs of $N3$ and $N4$ are 1. By the argument given in the preceding paragraph, the state Q_n of the FLIP-FLOP does not change. Hence, after the pulse passes (in the bit time $n + 1$), the state Q_{n+1} is identical with Q_n. If we denote the values of R and S in the interval just before $t = nT$ by R_n and S_n, then $Q_{n+1} = Q_n$ if $S_n = 0$ and $R_n = 0$. This relationship is indicated in the first row of the truth table of Fig. 8-3b.

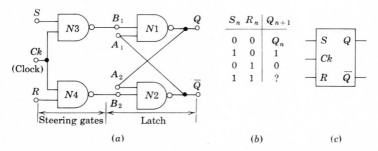

Fig. 8-3 (a) An S-R clocked FLIP-FLOP; (b) the truth table (the question mark in the last row indicates that this state cannot be predicted); (c) the logic symbol.

If $Ck = 1$, $S_n = 1$, and $R_n = 0$, then the situation is that pictured in Fig. 8-2b and the output state is 1. Hence, after the clock pulse passes (at bit time $n + 1$), we find $Q_{n+1} = 1$, confirming the second row of the truth table. If R and S are interchanged and if simultaneously Q is interchanged with $\bar{Q}$, then the logic diagram of Fig. 8-3a is unaltered. Hence the third row of Fig. 8-3b follows from the second row.

If $Ck = 1$, $S_n = 1$, and $R_n = 1$, then the outputs of the NAND gates $N3$ and $N4$ are both 0. Hence the input B_1 of $N1$ as well as B_2 of $N2$ is 0, so that the outputs of *both* $N1$ and $N2$ must be 1. This condition is logically inconsistent with our labeling the two outputs Q and $\bar{Q}$. We must conclude that the output transistor ($Q3$ of Fig. 6-22) of each gate $N1$ and $N2$ is cut off, resulting in both outputs being high (1). At the end of the pulse the inputs at B_1 and B_2 rise from 0 toward 1. Depending upon which input increases faster and upon circuit parameter asymmetries, either the stable state $Q = 1$ ($\bar{Q} = 0$) or $Q = 0$ ($\bar{Q} = 1$) will result. Therefore we have indicated a question mark for Q_{n+1} in the fourth row of the truth table of Fig. 8-3b. This state is said to be *indeterminate, ambiguous,* or *undefined,* and the condition $S_n = 1$ and $R_n = 1$ is forbidden; it must be prevented from taking place.

8-2   FLIP–FLOPS

In addition to the *S-R* FLIP-FLOP, three other variations of this basic 1-bit memory are commercially available: the *J-K*, *T*, and *D* types. The *J-K* FLIP-FLOP removes the ambiguity in the truth table of Fig. 8-3b. The *T* FLIP-FLOP acts as a toggle switch and changes the output state with each clock pulse; $Q_{n+1} = \bar{Q}_n$. The *D* type acts as a delay unit which causes the output Q to follow the input D, but delayed by 1 bit time; $Q_{n+1} = D_n$. We now discuss each of these three FLIP-FLOP types.

The *J-K* FLIP-FLOP   This building block is obtained by augmenting the *S-R* FLIP-FLOP with two AND gates $A1$ and $A2$ (Fig. 8-4a). Data input J and the output $\bar{Q}$ are applied to $A1$. Since its output feeds S, then $S = J\bar{Q}$. Similarly, data input K and the output Q are applied to $A2$, and hence $R = KQ$. The logic followed by this system is given in the truth table of Fig. 8-4b. This logic can be verified by referring to Table 8-1. There are four possible combinations for the two data inputs J and K. For each of these there are two possible states for Q, and hence Table 8-1 has eight rows. From the J_n, K_n, Q_n, and $\bar{Q}_n$ bits in each row, $S_n = J_n\bar{Q}_n$ and $R_n = K_nQ_n$ are calculated and are entered into the fifth and sixth columns of the table. Using these values of S_n and R_n and referring to the *S-R* FLIP-FLOP truth table of Fig. 8-3b, the seventh column is obtained. Finally, column 8 follows from column 7 because $Q_n = 1$ in row 4, $Q_n = 0$ in row 5, $\bar{Q}_n = 1$ in row 7, and $\bar{Q}_n = 0$ in row 8.

Columns 1, 2, and 8 of Table 8-1 form the *J-K* FLIP-FLOP truth table of Fig. 8-4b. Note that *the first three rows of a J-K are identical with the cor-*

TABLE 8-1   Truth table for Fig. 8-4a

Column	1	2	3	4	5	6	7	8
Row	J_n	K_n	Q_n	$\bar{Q}_n$	S_n	R_n	Q_{n+1}	
1	0	0	0	1	0	0	Q_n	Q_n
2	0	0	1	0	0	0	Q_n	
3	1	0	0	1	1	0	1	1
4	1	0	1	0	0	0	Q_n	
5	0	1	0	1	0	0	Q_n	0
6	0	1	1	0	0	1	0	
7	1	1	0	1	1	0	1	$\bar{Q}_n$
8	1	1	1	0	0	1	0	

responding rows for an S-R truth table (Fig. 8-3b).   However, the ambiguity of the state $S_n = 1 = R_n$ is now replaced by $Q_{n+1} = \bar{Q}_n$ for $J_n = 1 = K_n$. *If the two data inputs in the J-K* FLIP-FLOP *are high, the output will be complemented by the clock pulse.*

It is really not necessary to use the AND gates $A1$ and $A2$ of Fig. 8-4a since the same function can be performed by adding an extra input terminal to each NAND gate $N3$ and $N4$ of Fig. 8-3a.   This simplification is indicated in Fig. 8-5.   (Ignore the dashed inputs; i.e., assume that they are both 1.) Note that Q and $\bar{Q}$ at the inputs are obtained by the feedback connections (drawn heavy) from the outputs.

Preset and Clear   The truth table of Fig. 8-4b tells what happens to the output with the application of a clock pulse, as a function of the data inputs J and K.   However, the value of the output before the pulse is applied is arbitrary.   The addition of the dashed inputs in Fig. 8-5 allows the initial state of the FLIP-FLOP to be assigned.   For example, it may be required to *clear* the latch, i.e., to specify that $Q = 0$ when $Ck = 0$.

The clear operation may be accomplished by programming the clear input to 0 and the *preset* input to 1; $Cr = 0$, $Pr = 1$, $Ck = 0$.   Since $Cr = 0$, the output of $N2$ (Fig. 8-5) is $\bar{Q} = 1$.   Since $Ck = 0$, the output of $N3$ is 1, and hence all inputs to $N1$ are 1 and $Q = 0$, as desired.   Similarly, if it is required to preset the latch into the 1 state, it is necessary to choose $Pr = 0$, $Cr = 1$, $Ck = 0$.   The preset and clear data are called *direct*, or *asynchronous*,

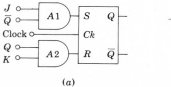

J_n	K_n	Q_{n+1}
0	0	Q_n
1	0	1
0	1	0
1	1	$\bar{Q}_n$

(a)   (b)

Fig. 8-4   (a) An S-R FLIP-FLOP is converted into a J-K FLIP-FLOP; (b) the truth table.

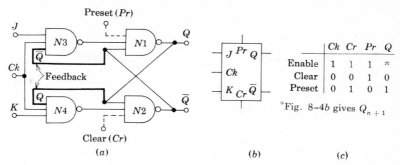

Fig. 8-5 (a) A J-K FLIP-FLOP; (b) the logic symbol; (c) the necessary conditions for synchronous operation (row 1) or for asynchronous clearing (row 2) or presetting (row 3).

inputs; i.e., they are not in synchronism with the clock, but may be applied at any time in between clock pulses. Once the state of the FLIP-FLOP is established asynchronously, the direct inputs must be maintained at $Pr = 1$, $Cr = 1$, before the next pulse arrives in order to *enable* the FLIP-FLOP. The data $Pr = 0$, $Cr = 0$, must not be used since they lead to an ambiguous state. Why?

The logic symbol for the J-K FLIP-FLOP is indicated in Fig. 8-5b, and the inputs for proper operation are given in Fig. 8-5c.

The Race-around Condition   There is a possible physical difficulty with the J-K FLIP-FLOP constructed as in Fig. 8-5. Truth table 8-1 is based upon combinational logic, which assumes that the inputs are independent of the outputs. However, because of the feedback connection $Q(\bar{Q})$ at the input to $K(J)$, the input will change during the clock pulse ($Ck = 1$) if the output changes state. Consider, for example, that the inputs to Fig. 8-5 are $J = 1$, $K = 1$, and $Q = 0$. When the pulse is applied, the output becomes $Q = 1$ (according to row 7 of Table 8-1), this change taking place after a time interval Δt equal to the propagation delay (Sec. 6-13) through two NAND gates in series in Fig. 8-5. Now $J = 1$, $K = 1$, and $Q = 1$, and from row 8 of Table 8-1, we find that the input changes back to $Q = 0$. Hence we must conclude that for the duration t_p (Fig. 8-1) of the pulse (while $Ck = 1$), the output will oscillate back and forth between 0 and 1. At the end of the pulse ($Ck = 0$), the value of Q is ambiguous.

The situation just described is called a *race-around condition*. It can be avoided if $t_p < \Delta t < T$. However, with IC components the propagation delay is very small, usually much less than the pulse width t_p. Hence the above inequality is *not* satisfied, and the output is indeterminate. Lumped delay lines can be used in series with the feedback connections of Fig. 8-5 in order to increase the loop delay beyond t_p, and hence to prevent the race-around difficulty. A more practical IC solution is now to be described.

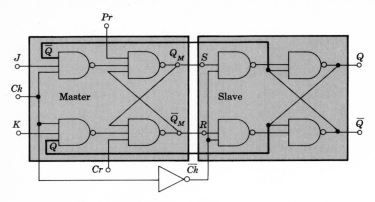

Fig. 8-6 A J-K master-slave FLIP-FLOP.

The Master-Slave J-K FLIP-FLOP In Fig. 8-6 is shown a cascade of two S-R FLIP-FLOPS with feedback from the output of the second (called the *slave*) to the input of the first (called the *master*). Positive clock pulses are applied to the master, and these are inverted before being used to excite the slave. For $Pr = 1$, $Cr = 1$, and $Ck = 1$, the master is enabled and its operation follows the J-K truth table of Fig. 8-4b. Furthermore, since $\overline{Ck} = 0$, the slave S-R FLIP-FLOP is inhibited (cannot change state), so that Q_n is invariant for the pulse duration t_p. Clearly, the race-around difficulty is circumvented with the master-slave topology. After the pulse passes, $Ck = 0$, so that the master is inhibited and $\overline{Ck} = 1$, which causes the slave to be enabled. The slave is an S-R FLIP-FLOP, which follows the logic in Fig. 8-3b. If $S = Q_M = 1$ and $R = \bar{Q}_M = 0$, then $Q = 1$ and $\bar{Q} = 0$. Similarly, if $S = Q_M = 0$ and $R = \bar{Q}_M = 1$, then $Q = 0$ and $\bar{Q} = 1$. In other words, in the interval between clock pulses, the value of Q_M is transferred to the output Q. In summary, during a clock pulse the output Q does not change but Q_M follows J-K logic; at the end of the pulse, the value of Q_M is transferred to Q.

It should be emphasized that the data J and K must remain constant for the pulse duration or an erroneous output may result (Prob. 8-7). Also note that some commercially available FLIP-FLOPS have internal AND gates at the inputs to provide multiple J and K inputs, thereby avoiding the necessity of external gates in applications where these may be required.

The D-Type FLIP-FLOP If a J-K FLIP-FLOP is modified by the addition of an inverter as in Fig. 8-7a, so that K is the complement of J, the unit is called a D (*delay*) FLIP-FLOP. From the J-K truth table of Fig. 8-4b, $Q_{n+1} = 1$ for $D_n = J_n = \bar{K}_n = 1$ and $Q_{n+1} = 0$ for $D_n = J_n = \bar{K}_n = 0$. Hence $Q_{n+1} = D_n$. The output Q_{n+1} after the pulse (bit time $n + 1$) equals the input D_n before the pulse (bit time n), as indicated in the truth table of Fig.

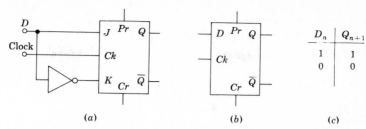

Fig. 8-7 (a) A J-K FLIP-FLOP is converted into a D-type latch; (b) the logic symbol; and (c) the truth table.

8-7c. If the FLIP-FLOP in Fig. 8-7a is of the S-R type, the unit also functions as a D-type latch. There is no ambiguous state because $J = K = 1$ is not possible.

The D-type FLIP-FLOP is a binary used to provide delay. The bit on the D line is transferred to the output at the next clock pulse, and hence this unit functions as a 1-bit delay device. There is available an MSI package (TI 74100) with two independent quadruple D-type latches, so that a total of 8 bits can be stored and transferred.

The T-Type FLIP-FLOP This unit changes state with each clock pulse, and hence it acts as a toggle switch. If $J = K = 1$, then $Q_{n+1} = \bar{Q}_n$, so that the J-K FLIP-FLOP is converted into a T-type FLIP-FLOP. In Fig. 8-8a such a system is indicated with a data input T. The logic symbol is shown in Fig. 8-8b, and the truth table in Fig. 8-8c. The S-R- and the D-type latches can also be converted into toggle, or complementing, FLIP-FLOPS (Prob. 8-9).

Summary Four FLIP-FLOP configurations S-R, J-K, D, and T are important. The logic satisfied by each type is repeated for easy reference in Table 8-2. An IC FLIP-FLOP is driven synchronously by a clock, and in addition it may (or may not) have direct inputs for asynchronous operation, preset

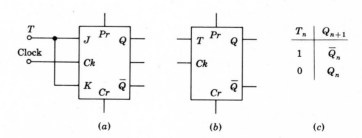

Fig. 8-8 A J-K FLIP-FLOP is converted into a T-type FLIP-FLOP, with a data input T; (b) the logic symbol; and (c) the truth table.

TABLE 8-2   FLIP-FLOP truth tables

S-R			J-K			D		T		Direct inputs			
S_n	R_n	Q_{n+1}	J_n	K_n	Q_{n+1}	D_n	Q_{n+1}	T_n	Q_{n+1}	Ck	Cr	Pr	Q
0	0	Q_n	0	0	Q_n	1	1	1	$\bar{Q}_n$	0	1	0	1
1	0	1	1	0	1	0	0	0	Q_n	0	0	1	0
0	1	0	0	1	0					1	1	1	†
1	1	?	1	1	$\bar{Q}_n$								
Fig. 8-3			Fig. 8-6			Fig. 8-7		Fig. 8-8					

† Refer to truth table S-R, J-K, D, or T for Q_{n+1} as a function of the inputs.

(Pr) and clear (Cr). A direct input can be 0 only in the interval between clock pulses when $Ck = 0$. When $Ck = 1$, both asynchronous inputs must be high; $Pr = 1$ and Cr = 1. The inputs must remain constant during a pulse width, $Ck = 1$. For a master-slave FLIP-FLOP the output Q remains constant for the pulse duration and changes only after Ck changes from 1 to 0, at the negative-going edge of the pulse.

The toggle, or complementing, FLIP-FLOP is not available commercially because a J-K can be used as a T type by connecting the J and K inputs together (Fig. 8-8).

8-3   SHIFT REGISTERS

Since a binary is a 1-bit memory, then n FLIP-FLOPS can store an n-bit word. This combination is referred to as a *register*. To allow the data in the word to be read into the register serially, the output of one FLIP-FLOP is connected to the input of the following binary. Such a configuration, called a *shift register*, is indicated in Fig. 8-9. Each FLIP-FLOP is of the S-R (or J-K) master-slave type. Note that the stage which is to store the most significant bit (MSB) is converted into a D-type latch (Fig. 8-7) by connecting S and R through an inverter. The 5-bit shift register indicated in Fig. 8-9 is available on a single chip in a 16-pin package (medium-scale integration). We shall now explain the operation of this system by assuming that the serial data 01011 (LSB) is to be registered. (The least significant bit is the rightmost digit, which in this case is a 1.)

Serial-to-Parallel Converter The FLIP-FLOPS are cleared by applying a 0 to the *clear* input so that every output Q_0, Q_1, . . . , Q_4 is 0. Then Cr is set to 1 and Pr is held constant at 1 (by keeping the preset enable at 0). The serial data train and the synchronous clock are now applied. The least significant bit (LSB) is entered into FF4 when Ck changes from a 0 to a 1

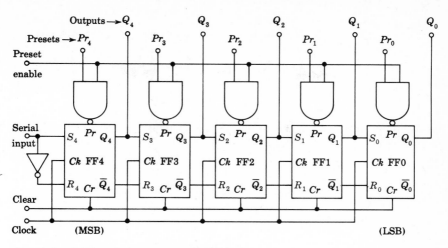

Fig. 8-9   A 5-bit shift register (TI 7496).

by the action of a D-type FLIP-FLOP. After the clock pulse, $Q_4 = 1$, while all other outputs remain at 0.

At the second clock pulse the state of Q_4 is transferred to the master latch of FF3 by the action of an S-R FLIP-FLOP. Simultaneously, the next bit (a 1 in the 01011 word) enters the master of FF4. After the second clock pulse the bit in each master transfers to its slave and $Q_4 = 1$, $Q_3 = 1$, and the other outputs remain 0. The readings of the register *after* each pulse are given in Table 8-3. For example, after the third pulse, Q_3 has shifted to Q_2, Q_4 to Q_3, and the third input bit (0) has entered FF4, so that $Q_4 = 0$. We may easily follow this procedure and see that by registering each bit in the MSB FLIP-FLOP and then shifting to the right to make room for the next digit, the input word becomes installed in the register after the nth clock pulse (for an n-bit code). Of course, the clock pulses must stop at the moment the word is registered. Each output is available on a separate line, and they may be read simultaneously. Since the data entered the system serially and came out in parallel, this shift register is a *serial-to-parallel* converter. It is also

TABLE 8-3   Reading of shift register after each clock pulse

Clock pulse	Word bit	Q_4	Q_3	Q_2	Q_1	Q_0
1	1 ⟶ 1		0	0	0	0
2	1 ⟶ 1	⟶ 1	0	0	0	
3	0 ⟶ 0	⟶ 1	1	0	0	
4	1 ⟶ 1	0	1	1	0	
5	0 ⟶ 0	⟶ 1	0	1	1	

referred to as a *series-in, parallel-out register*. A *temporal code* (a time arrangement of bits) has been changed to a *spacial code* (information stored in a static memory).

Master-slave FLIP-FLOPS are required because of the race problem between stages (Sec. 8-2). If all FLIP-FLOPS were to change states simultaneously, there would be an ambiguity as to what data would transfer from the preceding stage. For example, at the third clock pulse, Q_4 changes from 1 to 0, and it would be questionable as to whether Q_3 would become a 1 or a 0. Hence it is necessary that Q_4 remain a 1 until this bit is entered into FF3, and only then may it change to 0. The master-slave configuration provides just this action. If in Fig. 8-6, the $J(K)$ input is called $S(R)$ and if the (heavy) feedback connections are omitted an S-R master-slave FLIP-FLOP results.

Series-In, Series-Out Register We may take the output at Q_0 and read the register serially if we apply n clock pulses, for an n-bit word. After the nth pulse each FLIP-FLOP reads 0. Note that the shift-out clock rate may be for changing the spacing in time of a binary code, a process referred to as greater or smaller than the original pulse frequency. Hence here is a method *buffering*.

Parallel-to-Serial Converter Consider the situation where the word bits are available in parallel, e.g., at the outputs from an ROM (Sec. 7-6). It is desired to present this code, say 01011, in serial form.

The LSB is applied to Pr_0 the 2^1 bit to Pr_1 . . . , so that $Pr_0 = 1$, $Pr_1 = 1$, $Pr_2 = 0$, $Pr_3 = 1$, and $Pr_4 = 0$. The register is first cleared by $Cr = 0$, and then $Cr = 1$ is maintained. A 1 at the *preset enable* input activates all kth input NAND gates for which $Pr_k = 1$. The preset of the corresponding kth FLIP-FLOP is $Pr = 0$, and this stage is therefore preset to 1 (Table 8-2). In the present illustration FF0, FF1, and FF3 are preset and the input word 01011 is written into the register, all bits in parallel, by the preset enable pulse.

As explained above, the stored word may be read serially at Q_0 by applying five clock pulses. This is a *parallel-to-serial*, or a *spacial-to-temporal*, *converter*.

Parallel In, Parallel Out The data are entered as explained above by applying a 1 at the preset enable, or *write*, terminal. It is then available in parallel form at the outputs Q_0, Q_1, ... If it is desired to *read* the register during a selected time, each output Q_k is applied to one input of a two-input AND gate N_k, and the second input of each AND is excited by a read pulse. The output of N_k is 0 except for the pulse duration, when it reads 1 if $Q_k = 1$. (The gates N_k are not shown in Fig. 8-9.)

Note that in this application the system is not operating as a shift register since there is no clock required (and no serial input). Each FLIP-FLOP is simply used as an isolated 1-bit read/write memory.

Right-Shift, Left-Shift Register Some commercial shift registers are equipped with gates which allow shifting the data from right to left as well as in the reverse direction. One application for such a system is to perform multiplication or division by multiples of 2, as will now be explained. Consider first a right-shift register as in Fig. 8-9 and that the serial input is held low.

Assume that a binary number is stored in a shift register, with the least-significant bit stored in FF0. Now apply one clock pulse. Each bit then moves to the next lower significant place, and hence is divided by 2. The number now held in the register is half the original number, provided that FF0 was originally 0. Since the 2^0 bit is lost in the shift to the right, then if FF0 was originally in the 1 state, corresponding to the decimal number 1, after the shift the register is in error by the decimal number 0.5. The next clock pulse causes another division by 2, etc.

Consider now that the system is wired so that each clock pulse causes a shift to the left. Each bit now moves to the next higher significant digit, and the number stored is multiplied by 2.

Digital Delay Line A shift register may be used to introduce a time delay Δ into a system, where Δ is an integral multiple of the clock period T. Thus an input pulse train appears at the output of an n-stage register delayed by a time $(n-1)T = \Delta$

Sequence Generator An important application of a shift register is to generate a binary sequence. This system is also called a *word, code,* or *character, generator.* The shift register FLIP-FLOPS are preset to give the desired code. Then the clock applies shift pulses, and the output of the shift register gives the temporal pattern corresponding to the specified sequence. Clearly, we have just described a parallel-in, series-out register. For test purposes it is often necessary that the code be repeated continuously. This mode of operation is easily obtained by feeding the output Q_0 of the register back into the serial input to form a "reentrant shift register." Such a configuration is called a *dynamic,* or *circulating, memory,* or a *shift-register read-only memory.*

Shift-Register Ring Counter Consider the 5-bit shift register (Fig. 8-9) with Q_0 connected to the serial input. Such a circulating memory forms a *ring counter.* Assume that all FLIP-FLOPS are cleared and then that FF0 is preset so that $Q_0 = 1$ and $Q_4 = Q_3 = Q_2 = Q_1 = 0$. The first clock pulse transfers the state of FF0 to FF4, so that after the pulse $Q_4 = 1$ and

$$Q_3 = Q_2 = Q_1 = Q_0 = 0$$

Succeeding pulses will transfer the state 1 progressively around the ring. The count is read by noting which FLIP-FLOP is in state 1; no decoding is necessary.

Consider a ring counter with N stages. If the interval between triggers is T, then the output from any binary stage is a pulse train of period NT, with each pulse of duration T. The output pulse of one stage is delayed by a time T from a pulse in the preceding stage. These pulses may be used where a set of sequential gating waveforms is required. Thus a ring counter is analogous to a stepping switch, where each triggering pulse causes an advance of the switch by one step.

Since there is one output pulse for each N clock pulses, the counter is also a *divide-by-N* unit, or an $N:1$ *scaler*. Typically, TTL shift-register counters operate at frequencies as high as 25 MHz.

Twisted-Ring Counter　The topology where $\bar{Q}_0$ (rather than Q_0) is fed back to the input of the shift register is called a *twisted-ring*, or *Johnson, counter*. This system is a $2N:1$ scaler. To verify this statement consider that initially all stages in Fig. 8-9 are in the 0 state. Since $S_4 = \bar{Q}_0 = 1$, the first pulse puts FF4 into the 1 state; $Q_4 = 1$, and all other FLIP-FLOPS remain in the 0 state. Since now $S_3 = Q_4 = 1$ and S_4 remains in the 1 state, then after the next pulse there results $Q_4 = 1$, $Q_3 = 1$, $Q_2 = 0$, $Q_1 = 0$, and $Q_0 = 0$. In other words, pulse 1 causes only Q_4 to change state, and pulse 2 causes only Q_3 to change from 0 to 1. Continuing the analysis, we see that pulses 3, 4, and 5 cause Q_2, Q_1, and Q_0, in turn, to switch from the 0 to the 1 state. At the end of five pulses all FLIP-FLOPS are in the 1 state.

After pulse 5, $S_4 = \bar{Q}_0$ changes from 1 to 0. Hence the sixth pulse causes Q_4 to change to 0. The seventh pulse resets Q_3 to 0, and so on, until, at the tenth pulse, all stages have been returned to the 0 state, and the counting cycle is complete. We have demonstrated that this five-stage twisted-ring configuration is a 10:1 counter. To read the count requires a 5-to-10-line decoder, but because of the unique waveforms generated, only two-input AND gates are required (Prob. 8-11).

MOS shift registers are considered in Sec. 9-5.

8-4　COUNTERS

The ring counters discussed in the preceding section do not make efficient use of the FLIP-FLOPS. A 5:1 counter (or 10:1 with the Johnson ring) is obtained with five stages, whereas five FLIP-FLOPS define $2^5 = 32$ states. By modifying the interconnections between stages (*not* using the shift-register topology), we now demonstrate that n binaries can function as a $2^n:1$ counter.

Ripple Counter　Consider a chain of four J-K master-slave FLIP-FLOPS with the output Q of each stage connected to the clock input of the following binary, as in Fig. 8-10. The pulses to be counted are applied to the clock input of FF0. For all stages J and K are tied to the supply voltage, so that $J = K = 1$. This connection converts each stage to a T-type FLIP-FLOP (Fig. 8-8), with $T = 1$.

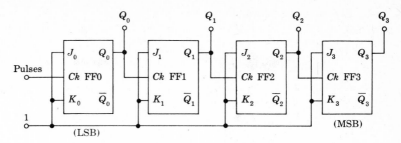

Fig. 8-10 A chain of FLIP-FLOPS connected as a ripple counter.

It should be recalled that, for a T-type binary with $T = 1$, the master changes state every time the waveform at its clock input changes from 0 to 1 and that the new state of the master is transferred to the slave when the clock falls from 1 to 0.   This operation requires that

1. Q_0 changes state at the *falling* edge of each pulse.
2. All other Q's make a transition when and only when the output of the preceding FLIP-FLOP changes from 1 to 0.   This negative transition "ripples" through the counter from the LSB to the MSB.

Following these two rules, the waveforms in Fig. 8-11 are obtained. Table 8-4 lists the states of all the binaries of the chain as a function of the number of externally applied pulses.   This table may be verified directly by comparison with the waveform chart of Fig. 8-11.   Note that in Table 8-4 the FLIP-FLOPS have been ordered in the reverse direction from their order in Fig. 8-10.   We observe that the ordered array of states 0 and 1 in any row

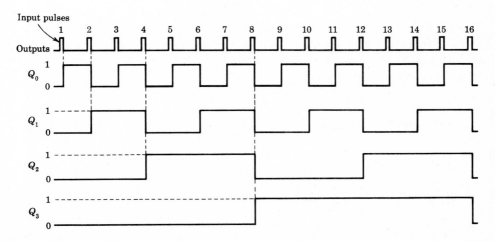

Fig. 8-11 Waveform chart for the four-stage ripple counter.

TABLE 8-4   States of the FLIP-FLOPS in
Fig. 8-10

Number of input pulses	FLIP-FLOP outputs			
	Q_3	Q_2	Q_1	Q_0
0	0	0	0	0
1	0	0	0	1
2	0	0	1	0
3	0	0	1	1
4	0	1	0	0
5	0	1	0	1
6	0	1	1	0
7	0	1	1	1
8	1	0	0	0
9	1	0	0	1
10	1	0	1	0
11	1	0	1	1
12	1	1	0	0
13	1	1	0	1
14	1	1	1	0
15	1	1	1	1
16	0	0	0	0

in Table 8-4 is precisely the binary representation of the number of input pulses as given in Table 6-2. Thus the chain of FLIP-FLOPS *counts* in the binary system.

A chain of n binaries will count up to the number 2^n before it resets itself into its original state. Such a chain is referred to as a counter *modulo* 2^n. To read the counter, the 4-bit words (numbers) in Table 8-4 are recognized with a decoder, which in turn drives visible numerical indicators (Sec. 7-3). Spikes are possible in any counter unless all FLIP-FLOPS change state simultaneously. To eliminate the spikes at the decoder output, a strobe pulse is used (S in Fig. 7-14) so that the counter is read only after the spikes have decayed and a steady state is reached.

Up-Down Counter   A counter which can be made to count in either the forward or reverse direction is called an *up-down*, a *reversible*, or a *forward-backward*, counter. Forward counting is accomplished, as we have seen, when the trigger input of a succeeding binary is coupled to the Q output of a preceding binary. The count will proceed in the reverse direction if the coupling is made instead to the $\bar{Q}$ output, as we shall now verify.

If a binary makes a transition from state 0 to 1, the output $\bar{Q}$ will make a transition from state 1 to 0. This negative-going transition in $\bar{Q}$ will induce a change in state in the succeeding binary. Hence, for the reversing connection, the following rules apply:

1. FLIP-FLOP FF0 makes a transition at each externally applied pulse.

2. Each of the other binaries makes a transition when and only when the preceding FLIP-FLOP goes from state 0 to state 1.

If these rules are applied to any of the numbers in Table 8-4, the next smaller number in the table results. For example, consider the number 12, which is 1100 in binary form. At the next pulse, the rightmost 0 (corresponding to Q_0) becomes 1. This change of state from 0 to 1 causes Q_1 to change state from 0 to 1, which in turn causes Q_2 to change state from 1 to 0. This last transition is in the direction not to affect the following binary, and hence Q_3 remains in state 1. The net result is that the counter reads 1011, which represents the number 11. Since we started with 12 and ended with 11, a reverse count has taken place.

The logic block diagram for an up-down counter is indicated in Fig. 8-12. For simplicity in drawing, no connections to J and K are indicated. For a ripple counter it is always to be understood that $J = K = 1$ as in Fig. 8-10. The two-level AND-OR gates CG1 and CG2 between stages control the direction of the counter. Note that this logic combination is equivalent to a NAND-NAND configuration (Fig. 6-18). If the input X is a 1 (0), then Q ($\bar{Q}$) is effectively connected to the following FLIP-FLOP and pulses are added (subtracted). In other words, $X = 1$ converts the system to an *up* counter and $X = 0$ to a *down* counter. The control X may not be changed from 1 to 0 (or 0 to 1) between input pulses, because a spurious count may be introduced by this transition. (The synchronous counter does not have this difficulty.)

Divide-by-N Counter It may be desired to count to a base N which is not a power of 2. We may prefer, for example, to count to the base 10, since the decimal system is the one with which we are most familiar. To construct such a counter, start with a ripple chain of n FLIP-FLOPS such that n is the smallest number for which $2^n > N$. Add a feedback gate so that at count N all binaries are reset to zero. This feedback circuit is simply a NAND gate whose output feeds all *clear* inputs in parallel. Each input to the NAND gate is a FLIP-FLOP output Q which becomes 1 at the count N.

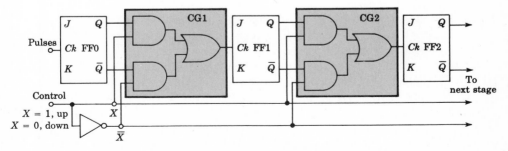

Fig. 8-12 An up-down ripple counter. (It is understood that $J = K = 1$.)

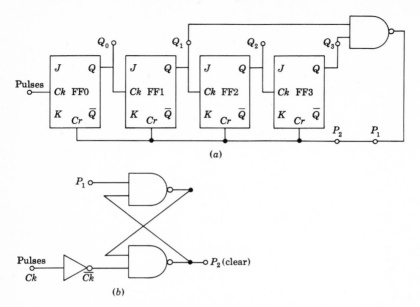

Fig. 8-13 (a) A decade counter $(J = K = 1)$; (b) a latch to eliminate resetting difficulties (due to unequal internal delays).

Let us illustrate the above procedure for a decade counter. Since the smallest value of n for which $2^n > 10$ is $n = 4$, then four FLIP-FLOPS are required. The decimal number 10 is the binary number 1010 (LSB), and hence $Q_0 = 0$, $Q_1 = 1$, $Q_2 = 0$, and $Q_3 = 1$. The inputs to the feedback NAND gate are therefore Q_1 and Q_3, and the complete circuit is shown in Fig. 8-13a. Note that after the tenth pulse Q_1 and Q_3 both go to 1, the output of the NAND gate becomes 0, and all FLIP-FLOPS are cleared (reset to 0). (Note that Q_1 and Q_3 first become 1 and then return to 0 after pulse 10, thereby generating a narrow spike.)

If the propagation delay from the clear input to the FLIP-FLOP output varies from stage to stage, the clear operation may not be reliable. In the above example, if FF3 takes appreciably longer time to reset than FF1, then when Q_1 returns to 0, the output of the NAND gate goes to 1, so that $Cr = 1$ and Q_3 will not reset. Wide variations in reset propagation time may occur if the counter outputs are unevenly loaded. A method of eliminating the difficulty with resetting is to use a latch to memorize the output of the NAND gate at the Nth pulse. The lead in Fig. 8-13a between the NAND output P_1 and the clear input P_2 is opened, and the circuit drawn in Fig. 8-13b is inserted between these two points. The operation of the latch is considered in detail in Prob. 8-15.

A divide-by-6 counter is obtained using a 3-bit ripple counter, and since for $N = 6$, $Q_1 = 1 = Q_2$, then Q_1 and Q_2 are the inputs to the feedback NAND gate. Similarly, a divide-by-7 counter requires a three-input NAND gate with inputs Q_0, Q_1, and Q_2.

In some applications it is important to be able to program the count (the value of N) of a divide-by-N counter, either by means of switches or through control data inputs at the preset terminals. Such a *programmable* or *preset-table* counter is indicated in the figure of Prob. 8-16.

Consider that it is required to count up to 10,000 and to indicate the count visually in the decimal system. Since $10,000 = 10^4$, then four decade-counter units, such as in Fig. 8-13, are cascaded. A BCD-to-decimal decoder/lamp driver (Sec. 7-3) or a BCD–to–seven-segment display decoder (Sec. 7-7) is used with each unit to make visible the four decimal digits giving the count.

Synchronous Counters   The *carry propagation delay* is the time required for a counter to complete its response to an input pulse. The carry time of a ripple counter is longest when each stage is in the 1 state. For in this situation, the next pulse must cause all previous FLIP-FLOPS to change state. Any particular binary will not respond until the preceding stage has nominally completed its transition. The clock pulse effectively "ripples" through the chain. Hence the carry time will be of the order of magnitude of the sum of the propagation delay times (Sec. 6-13) of all the binaries. If the chain is long, the carry time may well be longer than the interval between input pulses. In such a case, it will not be possible to read the counter between pulses.

If the asynchronous operation of a counter is changed so that all FLIP-FLOPS are clocked simultaneously (synchronously) by the input pulses, the propagation delay time may be reduced considerably. Control (AND) gates are interposed between T-type FLIP-FLOP stages in a synchronous counter. The gate output is used to trigger the following stage. The gate inputs must be properly chosen from the outputs of the preceding stage to satisfy the required counter logic. Typically, the maximum frequency of operation of a 4-bit synchronous counter using TTL logic is 32 MHz, which is about twice that of a ripple counter. Another advantage of the synchronous counter is that no decoding spikes appear at the output since all FLIP-FLOPS change state at the same time. Hence no strobe pulse is required when decoding a synchronous counter.

Synchronous up-down decade counters are available commercially (for example, TI 74192) on a single chip. Such a counter has the complexity of 55 equivalent gates, and hence is an example of MSI. The FLIP-FLOPS are provided with *preset* and *clear* inputs. Division by a number other than 10 (or a multiple of 2) is usually not commercially available.

8-5   APPLICATIONS OF COUNTERS

Many systems, including digital computers, data handling, and industrial control systems, use counters. We describe briefly some of the fundamental applications.

Direct Counting Direct counting finds application in many industrial processes. Counters will operate with reliability where human counters fail because of fatigue or limitations of speed. It is required, of course, that the event which is to be counted first be converted into an electric signal, but this requirement usually imposes no important limitation. For example, objects may be counted by passing them single-file on a conveyor belt between a photoelectric cell and a light source.

The *preset* input allows control of industrial processes. The counter may be preset so that it will deliver an output pulse when the count reaches a predetermined number. Such a counter may be used, for example, to count the number of pills dropped into a bottle. When the preset count is attained, the output pulse is used to divert the pills to the next container and at the same time to reset the counter for counting the next batch.

Divide-by-N There are many applications where it is desired to change the frequency of a square wave from f to f/N, where N is some multiple of 2. From the waveforms of Fig. 8-11 it is seen that a counter performs this function.

If instead of square waves it is required to use narrow pulses or spikes for system synchronization, these may be obtained from the waveforms of Fig. 8-11. A small RC coupling combination at the counter output, as in Fig. 14-2a, causes a positive pulse to appear at each transition from 0 to 1 and a negative pulse at each transition from 1 to 0 (Fig. 14-2d). If we count only the positive pulses (the negative pulses may be eliminated by using a diode), it appears (Fig. 14-2e) that each binary divides by 2 the number of positive pulses applied to it. The four binaries together accomplish a division by a factor $2^4 = 16$. A single negative pulse will appear at the output for each 16 pulses applied at the input. A chain of n binaries used for this purpose of dividing or scaling down the number of pulses is referred to as a *scaler*. Thus a chain of four FLIP-FLOPS constitutes a scale-of-16 circuit, etc.

Measurement of Frequency The basic principle by which counters are used for the precise determination of frequency is illustrated in Fig. 8-14. The input signal whose frequency is to be measured is converted into pulses by a zero-crossing detector (Fig. 14-2) and applied through an AND gate to a counter. To determine the frequency, it is now only required to keep the gate open for transmission for a known time interval. If, say, the gating time is 1 s, the counter will yield the frequency directly in cycles per second (hertz). The *clock* for timing the gate interval is an accurate crystal oscillator whose frequency, is, say, 1 MHz. The crystal oscillator drives a scale-of-10^6 circuit which divides the crystal frequency by a factor of 1 million. The divider output consists of a 1-Hz signal whose period is as accurately maintained as the crystal frequency. This divider output signal controls the gating time by setting a toggle FLIP-FLOP to the 1 state for 1 s. The system is susceptible to only slight errors. One source of error results from the fact that a variation

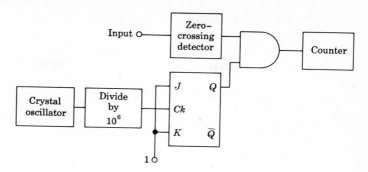

Fig. 8-14   A system for measuring frequency by means of a counter.

of ± 1 count may be obtained, depending on the instant when the first and last pulses occur in relation to the sampling time. Beyond these, of course, the accuracy depends on the accuracy of the crystal oscillator.

Measurement of Time   The time interval between two pulses may also be measured with the circuit of Fig. 8-14. The FLIP-FLOP is now converted into a set-reset type, with the first input pulse applied to the S terminal, the second pulse to the R terminal, and no connection made to Ck. With this configuration the first pulse opens the AND gate for transmission and the second pulse closes it. The crystal-oscillator signal (or some lower frequency from the divider chain) is converted into pulses, and these are passed through the gate into the counter. The number of counts recorded is proportional to the length of time the gate is open and hence gives the desired time interval.

Measurement of Distance   In radar or sonar systems a pulse is transmitted and a reflected pulse is received delayed by a time T. Since the velocity of light (or sound) is known, a measurement of the interval T, as outlined above, gives the distance from the transmitter to the object from which the reflection was received.

Measurement of Speed   A speed determination may be converted into a time measurement. For example, if two photocell–light-source combinations are set a fixed distance apart, the average speed of an object passing between these points is inversely proportional to the time interval between the generated pulses. Projectile velocities have been measured in this manner.

Digital Computer   In a digital computer a problem is solved by subjecting data to a sequence of operations in accordance with the program of instructions introduced into the computer. Counters may be used to count the operations as they are performed and to call forth the next operation from the memory when the preceding one has been completed.

Waveform Generation The waveforms which occur at the collectors or bases of binary counters may be combined either directly or in connection with logic gates to generate desirable pulse-type waveforms. Such waveforms are used for sequential data selection and parallel-to-serial conversion, as described in Sec. 7-4.

Conversion between Analog and Digital Information These systems are considered in Secs. 14-7 and 14-8.

REVIEW QUESTIONS

8-1 (a) Define a *sequential* system. (b) How does it differ from a *combinational* system?

8-2 (a) Define a *latch*. (b) Show how to construct this unit from NOT gates. (c) Verify that the circuit of part b has two stable states.

8-3 (a) Sketch the logic system for a latch with set S (preset) and reset R (clear) inputs. (b) Verify that if $S = 1$ and $R = 0$, the FLIP-FLOP is set to $Q = 1$.

8-4 (a) Sketch the logic system for a clocked S-R FLIP-FLOP. (b) Verify that the state of the system does not change in between clock pulses. (c) Give the truth table. (d) Justify the entries in the truth table.

8-5 (a) Augment an S-R FLIP-FLOP with two AND gates to form a J-K FLIP-FLOP. (b) Give the truth table. (c) Verify part b by making a table of J_n, K_n, Q_n, $\bar{Q}_n$, S_n, R_n, and Q_{n+1}.

8-6 Explain what is meant by a *race-around* condition in connection with the J-K FLIP-FLOP of Rev. 8-5.

8-7 Draw a clocked J-K FLIP-FLOP system and include *preset* (Pr) and *clear* (Cr) inputs. (b) Explain the clear operation.

8-8 (a) Draw a *master-slave* J-K FLIP-FLOP system. (b) Explain its operation and show that the race-around condition is eliminated.

8-9 (a) Show how to convert a J-K FLIP-FLOP into a *delay* (D-type) unit. (b) Give the truth table. (c) Verify this table.

8-10 Repeat Rev. 8-9 for a *toggle* (T-type) FLIP-FLOP.

8-11 Give the truth tables for each FLIP-FLOP type: (a) S-R, (b) J-K, (c) D, and (d) T. What are the direct inputs Pr and Cr and the clock Ck for (e) presetting, (f) clearing, and (g) normal clocked operation?

8-12 (a) Define a *register*. (b) Construct a shift register from S-R FLIP-FLOPS. (c) Explain its operation.

8-13 (a) Explain why there may be a race condition in a shift register. (b) How is this difficulty bypassed?

8-14 Explain how a shift register is used as a converter from (a) *serial-to parallel* data and (b) *parallel-to-serial* data.

8-15 Explain how a shift register is used as a *sequence generator*.

8-16 Explain how a shift register is used as a *read-only memory*.

8-17 (a) Explain how a shift register is used as a *ring counter*. (b) Draw the output waveform from each FLIP-FLOP of a three-stage unit.

8-18 (a) Sketch the block diagram for a *Johnson* (*twisted ring*) counter. (b)

Draw the output waveform from each FLIP-FLOP of a three-stage unit. (*c*) By what number N does this system divide?

8-19 (*a*) Draw the block diagram of a *ripple counter*. (*b*) Sketch the waveform at the output of each FLIP-FLOP for a three-stage counter. (*c*) Explain how this waveform chart is obtained. (*d*) By what number N does this system divide?

8-20 (*a*) Draw the block diagram for an *up-down counter*. (*b*) Explain its operation.

8-21 Explain how to modify a ripple counter so that it divides by N, where N is *not* a power of 2.

8-22 (*a*) Draw the block diagram of a decade ripple counter. (*b*) Explain its operation.

8-23 Repeat Rev. 8-22 for a divide-by-6 ripple counter.

8-24 What is the advantage of a *synchronous counter* over a ripple counter?

8-25 Explain how to measure frequency by means of a counter.

8-26 List six applications of counters. Give no explanations.

9 / MOS DIGITAL CIRCUITS AND LSI SYSTEMS

The field-effect transistor is a semiconductor device which depends for its operation on the control of current by an electric field. There are two types of field-effect transistors, the *junction field-effect transistor* (abbreviated JFET, or simply FET) and the *insulated-gate field-effect transistor* (IGFET), more commonly called the *metal-oxide-semiconductor (MOS) transistor* (MOST, or MOSFET).

The principles on which these devices operate, as well as the differences in their characteristics, are examined in this chapter. The principal applications of MOSFETs are as LSI digital arrays.

The field-effect transistor differs from the bipolar junction transistor in the following important characteristics:

1. Its operation depends upon the flow of majority carriers only. It is therefore a *unipolar* (one type of carrier) device.

2. It is simpler to fabricate and occupies less space in integrated form than a bipolar transistor.

3. It exhibits a high input resistance, typically many megohms.

4. It is less noisy than a bipolar transistor.

5. It exhibits no offset voltage at zero drain current, and hence makes an excellent signal chopper.

9-1   THE JUNCTION FIELD–EFFECT TRANSISTOR

The structure of an *n-channel* field-effect transistor is shown in Fig. 9-1. Ohmic contacts are made to the two ends of a semiconductor bar of *n*-type material (if *p*-type silicon is used, the device is referred to as a *p-channel* FET). Current is caused to flow along the length of the bar because of the voltage supply connected between the ends. This current consists of majority carriers, which in this case are elec-

190

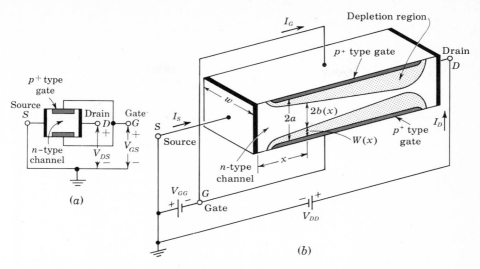

Fig. 9-1 The basic structure of an n-channel field-effect transistor.   (a) Simplified view.   (b) More detailed drawing.   The normal polarities of the drain-to-source (V_{DD}) and gate-to-source (V_{GG}) supply voltages are shown.   In a p-channel FET the voltages would be reversed.

trons.   A simple side view of a JFET is indicated in Fig. 9-1a and a more detailed sketch is shown in Fig. 9-1b.   The circuit symbol with current and voltage polarities marked is given in Fig. 9-2.   The following FET notation is standard.

Source   The *source* S is the terminal through which the majority carriers enter the bar.   Conventional current entering the bar at S is designated by I_S.

Drain   The *drain* D is the terminal through which the majority carriers leave the bar.   Conventional current entering the bar at D is designated by I_D. The drain-to-source voltage is called V_{DS}, and is positive if D is more positive than S.   In Fig. 9-1, $V_{DS} = V_{DD}$ = drain supply voltage.

Gate   On both sides of the n-type bar of Fig. 9-1, heavily doped (p^+) regions of acceptor impurities have been formed by alloying, by diffusion, or by any other procedure available for creating p-n junctions.   These impurity regions are called the *gate* G.   Between the gate and source a voltage $V_{GS} = -V_{GG}$ is applied in the direction to reverse-bias the p-n junction. Conventional current entering the bar at G is designated I_G.

Channel   The region in Fig. 9-1 of n-type material between the two gate regions is the *channel* through which the majority carriers move from source to drain.

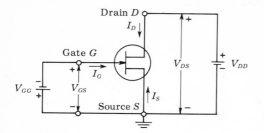

Fig. 9-2 Circuit symbol for an *n*-channel FET. (For a *p*-channel FET the arrow at the gate junction points in the opposite direction.) For an *n*-channel FET, I_D and V_{DS} are positive and V_{GS} is negative. For a *p*-channel FET, I_D and V_{DS} are negative and V_{GS} is positive.

FET Operation It is necessary to recall that on the two sides of the reverse-biased *p-n* junction (the transition region) there are space-charge regions (Sec. 2-6). The current carriers have diffused across the junction, leaving only uncovered positive ions on the *n* side and negative ions on the *p* side. The electric lines of field intensity which now originate on the positive ions and terminate on the negative ions are precisely the source of the voltage drop across the junction. As the reverse bias across the junction increases, so also does the thickness of the region of immobile uncovered charges. The conductivity of this region is nominally zero because of the unavailability of current carriers. Hence we see that the effective width of the *channel* in Fig. **9-1** will become progressively decreased with increasing reverse bias. Accordingly, for a fixed drain-to-source voltage, the drain current will be a function of the reverse-biasing voltage across the gate junction. The term *field effect* is used to describe this device because the mechanism of current control is the *effect* of the extension, with increasing reverse bias, of the *field* associated with the region of uncovered charges.

FET Static Characteristics The circuit, symbol, and polarity conventions for an FET are indicated in Fig. 9-2. The direction of the arrow at the gate of the junction FET in Fig. 9-2 indicates the direction in which gate current would flow if the gate junction were forward-biased. The common-source drain characteristics for a typical *n*-channel FET shown in Fig. 9-3 give I_D against V_{DS}, with V_{GS} as a parameter. To see qualitatively why the characteristics have the form shown, consider, say, the case for which $V_{GS} = 0$. For $I_D = 0$, the channel between the gate junctions is entirely open. In response to a small applied voltage V_{DS}, the *n*-type bar acts as a simple semiconductor resistor, and the current I_D increases linearly with V_{DS}. With increasing current, the ohmic voltage drop between the source and the channel region reverse-biases the junction, and the conducting portion of the channel begins to constrict. Because of the ohmic drop along the length of the channel itself, the constriction is not uniform, but is more pronounced at distances farther from the source, as indicated in Fig. 9-1. Eventually, a voltage V_{DS} is reached at which the channel is "pinched off." This is the voltage, not too sharply defined in Fig. 9-3, where the current I_D begins to level off and approach a constant value. It is, of course, in principle not possible for the channel to close completely and thereby reduce the current I_D to zero. For if such,

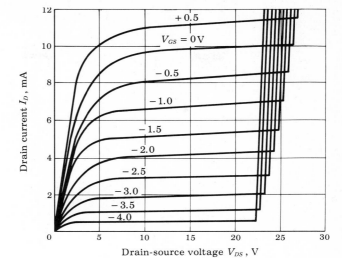

Fig. 9-3 Common-source drain characteristics of an n-channel field-effect transistor.

indeed, could be the case, the ohmic drop required to provide the necessary back bias would itself be lacking. Note that each characteristic curve has an ohmic region for small values of V_{DS}, where I_D is proportional to V_{DS}. Each also has a constant-current region for large values of V_{DS}, where I_D responds very slightly to V_{DS}.

If now a gate voltage V_{GS} is applied in the direction to provide additional reverse bias, pinch-off will occur for smaller values of $|V_{DS}|$, and the maximum drain current will be smaller. This feature is brought out in Fig. 9-3. Note that a plot for a silicon FET is given even for $V_{GS} = +0.5$ V, which is in the direction of forward bias. We note from Table 4-1 that, actually, the gate current will be very small, because at this gate voltage the Si junction is barely at the cutin voltage V_γ.

The maximum voltage that can be applied between any two terminals of the FET is the lowest voltage that will cause avalanche breakdown (Sec. 2-9) across the gate junction. From Fig. 9-3 it is seen that avalanche occurs at a lower value of $|V_{DS}|$ when the gate is reverse-biased than for $V_{GS} = 0$. This is caused by the fact that the reverse-bias gate voltage adds to the drain voltage, and hence increases the effective voltage across the gate junction.

We note from Fig. 9-2 that the n-channel FET requires zero or negative gate bias and positive drain voltage. The p-channel FET requires opposite voltage polarities. Either end of the channel may be used as a source. We can remember supply polarities by using the channel type, p or n, to designate the polarity of the *source* side of the drain supply.

A Practical FET Structure The structure shown in Fig. 9-1 is not practical because of the difficulties involved in diffusing impurities into both sides of a semiconductor wafer. Figure 9-4 shows a single-ended-geometry junction

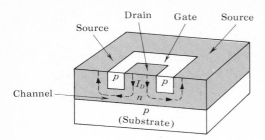

Fig. 9-4 Single-ended-geometry junction FET.

FET where diffusion is from one side only. The substrate is of p-type material onto which an n-type channel is epitaxially grown (Sec. 5-3). A p-type gate is then diffused into the n-type channel. The substrate which may function as a second gate is of relatively low resistivity material. The diffused gate is also of very low resistivity material, allowing the depletion region to spread mostly into the n-type channel.

9-2   THE METAL–OXIDE–SEMICONDUCTOR FET (MOSFET)

We now turn our attention to the insulated-gate FET, or metal-oxide-semiconductor FET, which is of much greater commercial importance than the junction FET.

The p-channel MOSFET consists of a lightly doped n-type substrate into which two highly doped p^+ regions are diffused, as shown in Fig. 9-5. These n^+ sections, which will act as the source and drain, are separated by about 5 to 10 μm. A thin (1000 to 2000 Å) layer of insulating silicon dioxide (SiO_2) is grown over the surface of the structure, and holes are cut into the oxide layer, allowing contact with the source and drain. Then the gate-metal area is overlaid on the oxide, covering the entire channel region. Simultaneously, metal contacts are made to the drain and source, as shown in Fig. 9-5. The contact to the metal over the channel area is the gate terminal. The chip area of a MOSFET is 3 square mils or less, which is only about 5 percent of that required by a bipolar junction transistor.

The metal area of the gate, in conjunction with the insulating dielectric oxide layer and the semiconductor channel, form a parallel-plate capacitor. The insulating layer of silicon dioxide is the reason why this device is called the insulated-gate field-effect transistor. This layer results in an extremely high input resistance (10^{10} to 10^{15} Ω) for the MOSFET. The p-channel enhancement MOSFET characteristic will now be described.

The Enhancement MOSFET   If we ground the substrate for the structure of Fig. 9-5 and apply a negative voltage at the gate, an electric field will be directed perpendicularly through the oxide. This field will end on "induced" positive charges on the semiconductor site, as shown in Fig. 9-5. The positive

**Fig. 9-5 Enhancement in a
p-channel MOSFET. (Courtesy of
Motorola Semiconductor Products,
Inc.)**

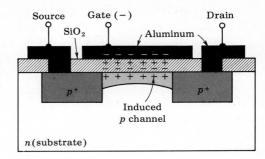

charges, which are minority carriers in the *n*-type substrate, form an "inversion
layer." As the magnitude of the negative voltage on the gate increases, the induced positive charge in the semiconductor increases. The region beneath the
oxide now has *p*-type carriers, the conductivity increases, and current flows from
source to drain through the induced channel. Thus the drain current is "en-
hanced" by the negative gate voltage, and such a device is called an *enhancement-type* MOS.

Threshold Voltage The volt-ampere drain characteristics of a *p*-channel
enhancement-mode MOSFET are given in Fig. 9-6*a*, and its transfer curve in
Fig. 9-6*b*. The current I_D at $V_{GS} \geq 0$ is very small, of the order of a few
nanoamperes. This current is labeled I_{DSS} in Fig. 9-6*b*. As V_{GS} is made
negative, the current $|I_D|$ increases slowly at first, and then much more rapidly
with an increase in $|V_{GS}|$. The manufacturer often indicates the *gate-source
threshold voltage* V_{GST}, or V_T,† at which $|I_D|$ reaches some defined small value,
say 10 μA. A current $I_{D,\text{ON}}$ corresponding approximately to the maximum
value given on the drain characteristics, and the value of V_{GS} needed to obtain
this current are also usually given on the manufacturer's specification sheets.

† In this chapter the threshold voltage should not be confused with the volt equivalent
of temperature V_T of Sec. 1-9.

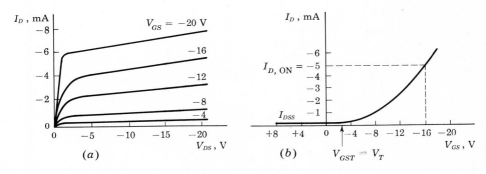

Fig. 9-6 (a) The drain characteristics and (b) the transfer curve for ($V_{DS} = 10$ V)
of a *p*-channel enhancement-type MOSFET.

The value of V_T for the p-channel standard MOSFET is typically -4 V, and it is common to use a power-supply voltage of -12 V for the drain supply. This large voltage is incompatible with the power-supply voltage of typically 5 V used in bipolar integrated circuits. Thus various manufacturing techniques have been developed to reduce V_T. In general, a low threshold voltage allows (1) the use of a small power-supply voltage, (2) compatible operation with bipolar devices, (3) smaller switching time due to the smaller voltage swing during switching, and (4) higher packing densities.

Three methods are used to lower the magnitude of V_T.

1. The high-threshold MOSFET described above uses a silicon crystal with $\langle 111 \rangle$ orientation. If a crystal is utilized in the $\langle 100 \rangle$ direction it is found that a value of V_T results which is about one-half that obtained with $\langle 111 \rangle$ orientation.

2. The silicon nitride approach makes use of a layer of Si_3N_4 and SiO_2, whose dielectric constant is about twice that of SiO_2 alone. A FET constructed in this manner (designated an MNOS device) decreases V_T to approximately 2 V.

3. Polycrystalline silicon doped with boron is used as the gate electrode instead of aluminum. This reduction in the difference in contact potential between the gate electrode and the gate dielectric reduces V_T. Such devices are called *silicon gate* MOS transistors. All three of the fabrication methods described above result in a low-threshold device with V_T in the range 1.5 to 2.5 V, whereas the standard high-threshold MOS has a V_T of approximately 4 to 6 V.

Power Supply Requirements   Table 9-1 gives the voltages customarily used with high-threshold and low-threshold p-channel MOSFETs. Note that V_{SS} refers to the source, V_{DD} to the drain, and V_{GG} to the gate supply voltages. The subscript 1 denotes that the source is grounded and the subscript 2 designates that the drain is at ground potential.

The low-threshold MOS circuits require lower power supply voltages and this means less expensive system power supplies. In addition, the input voltage swing for turning the device ON and OFF is smaller for the lower-threshold voltage, and this means faster operation. Another very desirable feature of low-threshold MOS circuits is that they are directly compatible with bipolar ICs. They require and produce essentially the same input and output signal

TABLE 9-1   Power supply voltages for p-channel MOSFETs, in volts

	V_{SS1}	$-V_{DD1}$	$-V_{GG1}$	V_{SS2}	V_{DD2}	$-V_{GG2}$
High threshold	0	-12	-24	$+12$	0	-12
Low threshold	0	-5	-17	$+5$	0	-12

Fig. 9-7 Ion implantation in
MOS devices.

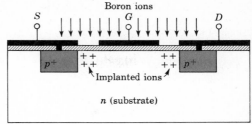

swings and the system designer has the flexibility of using MOS and bipolar circuits in the same system.

Ion Implantation The ion-implantation technique demonstrated in Fig. 9-7 provides very precise control of doping. Ions of the proper dopant such as phosphorus or boron are accelerated to a high energy of up to 300,000 eV and are used to bombard the silicon wafer target. The energy of the ions determines the depth of penetration into the target. In those areas where ion implantation is not desired, an aluminum mask or a thick (12,000 Å) oxide layer absorbs the ion. Virtually any value of V_T can be obtained using ion implantation. In addition, we see from Fig. 9-7 that there is no overlap between the gate and drain or gate and source electrodes (compare Fig. 9-7 with Fig. 9-5). Consequently, due to ion implantation, there is a drastic reduction in C_{gd}, the capacitance between gate and drain, and C_{gs}, the capacitance between gate and source.

The Depletion MOSFET A second type of MOSFET can be made if, to the basic structure of Fig. 9-5, a channel is diffused between the source and the drain, with the same type of impurity as used for the source and drain diffusion. Let us now consider such an n-channel structure, shown in Fig. 9-8a. With this device an appreciable drain current I_{DSS} flows for zero gate-to-source voltage $V_{GS} = 0$. If the gate voltage is made negative, positive charges are induced in the channel through the SiO_2 of the gate capacitor. Since the current in an FET is due to majority carriers (electrons for an n-type material), the induced positive charges make the channel less conductive, and the drain current drops as V_{GS} is made more negative. The redistribution of charge in the channel causes an effective depletion of majority carriers, which accounts for the designation *depletion* MOSFET. Note in Fig. 9-8b that, because of the voltage drop due to the drain current, the channel region nearest the drain is more depleted than is the volume near the source. This phenomenon is analogous to that of pinch-off occurring in a JFET at the drain end of the channel (Fig. 9-1). As a matter of fact, the volt-ampere characteristics of the depletion-mode MOS and the JFET are quite similar.

A MOSFET of the depletion type just described may also be operated in an enhancement mode. It is only necessary to apply a positive gate voltage

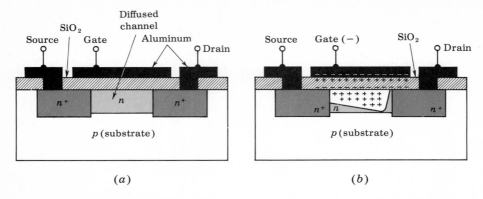

Fig. 9-8   (a) An n-channel depletion-type MOSFET.   (b) Channel depletion with the application of a negative gate voltage.   (Courtesy of Motorola Semiconductor Products, Inc.)

so that negative charges are induced into the n-type channel.   In this manner the conductivity of the channel increases and the current rises above I_{DSS}. The volt-ampere characteristics of this device are indicated in Fig. 9-9a, and the transfer curve is given in Fig. 9-9b.   The depletion and enhancement regions, corresponding to V_{GS} negative and positive, respectively, should be noted.   The manufacturer sometimes indicates the *gate-source cutoff voltage* $V_{GS,\text{OFF}}$, at which I_D is reduced to some specified negligible value at a recommended V_{DS}.   This gate voltage corresponds to the pinch-off voltage V_P of a JFET.

The foregoing discussion is applicable in principle also to the p-channel MOSFET.   For such a device the signs of all currents and voltages in the volt-ampere characteristics of Fig. 9-9 must be reversed.

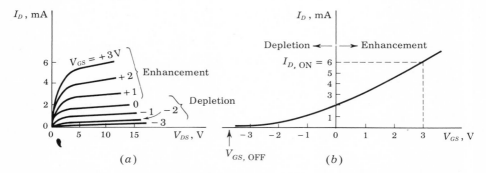

Fig. 9-9   (a) The drain characteristics and (b) the transfer curve (for $V_{DS} = 10$ V) for an n-channel MOSFET which may be used in either the enhancement or the depletion mode.

The ON Resistance $r_{d,\text{ON}}$ The volt-ampere characteristics of Fig. 9-6 suggest that, for very small values of V_{DS}, the MOSFET behaves like an ohmic resistance whose value is determined by V_{GS}. The ratio V_{DS}/I_D at the origin is called the ON *drain resistance $r_{d,\text{ON}}$*. This parameter is important in switching applications where the device is driven heavily ON.

Comparison of p- with n-Channel FETs   The p-channel enhancement FET, shown in Fig. 9-5, is very popular in MOS systems because it is much easier to produce than the n-channel device. Most of the contaminants in MOS fabrication are mobile ions which are positively charged and are trapped in the oxide layer between gate and substrate. In an n-channel enhancement device the gate is normally positive with respect to the substrate and, hence, the positively charged contaminants collect along the interface between the SiO_2 and the silicon substrate. The positive charge from this layer of ions attracts free electrons in the channel which tends to make the transistor turn on prematurely. In p-channel devices the positive contaminant ions are pulled to the opposite side of the oxide layer (to the aluminum-SiO_2 interface) by the negative gate voltage and there they cannot affect the channel.

The hole mobility in silicon and at normal field intensities is approximately 500 cm^2/V-s. On the other hand, electron mobility is about 1,300 cm^2/V-s. Thus the p-channel device will have more than twice the ON resistance of an equivalent n-channel of the same geometry and under the same operating conditions. In other words, the p-channel device must have more than twice the area of the n-channel device to achieve the same resistance. Therefore n-channel MOS circuits can be smaller for the same complexity than p-channel devices. The higher packing density of the n-channel MOS also makes it faster in switching applications due to the smaller junction areas. The operating speed is limited primarily by the internal RC time constants, and the capacitance is directly proportional to the junction cross sections. For all the above reasons it is clear that n-channel MOS circuits are more desirable than p-channel circuits. However, the more extensive process control needed for n-channel fabrication makes them more expensive than p-channel devices.

MOSFET Gate Protection   Since the SiO_2 layer of the gate is extremely thin, it may easily be damaged by excessive voltage. An accumulation of charge on an open-circuited gate may result in a large enough field to punch through the dielectric. To prevent this damage some MOS devices are fabricated with a Zener diode between gate and substrate. In normal operation this diode is open and has no effect upon the circuit. However, if the voltage at the gate becomes excessive, then the diode breaks down and the gate potential is limited to a maximum value equal to the Zener voltage.

Circuit Symbols   It is possible to bring out the connection to the substrate externally so as to have a tetrode device. Most MOSFETs, however, are triodes, with the substrate internally connected to the source. The circuit

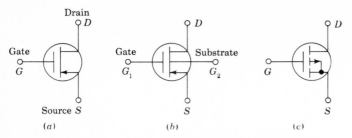

Fig. 9-10 Three circuit symbols for a p-channel MOSFET. (a) and (b) can be either depletion or enhancement types, whereas (c) represents specifically an enhancement device. In (a) the substrate is understood to be connected internally to the source. For an n-channel MOSFET the direction of the arrow is reversed.

symbols used by several manufacturers are indicated in Fig. 9-10. Often the substrate lead is omitted from the symbol as in (a), and is then understood to be connected to the source internally. For the enhancement-type MOSFET of Fig. 9-10c, G_2 is shown to be internally connected to S.

9-3 DIGITAL MOSFET CIRCUITS

The most common applications of MOS devices are digital, such as logic gates, registers, or memory arrays. Because of the gate-to-drain and gate-to-source and substrate parasitic capacitances, MOSFET circuits are slower than corresponding bipolar circuits. However, the lower power dissipation and higher density of fabrication make MOS devices attractive and economical for many applications of repetition frequencies below about 15 MHz.

Inverter MOSFET digital circuits consist *entirely* of FETs and no other devices such as diodes, resistors, or capacitors (except for parasitic capacitances). For example, consider the MOSFET inverter of Fig. 9-11a. Device $Q1$ is the *driver FET*, whereas $Q2$ acts as its load resistance and is called the *load FET*. The nonlinear character of the load is brought into evidence as follows: Since the gate is tied to the drain, $V_{GS2} = V_{DS2}$. The drain characteristics of Fig. 9-6 are reproduced in Fig. 9-12a, and the shaded curve represents the locus of the points $V_{GS2} = V_{DS2} = V_L$. This curve also gives I_{D2} versus V_L (for $V_{GS2} = V_{DS2}$), and its slope gives the incremental load conductance g_L of $Q2$ as a load. Clearly, the load resistance is nonlinear. Note that $Q2$ is always conducting, (for $|V_{DS2}| > |V_T|$), regardless of whether $Q1$ is ON or OFF.

The incremental resistance is not a very useful parameter when considering large-signal (ON-OFF) digital operation. It is necessary to draw the

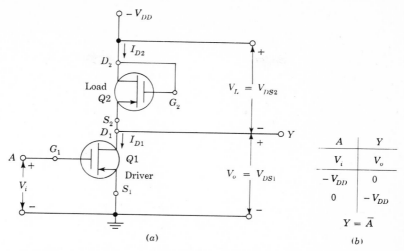

A	Y
V_i	V_o
$-V_{DD}$	0
0	$-V_{DD}$

$$Y = \bar{A}$$

(a)   (b)

Fig. 9-11 (a) MOS inverter; $Q2$ acts as the load for the driver $Q1$.
(b) The voltage truth table and Boolean expression.

load curve (corresponding to a *load line* with a constant resistance) on the volt-ampere characteristics of the driver FET $Q1$. The *load curve* is a plot of

$$I_D = I_{D1} \qquad \text{versus} \qquad V_{DS1} = V_o = -V_{DD} - V_L = -20 - V_{DS2}$$

where we have assumed a 20-V power supply. For a given value of $I_{D2} = I_{D1}$, we find $V_{DS2} = V_L$ from the shaded curve in Fig. 9-12a and then plot the locus of the values I_{D1} versus $V_o = V_{DS1}$ in Fig. 9-12b. For example, from Fig. 9-12a for $I_{D2} = 4$ mA, we find $V_{DS2} = -14$ V. Hence $I_{D1} = 4$ mA is located at $V_{DS1} = -20 + 14 = -6$ V in Fig. 9-12b.

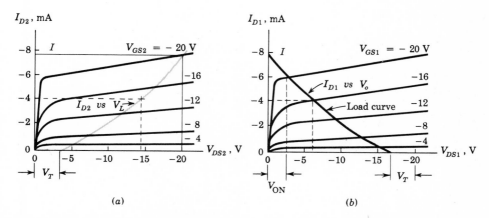

(a)   (b)

Fig. 9-12 (a) The load current I_{D2} versus $V_L = V_{DS2}$. (b) The load curve $I_{D1} = I_{D2}$ versus $V_o = V_{DS1}$.

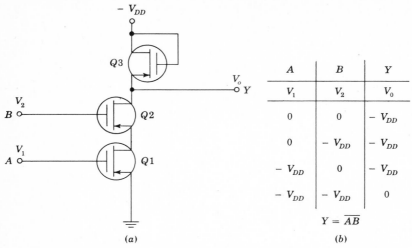

A	B	Y
V_1	V_2	V_o
0	0	$-V_{DD}$
0	$-V_{DD}$	$-V_{DD}$
$-V_{DD}$	0	$-V_{DD}$
$-V_{DD}$	$-V_{DD}$	0

$$Y = \overline{AB}$$

(a)   (b)

Fig. 9-13 (a) MOSFET (negative) NAND gate and (b) voltage truth table and Boolean expression. (Remember that 0 V is the zero state and $-V_{DD}$ is the 1 state.)

We now confirm that the circuit of Fig. 9-11 is an inverter, or NOT circuit. Let us assume *negative* logic (Sec. 6-1) with the 1, or low state, given by $V(1) \approx -V_{DD} = -20$ V and the 0, or high state, given by $V(0) \approx 0$. If $V_i = V_{GS1} = -20$ V, then from Fig. 9-12b, $V_o = V_{ON} \approx -2$ V. Hence, $V_i = V(1)$ gives $V_o = V(0)$. Similarly from Fig. 9-12b, if $V_i = 0$ V, then $V_o = -V_{DD} - V_T = -17$ V for $V_T \approx -3$ V. Hence, $V_i = V(0)$ gives $V_o = V(1)$, thus confirming the truth table of Fig. 9-11b.

We shall simplify the remainder of the discussion in this section by assuming that $|V_{ON}|$ and $|V_T|$ are small compared with $|V_{DD}|$ and shall take $V_{ON} = 0$ and $V_T = 0$. Hence, to a first approximation the load FET may be considered to be a constant resistance R_L and may be represented by a load line passing through $I_D = 0$, $V_{DS} = -V_{DD}$ and $I_D = I$, $V_{DS} = 0$, where I is the drain current for $V_{DS} = V_{GS} = -V_{DD}$. In other words, $R_L = -V_{DD}/I$. Most MOSFETs are p-channel enhancement-type devices, and negative logic is used with $V(0) = 0$ and $V(1) = -V_{DD}$.

NAND **Gate** The operation of the negative NAND gate of Fig. 9-13 can be understood if we realize that if either input V_1 or V_2 is at 0 V (the 0 state), the corresponding FET is OFF and the current is zero. Hence the voltage drop across the load FET is zero and the output $V_o = -V_{DD}$ (the 1 state). If both V_1 and V_2 are in the 1 state ($V_1 = V_2 = -V_{DD}$), then both $Q1$ and $Q2$ are ON and the output is 0 V, or at the 0 state. These values are in agreement with the voltage truth table of Fig. 9-13b. If 1 is substituted for $-V_{DD}$ in Fig. 9-13b, then this logic agrees with the truth table

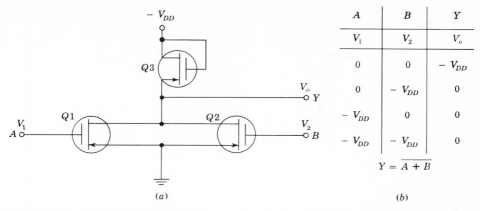

A	B	Y
V_1	V_2	V_o
0	0	$-V_{DD}$
0	$-V_{DD}$	0
$-V_{DD}$	0	0
$-V_{DD}$	$-V_{DD}$	0

$$Y = \overline{A + B}$$

(a)   (b)

Fig. 9-14 (a) MOSFET (negative) NOR gate and (b) voltage truth table and Boolean equation.

for a NAND gate, given in Fig. 6-14. We note that only during one of the four possible input states is power delivered by the power supply.

NOR **Gate**   The circuit of Fig. 9-14a is a negative NOR gate. When either one of the two inputs (or both) is at $-V_{DD}$, the corresponding FET is ON and the output is at 0 V. If both inputs are at 0 V, both transistors $Q1$ and $Q2$ are OFF and the output is at $-V_{DD}$. These values agree with the truth table of Fig. 9-14b. Note that power is drawn from the supply during three of the four possible input states. Because of the high density of MOS devices on the same chip, it is important to minimize power consumption in LSI MOSFET systems (Sec. 9-6).

The circuit of Fig. 9-13 may be considered to be a *positive* NOR gate, and that of Fig. 9-14 to be a *positive* NAND gate (Sec. 6-8). These MOSFET circuits are examples of direct-coupled transistor logic (DCTL). A FLIP-FLOP constructed from MOSFETs is indicated in Prob. 9-3. An AND (OR) gate is obtained by cascading a NAND (NOR) gate with a NOT gate. Typically, a three-input NAND gate uses about 16 mils² of chip area, whereas a single bipolar junction transistor may need about three times this area.

Complementary MOS (CMOS)   It is possible to reduce the power dissipation to very small (50-nW) levels by using complementary p-channel and n-channel enhancement MOS devices on the same chip. The basic complementary MOS inverter circuit is shown in Fig. 9-15. Transistor $Q1$ is the p-channel unit, and transistor $Q2$ is n-channel. The two devices are in series, with their drains tied together and their gates also connected together. The logic swing gate voltage V_i varies from 0 V to the power supply $-V_{DD}$. When $V_i = -V_{DD}$ (logic 1) transistor $Q1$ is turned ON (but draws no appreciable steady-state current) and $Q2$ is turned OFF, the output V_o is then at 0 V

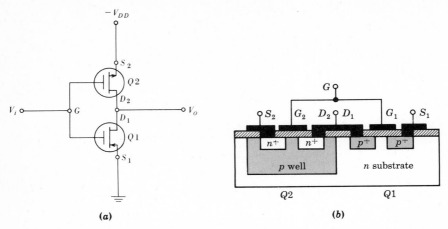

Fig. 9-15 (a) Complementary MOS inverter. (b) Cross section of a complementary MOSFET. Note that the p-type well is diffused into the n-type substrate and that the n-channel MOS $Q2$ is formed in this region.

(logic 0), and inversion has been accomplished. When zero voltage (logic 0) is applied at the input, the n-channel $Q2$ is turned ON (at no steady-state current) and $Q1$ is turned OFF. Thus the output is at $-V_{DD}$ (logic 1). In either logic state, $Q1$ or $Q2$ is OFF and the quiescent power dissipation for this simple inverter is the product of the OFF leakage current and $-V_{DD}$.

More complicated digital CMOS circuits (NAND, NOR, and FLIP-FLOPS) can be formed by combining simple inverter circuits (Probs. 9-5 and 9-6).

9-4 DYNAMIC MOS CIRCUITS

The first commercial MOSFET (abbreviated MOS) circuits appeared in 1964, and since then they have continued to increase in popularity because of their high packing density, small power consumption, and low cost. As many as 5,000 MOS devices are fabricated on a chip 150 by 150 mils square. Such LSI construction makes possible very large MOS shift registers and memories, described in Secs. 9-5 to 9-7. Many of these systems are operated synchronously. Hence a discussion of such clocked (called *dynamic* MOS) circuits follows. Note that p-channel devices are used with negative logic so that the 0 state is 0 V and the 1 state is −10 V.

Dynamic MOS circuits make use of the parasitic capacitance between gate and substrate to provide temporary storage. This storage can be made permanent using refreshing operations by means of clock waveforms. Since the leakage of the gate circuit is extremely low, the time constants are of the order of milliseconds, and to maintain the stored data, the refreshing or cycling must not be allowed to fall below some minimum rate, typically 1 kHz.

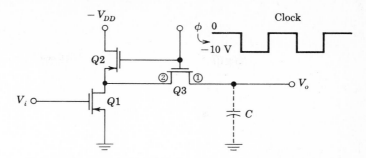

Fig. 9-16   Dynamic MOS inverter.   (V_o is read only when $\phi = -10$ V.)

Dynamic MOS Inverter   The circuit of Fig. 9-16 shows a dynamic MOS inverter which requires a train of clock pulses ϕ for proper operation.   Compare this circuit with the static MOS inverter of Fig. 9-11.   When the clock ϕ is at 0 V, transistors $Q2$ and $Q3$ are OFF and the power supply is disconnected from the circuit and delivers essentially no power.   When the clock pulse is at -10 V, both $Q2$ and $Q3$ are ON and inversion of the input V_i takes place.   Thus, if $V_i = -10$ V, $Q1$ is ON and the output is $V_o \approx 0$, whereas if $V_i = 0$ V, then $Q1$ is OFF and the output becomes $V_o \approx -V_{DD}$ (say -10 V).   Note that $Q3$ is a bidirectional switch; ① acts as the source when C charges to -10 V, whereas ② acts as the source when C discharges to 0 V.   During the time the clock is at 0 V, the output capacitor C retains its charge.   This parasitic capacitance of $Q3$ between source and ground has a typical value of 0.5 pF.

The inverter discussed above has been called a *ratio inverter*.   The name derives from the fact that when the input is low and the clock is low, transistors $Q1$ and $Q2$ form a voltage divider between $-V_{DD}$ and ground.   Therefore the output voltage V_o depends on the ratio of the ON resistance of $Q1$ and the effective load resistance of $Q2$ (typically, $<1:5$).   This ratio is related to the physical size of $Q1$ and $Q2$ and is often referred to as the *aspect ratio*.

Figure 9-17 shows a dynamic p-channel MOS NAND gate, corresponding to the static NAND of Fig. 9-13.   A dynamic MOS NOR is constructed by modifying the circuit of Fig. 9-14 in a similar manner.   The reader should verify that the dynamic circuits dissipate less power than the corresponding static circuits.

Two-Phase MOS   The inverter shown in Fig. 9-18a consists of three p-channel enhancement devices and dissipates almost no power due to the fact that two clock trains are used in the phase sequence shown in Fig. 9-18b.   During the clock pulse ϕ_1 (precharge clock), the parasitic capacitor C is charged to $-V_{DD}$.   Clock pulse ϕ_2, which comes after ϕ_1, performs the inversion.   If V_i is at -10 V, $Q3$ is turned ON, and since $Q2$ is also ON, the capacitor discharges to ground and the output becomes 0 V.   If $V_i = 0$ V, then $Q3$ is OFF,

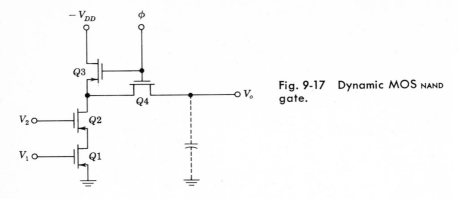

Fig. 9-17   Dynamic MOS NAND gate.

$Q2$ is ON, and there is no path to ground for C to discharge.   Thus V_o remains at $-V_{DD}$.   We note that, except during switching, all of the transistors $Q1$, $Q2$, and $Q3$ are always OFF and quiescent power dissipation is of the order of 10 nW for $-V_{DD} = -10$ V.   The circuit has no dc current path regardless of the state of the clocks or the data stored on the parasitic capacitor C.   Such a circuit has the significant advantage in that its output does not depend on the ratio of the resistances of any of its devices; therefore all devices can be of minimum geometry, reducing the chip size for a given number of gates.   Circuits that use the above feature are called *ratioless powerless*.

9-5   MOS SHIFT REGISTERS

Bipolar and MOS shift registers can be used for the same purpose.   However, if a large number of bits are involved, the MOS shift registers are preferred over bipolar circuits because they are more economical (in power and cost)

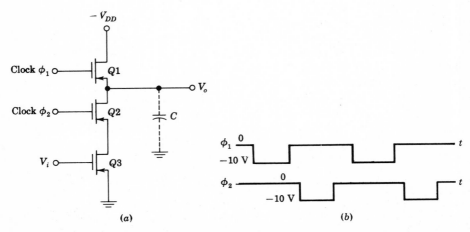

Fig. 9-18   (a) Two-phase MOS inverter.   (b) Timing of clock pulses ϕ_1 and ϕ_2.

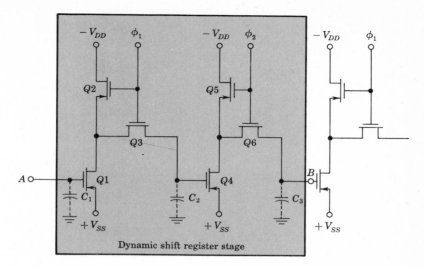

Fig. 9-19   A two-phase dynamic MOS shift register.   The clock pulses have the waveshapes indicated in Fig. 9-18, but the binary levels are +5 V and −12 V.

and are smaller in size.   Typical applications for MOS shift registers are cal-culators, display systems, refresh memories, scratch-pad memories, buffer memories, communication equipment, computer peripherals, and delay lines. MOS shift registers are available with from several hundred up to 1,024 bits of storage and shifting rates which run up to 5 MHz for p-channel devices and up to 15 MHz for n-channel MOS transistors.   Various configurations can be supplied, such as serial-in, serial-out or serial-in, parallel-out or paral-lel-in, serial-out.   For the sake of simplicity and because of their popularity, we shall treat only serial-in, serial-out registers in this section.

Dynamic MOS Shift Register   There are two types of MOS shift registers, dynamic and static (more properly called *dc stable*).   In a dynamic shift regis-ter each information bit is stored on the gate capacitance of a device and is transferred by pulsing the subsequent inverter.   A typical MOS dynamic shift register stage is shown in Fig. 9-19.   It uses the 2-phase clock pulses of Fig. 9-18b.   Each stage of the register requires six MOSFETs.   The input to the stage is the charge on the gate capacitance, such as C_1 of $Q1$, deposited there by the previous stage.   When clock ϕ_1 goes negative (for p-channel devices), transistors $Q1$ and $Q2$ form an inverter.   The common node of $Q1$, $Q2$, and $Q3$ approaches $+V_{SS}$ if the charge on C_1 is negative enough to turn $Q1$ on, or it approaches $-V_{DD}$ if the charge is positive and $Q1$ is off.   We should note that during the time ϕ_1 is negative, $Q3$ is on and the gate capacitance C_2 of $Q4$ gets charged to either $+V_{SS}$ or $-V_{DD}$.   When ϕ_1 goes high, C_2 retains the charge.   Pulsing ϕ_2 negative then shifts and inverts the data, depositing the

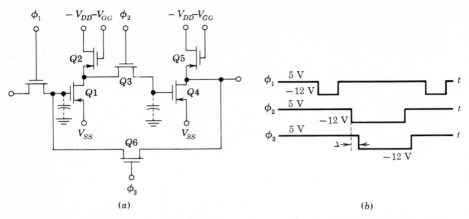

Fig. 9-20   (a) Basic MOS static shift-register cell; (b) timing diagram of static shift-register clocks.

charge on C_3.   The information at C_3 is identical with that on C_1 at A, but delayed by an amount predetermined by the clock period.   The combination $Q1Q2Q3$ can be called the *master* inverter, and $Q4Q5Q6$ the *slave* section.   To retain data stored in the register, the rate at which the data are clocked through the circuit must not fall below some minimum value.

A typical two-phase dynamic MOS shift register is the Texas Instruments TI 3401LC 512-bit register.   The entire device is constructed using MOS *p*-channel, low-threshold devices.   It operates at a minimum clock rate of 20 kHz and maximum rate of 5 MHz.   Both input and output are directly compatible with TTL integrated circuits, and the power dissipation is 0.2 mW/bit at 1 MHz.

Static MOS Shift Registers   A "static" shift register is dc stable and can operate without a minimum clock rate.   That is, it can store data indefinitely provided that power is supplied to the circuit.   However, static shift-register cells are larger than the dynamic cells and consume more power.   As shown in Fig. 9-20, the stage consists of a pair of static inverters with unclocked loads (Fig. 9-13) cross-connected through two transmission gates $Q3$ and $Q6$.   When both clocks ϕ_2 and ϕ_3 are low, the feedback loop is closed and the two inverters form a FLIP-FLOP, or *latch* (Fig. 8-2a).   In this condition the cell will hold information indefinitely.

Under normal conditions information is shifted by pulsing ϕ_1 and ϕ_2, very much as in the two-phase shift register of Fig. 9-19.   As long as the clocking frequency is high, the static shift register operates in the same manner as the dynamic shift register, with the feedback loop open (clock ϕ_3 at a high value).   When the frequency falls below a certain level, ϕ_3 is generated internally (on the chip).   Clock ϕ_3 is identical with ϕ_2, except that it is delayed (slightly) by Δ with respect to ϕ_2.   This signal ϕ_3 is used to close the feedback loop.   An example of a static shift register is the TI 3101LC, which is a dual 100-bit unit.   Each register has independent input and output terminals, common clocks, and power leads (nominal values $V_{SS} = +5$ V, $V_{DD} = 0$ V, and $-V_{GG} = -12$ V), and can operate from dc to 2.5 MHz.   The MOS gate inputs are protected with Zener diodes and can be driven directly from DTL/ TTL voltage levels, and the register outputs can drive DTL/TTL circuits without the addition of external components.   Two external clocks ϕ_1 and ϕ_2 are required for operation.   Data are transferred into the register when clock pulse ϕ_1 is at low level, and data are shifted when clock pulse ϕ_1 is returned to high level (typically, $+5$ V) and clock pulse ϕ_2 is pulsed to low level.   For long-term storage, clock pulses ϕ_1 must be held at a high level and clock pulses ϕ_2 and ϕ_3 at a low level.

Four-phase registers are available for very high density circuits operating at very high speeds.   For example, the TI3309JC consists of two 512-bit shift registers constructed on a monolithic chip using p-channel enhancement-mode transistors.   It contains 6.144 MOSFETs, exclusive of control circuitry. Operation at repetition rates from 10 kHz to 5 MHz is possible, and the power dissipation is less than 90 μW/bit at 1 MHz.

9-6   MOS READ–ONLY MEMORY

The ROM is discussed in Sec. 7-6, where it is seen (Fig. 7-18) to consist of a decoder, followed by an encoder (memory matrix).   Consider, for example, a 10-bit input code, resulting in $2^{10} = 1,024$ word lines, and with 4 bits per output code.   The memory matrix for this system consists of $1,024 \times 4$ intersections, as indicated schematically in Fig. 9-21.   The code conversion to be performed by the ROM is permanently programmed during the fabrication process by using a custom-designed mask so as to construct or omit an MOS transistor at each matrix intersection.   Such an encoder is indicated in Fig. 9-21, which shows how the memory FETs are connected between *word* and *bit* lines.

In Sec. 7-6 it is demonstrated that the relationship between the output bits Y and the word lines W is satisfied by the logic OR function.   Consider, for example, that it is required by the desired code conversion that

$$\bar{Y}_0 = W_0 + W_2 \qquad\qquad \bar{Y}_1 = W_1$$
$$\bar{Y}_2 = W_1 + W_2 + W_{500} \qquad \bar{Y}_3 = W_0 + W_{500} \tag{9-1}$$

These relationships are satisfied by the connections in Fig. 9-21.   The NOR gate for Y_0 of Eq. (9-1) is precisely that drawn in Fig. 9-14, with $Q3$ as the load FET and with signals W_0 and W_2 applied to the gates of $Q2$ and $Q1$, respectively.

The presence or absence of a MOS memory cell at a matrix intersection is determined during fabrication in the oxide-gate mask step.   If the MOSFET has a normal thin-oxide gate, its threshold voltage V_T is low; if the gate oxide

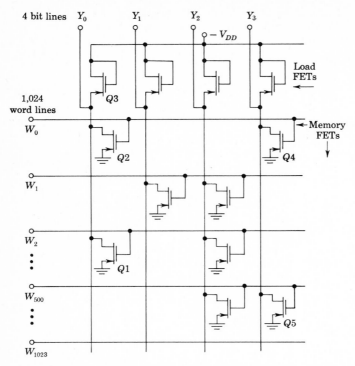

Fig. 9-21 MOS read-only-memory encoder. (Although there are a total of 1,024 word lines present, only 5 of these are indicated.)

is thick, then V_T is high. In response to a negative pulse on the word line, the low-threshold device will conduct and a logic 0 (because of inverter action) will be detected on the bit line. On the other hand, if a negative pulse is applied to the thick-oxide gate (high-threshold device), it does not conduct; it is effectively missing from the circuit. In other words, growing a thick-oxide gate at a matrix location is equivalent to *not* constructing an MOSFET at this position, as shown in Fig. 9-22.

In a static ROM no clocks are needed. The time required for a valid output to appear on the bit lines from the time an input address is applied to the memory is defined as the access time ($\sim$300 ns to 5 μs). In a static ROM the output is available as long as the input address remains valid. An example of a static MOS ROM is the EA 4900, Electronic Arrays (24-pin dual-in-line package) which consists of 2,048 words of 8 bits each for a total of 16,384 bits and with maximum access time of 950 ns.

The decoder in a static MOS ROM (Fig. 7-18) contains NAND gates which are static. Power dissipation as a result is relatively high. In the case of the EA 4900, power dissipation is typically 0.032 mW/bit. A dynamic

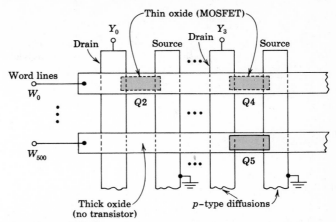

Fig. 9-22 MOS read-only-memory matrix. (Only word lines W_0 and W_{500} and bit lines Y_0 and Y_3 corresponding to Fig. 9-21 are indicated.)

ROM uses clocked or dynamic inverters in the decoder and requires a minimum clock rate, since otherwise the information is lost. However, its power dissipation is lower than for a static ROM. Most commercial ROMs are static because of the advantages of requiring no clocks and of giving an output which remains valid as long as the input address is applied.

9-7   RANDOM–ACCESS MEMORY (RAM)

The random-access memory, abbreviated RAM, is an array of storage cells that memorize information in binary form. In such a memory, as contrasted with an ROM, information can be randomly written into or read out of each storage element as required, and hence the name *random-access*, or *read/write*, *memory*. The basic monolithic storage cell is the latch, or FLIP-FLOP, discussed in Sec. 8-1.

Linear Selection To understand how the RAM operates we examine the simple 1-bit S-R FLIP-FLOP circuit shown in Fig. 9-23, with data input and output lines. From the figure we see that to read data out of or to write data into the cell, it is necessary to excite the *address line* ($X = 1$). To perform the write operation, the *write enable line* must also be excited. If the write input is a logic $1(0)$, then $S = 1(0)$ and $R = 0(1)$. Hence $Q = 1(0)$, and the data read out is $1(0)$, corresponding to that written in.

Suppose that we wish to read/write 16 words of 8 bits each. This system requires eight data inputs and eight data output lines. A total of $16 \times 8 = 128$ storage cells must be used. Of this number, 8 cells are arranged in a horizontal line, all excited by the same address line. There are 16 such

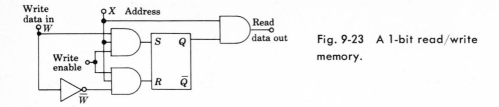

Fig. 9-23   A 1-bit read/write memory.

lines, each excited by a different address.   In other words, addressing is provided by exciting 1 of 16 lines.   This type of addressing is called *linear selection* (Prob. 9-11).

Coincident Selection   An RAM memory of sixteen 8-bit words has 16 lines with 8 storage cells per line, if linear addressing is used.   A more commonly used topology is to arrange 16 memory elements in a rectangular 4×4 array, each cell now storing one bit of one word.   Eight such matrix planes are required, one for each of the 8 bits in each word.

One plane of the above-described arrangement of cells is indicated in Fig. 9-24.   Each bit (indicated as a shaded rectangle) is located by addressing an X address line and a Y address line; the intersection of the two lines locates a point in the two-dimensional matrix, thus identifying the storage cell under consideration.   Such two-dimensional addressing is called X-Y, or *coincident*, *selection*.

Basic RAM Elements   In the 1-bit memory of Fig. 9-23 separate read and write leads are required.   For either the bipolar or the MOS RAM it is possible to construct a FLIP-FLOP (as we demonstrate in Fig. 9-27) which has a common terminal for both writing and reading, such as terminals 1 and 2 in Fig. 9-25.   This configuration requires the use not only of the write data W

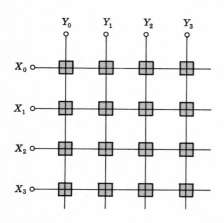

Fig. 9-24   Illustrating coincident addressing to locate 1 bit of a 16-word memory. The shaded squares represent schematically the storage cells, or memory units.

Fig. 9-25   A basic stor-
age cell can be con-
structed with complemen-
tary inputs and outputs
and with the write and
sense amplifiers meeting
at a common node (1) for
true data and (2) for
complementary data.

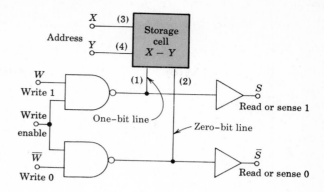

(write 1), but also of its complement $\bar{W}$ (write 0).   At the cell terminal to
which $W(\bar{W})$ is applied, there is obtained the read or sense data output $S(\bar{S})$.
Such a memory unit is indicated schematically in Fig. 9-25, where we note that
a total of four input/output leads to the storage cell are required, two for X-Y
addressing and two for read/write data (true and complemented).

The basic elements of which an RAM is constructed are indicated in Fig.
9-26.   These include the rectangular array of storage cells, the X and Y decoders, the write amplifiers to drive the memory, and the sense amplifiers to
detect (read) the stored digital information.   Some RAMs include a write enable input.   For such a unit, the write amplifiers of Fig. 9-26 are two-input
AND gates, as in Fig. 9-25.   Each word is identified by the matrix number
X-Y in the (shaded) memory cell.   For M-bit words there will be M planes,
like the one indicated in Fig. 9-26.

Magnetic cores have been used as storage elements for many years.
Semiconductor memories are now becoming increasingly popular.   Monolithic
RAMs are constructed using integrated-circuit technology and employing
either bipolar or MOS transistors for the storage and supporting circuits.
Some of the advantages of semiconductor over core memories are low cost, small
size, and nondestructive reading of the array.   On the other hand, the disadvantages include the *volatility of storage*, which means that all stored information is lost when the power supplies fail, and the power dissipation necessary
in order to retain the information stored in a FLIP-FLOP.

Since in Fig. 9-25 the output of the write amplifier is connected to the
input of the read amplifier, then clearly the sense amplifiers must not be used
to supply information on the state of a memory cell while a write amplifier
is excited.

An example of a 16-bit bipolar RAM having the pattern of Fig. 9-26 is
the TI 7481.   Average power dissipation is 275 mW, and reading propagation
delay is typically 20 ns.   An example of a larger RAM is the IM 5503 (Intersil
Memory Corporation) which has a $16 \times 16 = 256$-word by 1-bit organiza-
tion.   It has an access time of 75 ns.

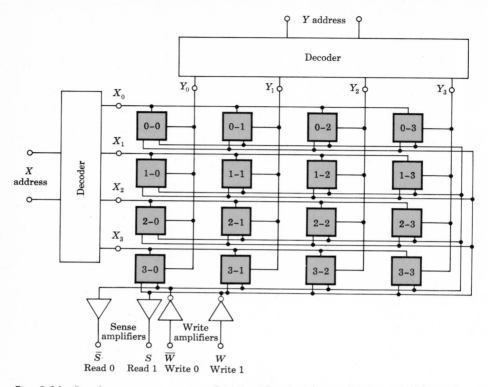

Fig. 9-26 Random-access memory (RAM) with coincident selection and 16 words by 1 bit. The shaded squares give the memory locations of one bit of each word.

Static MOS RAM The MOS FLIP-FLOP shown in Fig. 9-27 is used to store the binary information, and clocks are not needed. As long as power is maintained, the data will remain in storage. Devices $Q1$–$Q4$ form a bistable cross-coupled FLIP-FLOP circuit, whereas devices $Q5$–$Q8$ form the gating network through which the interior nodes $N1$ and $N2$ are connected to the ONE-bit line and the ZERO-bit line. Note that devices $Q5$ and $Q6$ or $Q7$ and $Q8$ form AND gates to which the X and Y address drives are applied for *coincident selection* of the storage cell. If the *linear-selection* scheme is used, devices $Q6$ and $Q8$ are omitted and the X address line represents the word line.

In the quiescent state both the X and the Y address lines are at ground potential, isolating the storage FLIP-FLOP from the bit lines. Assume that $Q2$ is ON and $Q1$ is OFF, so that node $N1$ is at $-V_{DD}$ and $N2$ at 0 V. To read the cell, both address lines are pulsed (negative for p-channel MOS devices), turning on gating devices $Q5$–$Q8$. Current will flow into the ONE-bit line, which is kept at $-V_{DD}$ though devices $Q7$, $Q8$, and the ON device $Q2$. Very little or essentially no current will flow through the ZERO-bit line, which is also kept at

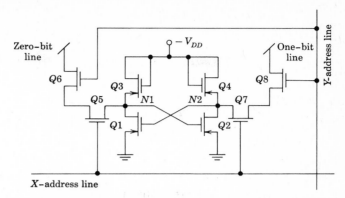

Fig. 9-27 Single MOS RAM cell including address-gating. The inverter $Q1$–$Q3$ is cross-coupled to the inverter $Q2$–$Q4$ to form a 1-bit latch. A logic 1 is stored if $Q2$ conducts.

$-V_{DD}$ since device $Q1$ is OFF. The state of the storage cell is thus determined by detecting on which bit line the sense current flows.

To write 1 into the cell, the address lines are again pulsed and the ONE-bit line is grounded. Since the ONE-bit line is pulsed from $-V_{DD}$ to ground, this is interpreted to mean that we desire to write a 1 in the cell. However, the cell is already in the state to which it was to be written, and no change occurs because $Q2$ is already ON. If the ZERO-bit line is pulsed by grounding it, node $N1$ is pulled toward ground, turning device $Q2$ OFF and device $Q1$ ON through the regenerative process of the FLIP-FLOP. Thus the cell changes state, and we have written the logic 0 into the FLIP-FLOP. We observe that the reading process is nondestructive.

Typical cycle times for full decoded bit-organized MOS chips generally lie in the range 500 ns to 1 μs. An example of a MOS RAM is the MK 4096 (Mostek, Inc.). The unit is a 4-K RAM. It is TTL/DTL compatible and the device is available in a 16-pin dual-in-line package.

Dynamic MOS RAM Instead of using an eight-device cell, it is possible to use a simpler single-transistor storage cell in which information is stored on the parasitic gate-to-substrate capacitance. Thus, at the expense of requiring a refresh operation to replenish the charge leaking off the storing capacitance, we achieve far greater density of storage cells on the same chip area.

An example of a dynamic MOS/LSI RAM is the TMS4050 manufactured by Texas Instruments. This 4-K memory is fully decoded. However, refreshing of all bits is required every 2 ms, and cycle time is 300 ns. The device is available in an 18-pin dual-in-line package. Power dissipation at room temperature is 400 mW.

REVIEW QUESTIONS

9-1 (*a*) Sketch the basic structure of an *n*-channel junction field-effect transistor. (*b*) Show the circuit symbol for the JFET.

9-2 (*a*) Draw a family of CS drain characteristics of an *n*-channel JFET. (*b*) Explain the shape of these curves qualitatively.

9-3 (*a*) Sketch the cross section of a *p*-channel enhancement MOSFET. (*b*) Show two circuit symbols for this MOSFET.

9-4 For the MOSFET in Rev. 9-3 draw (*a*) the drain characteristics and (*b*) the transfer curve.

9-5 Repeat Rev. 9-3 for an *n*-channel depletion MOSFET.

9-6 (*a*) Draw the circuit of a MOSFET NOT circuit. (*b*) Explain how it functions as an inverter.

9-7 (*a*) Explain how a MOSFET is used as a load. (*b*) Obtain the volt-ampere characteristic of this load graphically.

9-8 Sketch a two-input NAND gate and verify that it satisfies the Boolean NAND equation.

9-9 Repeat Rev. 9-8 for a two-input NOR gate.

9-10 Sketch a CMOS inverter and explain its operation.

9-11 (*a*) Draw the circuit of a *single-phase dynamic MOS inverter*. (*b*) Explain its operation.

9-12 Repeat Rev. 9-11 for a MOS NAND gate.

9-13 Repeat Rev. 9-11 for a *two-phase MOS inverter*. Draw the clocking waveforms.

9-14 (*a*) Draw the circuit of one stage of *a two-phase dynamic MOS shift register*. (*b*) Explain its operation. Draw the clocking waveforms.

9-15 (*a*) Draw the circuit of one stage of a *static MOS shift register*. (*b*) Draw the clocking waveforms. (*c*) Explain the operation of the circuit.

9-16 (*a*) Draw the circuit of the *encoder* of a MOS read-only memory. (*b*) Explain the operation of the circuit.

9-17 Explain how the MOS ROM is *programmed*.

9-18 (*a*) Draw the block diagram of a 1-bit *read/write memory*. (*b*) Explain its operation.

9-19 Explain *linear selection* in a *random-access memory* (RAM).

9-20 Repeat Rev. 9-19 for *coincident selection*.

9-21 (*a*) Draw in block-diagram form the basic elements of an RAM with coincident selection used to store four words of 1 bit each. (*b*) How is the system expanded to 3 bits/word? (*c*) How is the system expanded to 25 words of 3 bits/word?

9-22 List the advantages and disadvantages of a MOS RAM.

10 / LOW-FREQUENCY AMPLIFIERS

In a digital system a transistor operates in one of two states: it is either at cutoff or in saturation. On the other hand, when a transistor is used as an amplifier, it is biased so that it is operating in the active region. With no signal applied to the base, the collector current and voltage determine a point approximately in the center of the output characteristics. This zero-excitation operating point is called the *quiescent point* Q. In this chapter we are interested in finding the quiescent values of current and voltage, and then determining the output response when a signal is applied to the input of the transistor.

The large-signal response of a transistor is obtained graphically. For small signals the transistor operates with reasonable linearity, and we inquire into small-signal linear models which represent the operation of the transistor in the active region. A detailed study of the transistor amplifier in its various configurations is made.

Very often, in practice, a number of stages are used in cascade to amplify a signal from a source, such as a phonograph pickup, to a level which is suitable for the operation of another transducer, such as a loudspeaker. We consider the problem of cascading a number of transistor amplifier stages at low frequencies, where the transistor internal capacitances may be neglected.

10-1 GRAPHICAL ANALYSIS OF THE CE CONFIGURATION

It is our purpose in this section to analyze graphically the operation of the circuit of Fig. 10-1. In Fig. 10-2a the output characteristics of a *p-n-p* germanium transistor and in Fig. 10-2b the corresponding input characteristics are indicated. We have selected the CE configuration because, as we see in Sec. 10-3, it is the most generally useful configuration.

In Fig. 10-2a we have drawn a load line for a 250-Ω load with

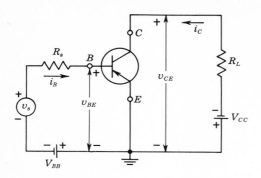

Fig. 10-1   The CE transistor configuration.

$V_{CC} = 15$ V.   If the input base-current signal is symmetric, the quiescent point Q is usually selected at about the center of the load line, as shown in Fig. 10-2a.   We postpone until Sec. 10-13 our discussion on biasing of transistors.

Notation   At this point it is important to make a few remarks on transistor symbols.   Specifically, instantaneous values of quantities which vary with time are represented by lowercase letters (i for current, v for voltage, and p for power).   Maximum, average (dc), and effective, or root-mean-square (rms), values are represented by the uppercase letter of the proper symbol (I, V, or P).   Average (dc) values and instantaneous total values are indicated by the uppercase subscript of the proper electrode symbol (B for base, C for collector, E for emitter).   Varying components from some quiescent value are indicated by the lowercase subscript of the proper electrode symbol.   A single subscript is used if the reference electrode is clearly understood.   If

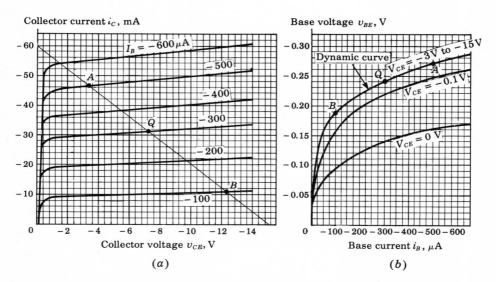

Fig. 10-2   (a) Output and (b) input characteristics of a p-n-p germanium transistor.

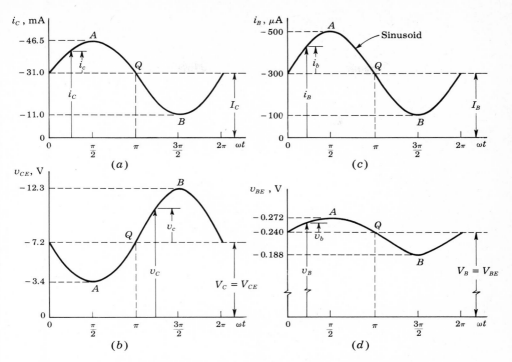

Fig. 10-3   (a,b) Collector and (c,d) base current and voltage waveforms.

there is any possibility of ambiguity, the conventional double-subscript notation should be used.   For example, in Figs. 10-3a to d and 10-1, we show collector and base currents and voltages in the common-emitter transistor configuration, employing the notation just described.   The collector and emitter
current and voltage component variations from the corresponding quiescent
values are

$$i_c = i_C - I_C = \Delta i_C \qquad v_c = v_C - V_C = \Delta v_C$$
$$i_b = i_B - I_B = \Delta i_B \qquad v_b = v_B - V_B = \Delta v_B$$

$(10\text{-}1)$

The *magnitude* of the supply voltage is indicated by repeating the electrode
subscript.   This notation is summarized in Table 10-1.

The Waveforms   Assume a 200-μA peak sinusoidally varying base current
around the quiescent point Q, where $I_B = -300$ μA.   Then the extreme·
points of the base waveform are A and B, where $i_B = -500$ μA and -100 μA,
respectively.   These points are located on the load line in Fig. 10-2a.   We
find i_C and v_{CE}, corresponding to any given value of i_B, at the intersection of
the load line and the collector characteristics corresponding to this value of i_B.
For example, at point A, $i_B = -500$ μA, $i_C = -46.5$ mA, and $v_{CE} = -3.4$ V.

TABLE 10-1   Notation summarized

	Base (collector) voltage with respect to emitter	Base (collector) current toward electrode from external circuit
Instantaneous total value..............................	v_B (v_C)	i_B (i_C)
Quiescent value................................	V_B (V_C)	I_B (I_C)
Instantaneous value of varying component........	v_b (v_c)	i_b (i_c)
Effective value of varying component (phasor, if a sinusoid)................................	V_b (V_c)	I_b (I_c)
Supply voltage (magnitude)....................	V_{BB} (V_{CC})	

The waveforms i_C and v_{CE} are plotted in Fig. 10-3a and b, respectively. We observe that the collector-current and collector-voltage waveforms are not the same as the base-current waveform (the sinusoid of Fig. 10-3c) because the collector characteristics in the neighborhood of the load line in Fig. 10-2a are not parallel lines equally spaced for equal increments in base current. This change in waveform is known as *output nonlinear distortion*.

The base-to-emitter voltage v_{BE} for any combination of base current and collector-to-be-emitter voltage can be obtained from the input characteristic curves. In Fig. 10-2b we show the *dynamic operating curve* drawn for the combinations of base current and collector voltage found along A-Q-B of the load line of Fig. 10-2a. The waveform v_{BE} can be obtained from the dynamic operating curve of Fig. 10-2b by reading the voltage v_{BE} corresponding to a given base current i_B. We now observe that, since the dynamic curve is not a straight line, the waveform of v_b (Fig. 10-3d) will not, in general, be the same as the waveform of i_b. This change in waveform is known as *input nonlinear distortion*. In some cases it is more reasonable to assume that v_b in Fig. 10-3d is sinusoidal, and then i_b will be distorted. The above condition will be true if the sinusoidal voltage source v_s driving the transistor has a small output resistance R_s in comparison with the input resistance R_i of the transistor, so that the transistor input-voltage waveform is essentially the same as the source waveform. However, if $R_s \gg R_i$, the variation in i_B is given by $i_b \approx v_s/R_s$, and hence the base-current waveform is also sinusoidal.

From Fig. 10-2b we see that *for a large sinusoidal base voltage v_b around the point Q, the base-current swing $|i_b|$ is smaller to the left of Q than to the right of Q*. This input distortion tends to cancel the output distortion because, in Fig. 10-2a, the collector-current swing $|i_c|$ for a given base-current swing is larger over the section BQ than over QA. Hence, if the amplifier is biased so that Q is near the center of the i_C-v_{CE} plane, there will be less distortion if the excitation is a sinusoidal base voltage than if it is a sinusoidal base current.

It should be noted here that the dynamic load curve can be approximated by a straight line over a sufficiently small line segment, and hence, if the input signal is small, there will be negligible input distortion under any conditions of operation (current-source or voltage-source driver).

Since the small-signal low-frequency response of a transistor is linear, it can be obtained analytically rather than graphically. As a matter of fact, for very small signals the graphical technique used in connection with Fig. 10-2 would require interpolation between the printed characteristics and would result in very poor accuracy. In the remainder of the chapter we assume small-signal (also called *incremental*) operation, obtain a linear circuit model for the transistor, and then analyze the network analytically.

10-2   TRANSISTOR HYBRID MODEL

The basic assumption in arriving at a transistor linear model or equivalent circuit is that the variations about the quiescent point are small, so that the transistor parameters can be considered constant over the signal excursion. The operating (quiescent) values of current and voltages are determined by the method employed to bias the transistor (Sec. 10-13). These values do not enter into an incremental model, which is used only to find small-signal variations about the Q point.

The transistor model presented in this chapter is given in terms of the h or hybrid parameters, which are *real numbers* at audio frequencies, are easy to measure, can also be obtained from the transistor static characteristic curves, and are particularly convenient to use in circuit analysis and design. Furthermore, a set of h parameters is specified for many transistors by the manufacturers.

To see how we can derive a hybrid model for a transistor, let us consider the common-emitter connection shown in Fig. 10-1. The variables i_B, i_C, $v_{BE} = v_B$, and $v_{CE} = v_C$ represent total instantaneous currents and voltages. From our discussion in Chap. 4 of transistor voltages and currents, we see that we may select the current i_B and voltage v_C as independent variables. Since v_B is some function of i_B and v_C and since i_C is another function of i_B and v_C, we may write the small-signal (incremental) voltage v_b and current i_c as a linear combination of i_b and v_c. Thus,

$$v_b = h_{ie}i_b + h_{re}v_c \tag{10-2}$$

$$i_c = h_{fe}i_b + h_{oe}v_c \tag{10-3}$$

From these equations it follows that

$$h_{ie} = \left.\frac{v_b}{i_b}\right|_{v_c=0} = \left.\frac{\Delta v_B}{\Delta I_B}\right|_{v_C=V_C} \tag{10-4}$$

$$h_{re} = \left.\frac{v_b}{v_c}\right|_{i_b=0} = \left.\frac{\cdot\,\Delta v_B}{\Delta v_C}\right|_{i_B=I_B} \tag{10-5}$$

$$h_{fe} = \left.\frac{i_c}{i_b}\right|_{v_c=0} = \left.\frac{\Delta i_C}{\Delta i_B}\right|_{v_C=V_C} \tag{10-6}$$

$$h_{oe} = \left.\frac{i_c}{v_c}\right|_{i_b=0} = \left.\frac{\Delta i_C}{\Delta v_C}\right|_{i_B=I_B} \tag{10-7}$$

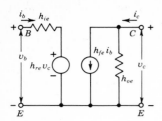

Fig. 10-4   The hybrid small-signal model for the common-emitter configuration.

Note that the ratios in the above equations are taken by keeping v_C = constant at the quiescent value V_C or i_B = constant at the quiescent value I_B. If a parameter is constant, its incremental change is zero. Hence, V_C = constant is equivalent to $v_c = 0$, and i_B = constant corresponds to $i_b = 0$.

Remembering that for the common-emitter configuration the input is the base and the output is the collector circuit, the above equations define the h parameters as follows:

$h_{ie} \equiv$ input resistance with output short-circuited (ohms).
$h_{re} \equiv$ fraction of output voltage at input with input open-circuited, or more simply, reverse-open-circuit voltage amplification (dimensionless).
$h_{fe} \equiv$ negative of current transfer ratio (or current gain) with output short-circuited. (Note that the current into a load across the output port would be the negative of i_c.) This parameter is usually referred to, simply, as the *short-circuit current gain* (dimensionless).
$h_{oe} \equiv$ output conductance with input open-circuited (mhos).

The Model   We may now use the four h parameters to construct a mathematical model of the transistor operating in the small-signal (incremental) mode. We can verify that the model of Fig. 10-4 satisfies Eqs. (10-2) and (10-3) by writing Kirchhoff's voltage and current laws for the input and output circuits respectively. Note that the input circuit contains a dependent voltage generator (a voltage $h_{re}v_c$ the value of which depends upon the output voltage), whereas the output circuit has a dependent current source (a current $h_{fe}i_b$ the value of which depends upon the input current). We note from Kirchhoff's current law that

$$i_b + i_e + i_c = 0 \qquad (10\text{-}8)$$

The circuit models and equations are valid for either an n-p-n or a p-n-p transistor and are independent of the type of load or the method of biasing.

Typical values of h parameters for a transistor operating at an emitter current $I_E = 1.3$ mA are

$$h_{ie} = 1.1 \text{ K} \qquad h_{re} = 2.5 \times 10^{-4} \qquad\qquad\qquad (10\text{-}9)$$
$$h_{fe} = 50 \qquad\quad h_{oe} = 2.5 \times 10^{-6} \text{ mho} \qquad \text{or} \qquad 1/h_{oe} = 40 \text{ K}$$

10-3   LINEAR ANALYSIS OF A TRANSISTOR CIRCUIT

To obtain the response of the amplifier of Fig. 10-1 to the small-signal excitation v_s proceed as follows:

1. Replace the transistor between points B, C, and E by the model of Fig. 10-4 between these corresponding nodes.
2. Transfer all circuit elements from the actual circuit to the equivalent circuit containing the h-parameter model. (For Fig. 10-1 the circuit components are the source v_s and the resistors R_s and R_L.)
3. Since we are interested only in changes from the quiescent values, replace each independent dc supply by its internal resistance. The ideal voltage sources V_{BB} and V_{CC} are each replaced by a short circuit (zero resistance).
4. Solve the resultant linear network for the currents and voltages by applying Kirchhoff's current and voltage laws (KCL and KVL).

It should be emphasized that the above procedure is not limited to the simple CE circuit of Fig. 10-1. It can be applied to a circuit such as the CE configuration with a resistor R_e added to the emitter node (Fig. 10-11), to a CC circuit (Fig. 10-7), to a cascade of amplifiers, or to any combination of transistors and circuit elements operating linearly at low frequencies.

Simplified h-Parameter Model In most practical cases, sufficiently accurate values of current and voltage can be obtained by using a simplified model that uses only two of the four h parameters in Fig. 10-4. It turns out that h_{re} and h_{oe} are small enough to be neglected completely, provided that the inequality

$$h_{oe}R_L \leq 0.1 \qquad\qquad (10\text{-}10)$$

is satisfied. This simplified model is shown in Fig. 10-5 and it may be used for any configuration by grounding the appropriate node. The signal is connected between the input terminal and ground, and the load R_L is placed between the output node and ground.

Consider the common-emitter configuration of Fig. 10-1. Figure 10-6 shows this CE stage with the transistor replaced by the approximate model of Fig. 10-5. Assuming sinusoidally varying voltages and currents, we can proceed with the analysis of this circuit, using the phasor notation to repre-

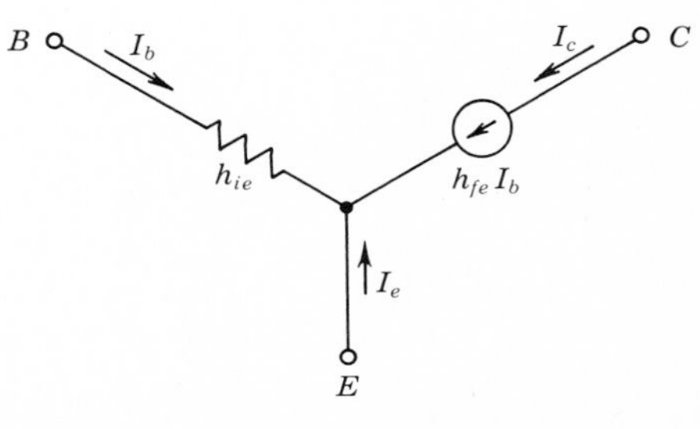

Fig. 10-5 Approximate hybrid model which may be used for all three configurations CE, CC, and CB.

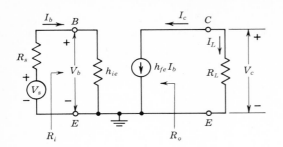

Fig. 10-6 Approximate CE model.

sent the sinusoidally varying quantities.   The quantities of interest are *the current gain, the input resistance, the voltage gain,* and *the output resistance.*

The Current Gain, or Current Amplification, A_I   For the transistor amplifier stage, A_I is defined as the ratio of output to input currents, or

$$A_I \equiv \frac{I_L}{I_b} = -\frac{I_c}{I_b} \tag{10-11}$$

From the circuit of Fig. 10-6 we have $I_c = h_{fe}I_b$.   Hence

$$A_I = -h_{fe} \tag{10-12}$$

Note that subject to the restriction of Eq. (10-10), A_I equals the short-circuit current gain, independent of the load R_L.

The Input Resistance R_i   The resistance R_s in Fig. 10-6 represents the signal-source resistance.   The resistance we see by looking into the transistor input terminals B and E is the amplifier *input resistance R_i*, or

$$R_i \equiv \frac{V_b}{I_b} = h_{ie} \tag{10-13}$$

Note that R_i equals the short-circuit input resistance, independent of R_L, if Eq. (10-10) is satisfied.

The Voltage Gain, or Voltage Amplification, A_V   The ratio of output voltage V_c to input voltage V_b gives the voltage gain of the transistor, or

$$A_V \equiv \frac{V_c}{V_b} = \frac{I_L R_L}{I_b h_{ie}} \tag{10-14}$$

Combining Eqs. (10-11), (10-13), and (10-14) we obtain

$$A_V = \frac{A_I R_L}{R_i} \tag{10-15}$$

The above derivation indicates that Eq. (10-15) is a general expression for the voltage gain in terms of the current gain and the input resistance, and *it is valid independent of the transistor configuration and also of the transistor model used.*

For the C E circuit Eq. (10-15) becomes

$$A_V = \frac{-h_{fe}R_L}{h_{ie}} \tag{10-16}$$

The Voltage Amplification A_{Vs} Taking into Account the Resistance R_s of the Source This overall voltage gain A_{Vs} is defined by

$$A_{Vs} \equiv \frac{V_c}{V_s} = \frac{V_c}{V_b}\frac{V_b}{V_s} = A_V\frac{V_b}{V_s} \tag{10-17}$$

From the equivalent input circuit of the amplifier, shown in Fig. 10-6,

$$V_b = \frac{V_s h_{ie}}{h_{ie} + R_s} \tag{10-18}$$

Then,

$$A_{Vs} = \frac{A_V h_{ie}}{h_{ie} + R_s} = \frac{A_I R_L}{h_{ie} + R_s} \tag{10-19}$$

where use has been made of Eq. (10-15). Note that, if $R_s = 0$, then $A_{Vs} = A_V$. Hence A_V is the voltage gain for an *ideal voltage source* (one with zero internal resistance). In practice, the quantity A_{Vs} is more meaningful than A_V since, usually, the source resistance has an appreciable effect on the overall voltage amplification. For example, if R_i is equal in magnitude to R_s, then $A_{Vs} = 0.5 A_V$.

The Output Resistance By definition, the output resistance R_o is obtained by setting the source voltage V_s to zero and the load resistance R_L to infinity and by driving the output terminals from a generator V_2. If the current drawn from V_2 is I_2, then

$$R_o \equiv \frac{V_2}{I_2} \quad \text{with} \quad V_s = 0 \text{ and } R_L = \infty \tag{10-20}$$

With $V_s = 0$ in Fig. 10-6, the input current I_b is zero. Hence, $I_2 = I_c = h_{fe}I_b = 0$ and

$$R_o = \frac{V_2}{0} = \infty$$

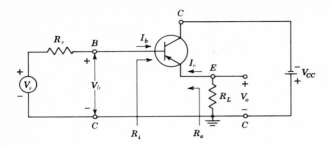

Fig. 10-7 A common-collector, or emitter-follower, configuration.

The CE stage, using the simplified h-parameter model, has infinite output resistance.

The Output Resistance R'_o Taking into Account the Resistance of the Load This resistance is clearly the parallel combination of R_o and R_L.

$$R'_o = \frac{R_o R_L}{R_o + R_L} \tag{10-21}$$

For the circuit of Fig. 10-6, $R'_o = R_L$ because $R_o = \infty$.

The approximate solution for the CE configuration is summarized in the first column of Table 10-2, and numerical values are given in Table 10-3.

10-4   THE EMITTER FOLLOWER

The circuit diagram of a common-collector (CC) transistor amplifier is given in Fig. 10-7. This configuration is also called the *emitter follower,* because its voltage gain is close to unity [Eq. (10-26)], and hence a change in base voltage appears as an equal change across the load at the emitter. In other words, the emitter *follows* the input signal. It is shown below that the input resistance R_i of an emitter follower is very high (hundreds of kilohms) and the output resistance R_o is very low (tens of ohms). Hence the most common use for the CC circuit is as a buffer stage which performs the function of resistance transformation (from high to low resistance) over a wide range of frequencies, with voltage gain close to unity. In addition, the emitter follower increases the power level of the signal.

Approximate Calculations for the Common-Collector Configuration
Figure 10-8 shows the simplified h-parameter model of Fig. 10-5 replacing the transistor in Fig. 10-7. Note that the collector is grounded with respect to

TABLE 10-2   Summary of approximate equations for $h_{oe}(R_e + R_L) \leq 0.1$†

	CE	CE with R_e	CC	CB
A_I	$-h_{fe}$	$-h_{fe}$	$1 + h_{fe}$	$\dfrac{h_{fe}}{1 + h_{fe}}$
R_i	h_{ie}	$h_{ie} + (1 + h_{fe})R_e$	$h_{ie} + (1 + h_{fe})R_L$	$\dfrac{h_{ie}}{1 + h_{fe}}$
A_V	$-\dfrac{h_{fe}R_L}{h_{ie}}$	$-\dfrac{h_{fe}R_L}{R_i} \approx -\dfrac{R_L}{R_e}$	$1 - \dfrac{h_{ie}}{R_i}$	$h_{fe}\dfrac{R_L}{h_{ie}}$
R_o	∞	∞	$\dfrac{R_s + h_{ie}}{1 + h_{fe}}$	∞
R'_o	R_L	R_L	$R_o \| R_L$	R_L

† $(R_i)_{CB}$ is an underestimation by less than 10 percent.   All other quantities except R_o are too large in magnitude by less than 10 percent.

the signal (because the supply V_{CC} must be replaced by a short-circuit, as indicated in rule 3, Sec. 10-3).

Current Gain   If KCL is applied at the emitter node E, then the load current $I_L = -I_e = (1 + h_{fe})I_b$.   Hence,

$$A_I = \frac{I_L}{I_b} = 1 + h_{fe} \tag{10-22}$$

Input Resistance   From Fig. 10-8 we obtain

$$V_b = I_b h_{ie} + (1 + h_{fe})I_b R_L \tag{10-23}$$

Hence,

$$R_i = \frac{V_b}{I_b} = h_{ie} + (1 + h_{fe})R_L \tag{10-24}$$

Note that $R_i \gg h_{ie}$ even if R_L is as small as 0.5 K because $h_{fe} \gg 1$.

Fig. 10-8   Simplified hybrid model for the CC circuit.

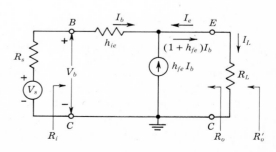

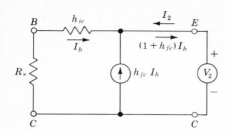

Fig. 10-9   Network used to find the output resistance of a CC stage.

Voltage Gain   From Eqs. (10-15) and (10-22),

$$A_V = \frac{A_I R_L}{R_i} = \frac{(1 + h_{fe})R_L}{R_i} \tag{10-25}$$

From Eq. (10-24), $(1 + h_{fe})R_L = R_i - h_{ie}$ and, hence, Eq. (10-25) becomes

$$A_V = 1 - \frac{h_{ie}}{R_i} \tag{10-26}$$

Since $R_i \gg h_{ie}$, Eq. (10-26) indicates that the voltage gain of a CC amplifier is approximately unity (but slightly less than unity).

Output Resistance   From the definition of R_o given in Eq. (10-20) we must set $V_s = 0$ and $R_L = \infty$ and apply an external generator V_2 to the output terminals, as indicated in Fig. 10-9.   From this figure we find

$$V_2 = -I_b(R_s + h_{ie}) \qquad \text{and} \qquad I_2 = -(1 + h_{fe})I_b$$

Hence,

$$R_o \equiv \frac{V_2}{I_2} = \frac{R_s + h_{ie}}{1 + h_{fe}} \tag{10-27}$$

Note that *the output resistance is a function of the source resistance R_s.*   Because $h_{fe} \gg 1$, the output resistance of an emitter follower is small (ohms) compared with the input resistance which is large (tens or hundreds of kilohms).

The output resistance R_o', taking the load into account, is obviously R_o in parallel with R_L.   The formulas derived in this section are summarized in the third column of Table 10-2.

10-5   APPROXIMATE COMMON–BASE AMPLIFIER SOLUTION

The circuit of Fig. 4-3 is a CB amplifier if a signal generator V_s (with internal resistance R_s) is added into the input (emitter circuit).   Using the simplified h-parameter model of Fig. 10-5, the equivalent circuit of Fig. 10-10 is obtained for the CB amplifier.   The current entering node E is I_e and that entering C is I_c.   Since, from KCL, $I_b = -(I_e + I_c)$, the current in h_{ie} must be I_b, as indi-

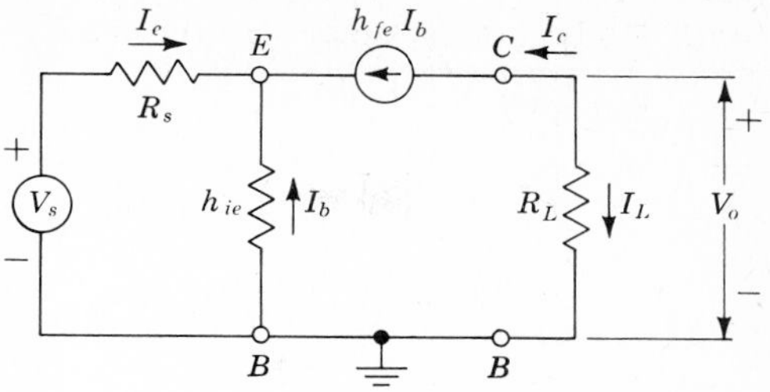

Fig. 10-10   The model for the CB amplifier.

cated in Fig. 10-10.   Applying the definitions of A_I, R_i, A_V, and R_o given in Sec. 10-3 to this figure, the formulas given in the fourth column of Table 10-2 are obtained (Prob. 10-8).

10-6   COMPARISON OF TRANSISTOR AMPLIFIER CONFIGURATIONS

Numerical values for A_I, A_V, R_i, and R_o for the CE, CC, and CB amplifiers are given in Table 10-3 for $R_L = 3$ K and $R_s = 3$ K, based upon the simplified model using the parameters $h_{ie} = 1.1$ K and $h_{fe} = 50$.   These agree within better than 10 percent with the exact values using the four h-parameter model of Fig. 10-4, except for the two values of R_o marked ∞ in Table 10-3.   The correct values for output impedance are 45.5 K and 1.72 M for the CE and CB configurations, respectively.   However, R_o', taking the 3-K load into account, is within 10 percent of 3 K (for the CE and CB stages), using either the two- or the four-parameter model.

   The CE Configuration   From Table 10-3 we see that only the CE stage is capable of both a voltage gain and a current gain greater than unity.   This configuration is the most versatile and useful of the three connections.   Note that the magnitudes of R_i and R_o lie between those for the CB and CC configurations (refer to the exact values of R_o quoted in the preceding paragraph).

   The CB Configuration   For the common-base stage, A_I is less than unity, A_V is high (equal to that of the CE stage), R_i is the lowest, and R_o is the highest of the three configurations.   The CB stage has few applications.   It is sometimes used to match a very low impedance source, to drive a high-impedance load, or as a noninverting amplifier with a voltage gain greater than unity.

   The CC Configuration   For the common-collector stage, A_I is high (approximately equal to that of the CE stage), A_V is less than unity, R_i is the highest, and R_o is the lowest of the three configurations.   This circuit finds wide application as a buffer stage between a high-impedance source and a low-impedance load.

TABLE 10-3   Comparison of transistor configurations ($R_L = 3$ K)

Quantity	CE	CC	CB
A_I...................	High (-50)	High (51)	Low (0.98)
A_V...................	High (-136)	Low (0.993)	High (136)
R_i...................	Medium ($1,100$ Ω)	High (154 K)	Low (21.6 Ω)
R_o ($R_s = 3$ K)........	High (∞)	Low (80.4 Ω)	High (∞)

10-7   THE CE AMPLIFIER WITH AN EMITTER RESISTOR

We see from Table 10-2 that the voltage gain of a CE stage depends upon h_{fe}. This transistor parameter depends upon temperature, aging, and operating point. Moreover, h_{fe} may vary widely from device to device, even for the same type of transistor. It is often necessary to stabilize the voltage amplification of each stage, so that A_V will become essentially independent of h_{fe}. A simple and effective way to obtain voltage-gain stabilization is to add an emitter resistor R_e to a CE stage, as indicated in the circuit of Fig. 10-11. This stabilization is a result of the feedback provided by the emitter resistor. The general concept of feedback is discussed in Chap. 12.

We show in this section that the presence of R_e has the following effects on the amplifier performance, in addition to the beneficial effect on bias stability discussed in Sec. 10-13. It leaves the current gain A_I essentially unchanged; it increases the input impedance by $(1 + h_{fe})R_e$; it increases the output impedance; and under the condition $(1 + h_{fe})R_e \gg h_{ie}$, it stabilizes the voltage gain, which becomes essentially equal to $-R_L/R_e$ (and thus is independent of the transistor).

The Approximate Solution   An approximate analysis of the circuit of Fig. 10-11a can be made using the simplified model of Fig. 10-5 as shown in Fig. 10-11b.

The current gain is, from Fig. 10-11b,

$$A_I = \frac{-I_c}{I_b} = \frac{-h_{fe}I_b}{I_b} = -h_{fe} \tag{10-28}$$

The current gain equals the short-circuit value, and is unaffected by the addition of R_e.

The input resistance, as obtained from inspection of Fig. 10-11b, is

$$R_i = \frac{V_i}{I_b} = h_{ie} + (1 + h_{fe})R_e \tag{10-29}$$

The input resistance is augmented by $(1 + h_{fe})R_e$, and may be very much larger than h_{ie}. Hence an emitter resistance greatly increases the input resistance.

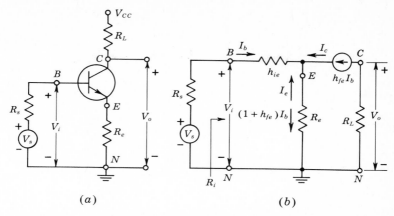

Fig. 10-11   (a) Common-emitter amplifier with an emitter resistor.
(b) Approximate small-signal equivalent circuit.

The voltage gain is

$$A_V = \frac{A_I R_L}{R_i} = \frac{-h_{fe} R_L}{h_{ie} + (1 + h_{fe}) R_e} \tag{10-30}$$

Clearly, the addition of an emitter resistance greatly reduces the voltage amplification. This reduction in gain is often a reasonable price to pay for the improvement in stability. We note that, if $(1 + h_{fe}) R_e \gg h_{ie}$, and since $h_{fe} \gg 1$, then

$$A_V \approx \frac{-h_{fe}}{1 + h_{fe}} \frac{R_L}{R_e} \approx \frac{-R_L}{R_e} \tag{10-31}$$

Subject to the above approximations, A_V is completely stable (if stable resistances are used for R_L and R_e), since it is independent of all transistor parameters.

The output resistance of the transistor alone (with R_L considered external) is infinite for the approximate circuit of Fig. 10-11, just as it was for the CE amplifier of Sec. 10-3 with $R_e = 0$. Hence the output resistance of the stage, including the load, is R_L.

The above expressions are included in the second column of Table 10-2. Incidentally, it can be shown† that the simplified h-parameter model of Fig. 10-5 is valid for the CE configuration with an emitter resistor R_e only if the inequality in Eq. (10-10) is generalized as follows:

$$h_{oe}(R_e + R_L) \leq 0.1 \tag{10-32}$$

---

† J. Millman, and C. C. Halkias: "Integrated Electronics; Analog and Digital Circuits and Systems," chap. 8, McGraw-Hill Book Company, New York, 1972.

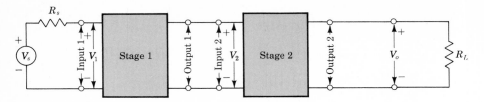

Fig. 10-12 Two cascaded stages.

10-8   CASCADING TRANSISTOR AMPLIFIERS

When the amplification of a single transistor is not sufficient for a particular purpose, or when the input or output impedance is not of the correct magnitude for the intended application, two or more stages may be connected in cascade; i.e., the output of a given stage is connected to the input of the next stage, as shown in Fig. 10-12. In the circuit of Fig. 10-13a the first stage is connected common-emitter, and the second is a common-collector stage. Figure 10-13b shows the small-signal circuit of the two-stage amplifier, with the biasing batteries omitted, since these do not affect the small-signal calculations.

To analyze a circuit such as the one of Fig. 10-13, we make use of the general expressions for A_I, R_i, A_V, and R_o from Table 10-2. It is necessary that we have available the h parameters for the specific transistors used in the circuit.

The voltage gain A_V of a multistage amplifier from the input of the first to the output of the last stage equals the product of the voltage gains of each stage. This statement is easily verified by reference to Fig. 10-12 or 10-13b. Thus,

$$A_V \equiv \frac{V_o}{V_1} = \frac{V_o}{V_2}\frac{V_2}{V_1} = A_{V2}A_{V1} \tag{10-33}$$

The overall voltage gain A_{Vs}, taking the source resistance into account, is given by

$$A_{Vs} \equiv \frac{V_o}{V_s} = \frac{V_o}{V_1}\frac{V_1}{V_s} = A_V\frac{R_{i1}}{R_{i1} + R_s} \tag{10-34}$$

because there is an effective resistance R_{i1} between base and emitter of the first stage.

The total current gain A_I is *not* equal to the product of the current gains of the individual stages because the output current of one stage is not equal to the input current of the following stage. For example, in Fig. 10-13b we see that $I_{b2} \neq -I_{c1}$. The correct expression for A_I is given by

$$A_I \equiv -\frac{I_{e2}}{I_{b1}} = -\frac{I_{e2}}{I_{b2}}\frac{I_{b2}}{I_{c1}}\frac{I_{c1}}{I_{b1}} = -A_{I2}\frac{I_{b2}}{I_{c1}}A_{I1} \tag{10-35}$$

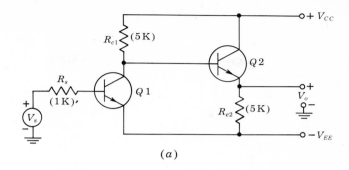

(a)

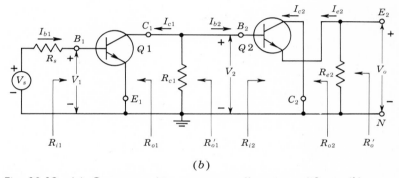

(b)

Fig. 10-13 (a) **Common-emitter–common-collector amplifier.** (b) **Small-signal circuit of the CE-CC amplifier.**

From Fig. 10-14 we can obtain the ratio I_{b2}/I_{c1}. Since $-I_{c1}$ passes through the parallel combination of R_{c1} and R_{i2},

$$V_{B_2N} = -\frac{I_{c2}R_{c1}R_{i2}}{R_{c1} + R_{i2}} = I_{b2}R_{i2}$$

or

$$\frac{I_{b2}}{I_{c2}} = -\frac{R_{c1}}{R_{c1} + R_{i2}} \tag{10-36}$$

Hence,

$$A_I = A_{I2}A_{I1}\frac{R_{c1}}{R_{i2} + R_{c1}} \tag{10-37}$$

Fig. 10-14 Relating to the calculation of overall current gain.

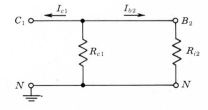

We observe that the effective load resistance R_{L1} of the first stage is the total resistance shunting C_1 to ground N. From Fig. 10-13 we find that R_{L1} is the parallel combination of R_{c1} and R_{i2}, or

$$R_{L1} = \frac{R_{c1}R_{i2}}{R_{c1} + R_{i2}} \tag{10-38}$$

EXAMPLE For the two-stage CE-CC configuration in Fig. 10-13 the hybrid parameters of each stage are $h_{ie} = 2$ K and $h_{fe} = 50$, whereas h_{re} and h_{oe} are small enough to be neglected. Find the input and output resistances and individual, as well as overall, voltage and current gains.

Solution We note that, in a cascade of stages, the collector resistance of one stage is shunted by the input resistance of the next stage. Hence it is advantageous to start the analysis with the last stage. In addition, it is convenient to compute, first, the current gain, then the input resistance and the voltage gain. Finally, the output resistance may be calculated if desired by starting this analysis with the first stage and proceeding toward the output stage.

For the CC output stage we have, from Table 10-2

$$A_{I2} = 1 + h_{fe} = 51$$

$$R_{i2} = h_{ie} + (1 + h_{fe})R_L = 2 + (51)(5) = 257 \text{ K}$$

$$A_{V2} = 1 - \frac{h_{ie}}{R_{i2}} = 1 - \frac{2}{257} = 0.992$$

Note the high input resistance of the CC stage and that its voltage gain is close to unity

For the CE stage we find from Table 10-2,

$$A_{I1} = -h_{fe} = -50 \qquad R_{i1} = h_{ie} = 2 \text{ K}$$

The input resistance R_i of the cascaded amplifier is the input resistance of the first stage, or $R_i = 2$ K. The effective load on the first stage, its voltage gain, and output resistance are

$$R_{L1} = \frac{R_{c1}R_{i2}}{R_{c1} + R_{i2}} = \frac{(5)(257)}{262} = 4.9 \text{ K}$$

$$A_{V1} = \frac{A_{I1}R_{L1}}{R_{i1}} = \frac{-(50)(4.9)}{2} = -123$$

$$R'_{o1} = R_{c1} = 5 \text{ K}$$

Since R'_{o1} is the effective source resistance for $Q2$, then, from Table 10-2,

$$R_{o2} = \frac{h_{ie} + R'_{o1}}{1 + h_{fe}} = \frac{2,000 + 5,000}{51} = 137 \text{ }\Omega$$

$$R'_{o2} = \frac{R_{o2}R_{L2}}{R_{o2} + R_{L2}} = \frac{(137)(5,000)}{5,137} = 134 \text{ }\Omega = R'_o$$

Finally, the overall voltage and current gains of the cascade are

$$A_V = A_{V1}A_{V2} = (-123)(0.992) = -122$$

$$A_I = A_{I1}A_{I2} \frac{R_{c1}}{R_{c1} + R_{i2}} = (-50)(51)\left(\frac{5}{5 + 257}\right) = -48.7$$

Alternatively, A_V may be computed from

$$A_V = A_I \frac{R_{L2}}{R_{i1}} = -\frac{(48.7)(5)}{2} = -122$$

The overall voltage gain, taking R_s into account, is given by Eq. (10-34):

$$A_{Vs} = A_V \frac{R_{i1}}{R_{i1} + R_s} = -\frac{(122)(2)}{2 + 1} = -81.3$$

Choice of the Transistor Configuration in a Cascade   It is important to note that the previous calculations of input and output resistances and voltage and current gains are applicable for any connection of the cascaded stages.   They could be CE, CC, CB, or combinations of all three possible connections.

Consider the following question: Which of the three possible connections must be used in cascade if maximum voltage gain is to be realized?   For the intermediate stages, the common-collector connection is not used because the voltage gain of such a stage is less than unity.   Hence it is not possible (without a transformer) to increase the overall voltage amplification by cascading common-collector stages.

Grounded-base (CB) coupled stages also are seldom cascaded because the voltage gain of such an arrangement is approximately the same as that of the output stage alone.   This statement may be verified as follows: The voltage gain of a stage equals its current gain times the effective load resistance R_L divided by the input resistance R_i.   The effective load resistance R_L is the parallel combination of the actual collector resistance R_c and (except for the last stage) the input resistance R_i of the following stage.   This parallel combination is certainly less than R_i, and hence, for identical stages, the effective load resistance is less than R_i.   The maximum current gain is less than unity (Table 10-2).   Hence the voltage gain of any stage (except the last, or output, stage) is less than unity.

Since the short-circuit current gain h_{fe} of a common-emitter stage is much greater than unity, it is possible to increase the voltage amplification by cascading such stages.   We may now state that *in a cascade the intermediate transistors should be connected in a common-emitter configuration.*

The choice of the input stage may be decided by criteria other than the maximization of voltage gain.   For example, the amplitude or the frequency response of the transducer V_s may depend upon the impedance into which it operates.   Some transducers require essentially open-circuit or short-circuit operation. In many cases the common-collector or common-base stage is used at the input because of impedance considerations, even at the expense of voltage or current gain.

The output stage is selected also on the basis of impedance considerations. Since a CC stage has a very low output resistance, it is often used for the last stage if it is required to drive low impedance (perhaps capacitive) load.

10-9   THE FIELD–EFFECT TRANSISTOR SMALL–SIGNAL MODEL

The linear incremental equivalent circuit for the FET can be obtained in a manner analogous to that used to derive the corresponding model for a bipolar transistor. We employ the same notation in labeling time-varying and dc currents and voltages as used in Secs. 10-1 and 10-2. The drain current i_D is a function of the gate voltage v_{GS} and the drain voltage v_{DS}. Hence, for small variations about the quiescent point, we may express the incremental current i_d as a linear combination of v_{gs} and v_{ds}. Thus,

$$i_d = g_m v_{gs} + \frac{1}{r_d} v_{ds} \qquad (10\text{-}39)$$

From this equation it follows that

$$g_m = \frac{i_d}{v_{gs}} \bigg|_{v_{ds}=0} \qquad (10\text{-}40)$$

Note that $v_{ds} = 0$ is equivalent to keeping $v_{DS} = V_{DS} = $ constant. The dimension of g_m is a reciprocal resistance (conductance) and is expressed in mhos or A/V. It is called the *mutual conductance* or *transconductance*. The second parameter r_d in Eq. (10-39) is the *drain*, or *output resistance*, and is defined by

$$r_d = \frac{v_{ds}}{i_d} \bigg|_{v_{gs}=0} \qquad (10\text{-}41)$$

The reciprocal of r_d (ohms) is designated by g_d and is called the *drain conductance*.

An *amplification factor* μ, of an FET may be defined by

$$\mu \equiv - \frac{v_{ds}}{v_{gs}} \bigg|_{i_d=0}$$

We may verify that μ, r_d, and g_m are related by

$$\mu = r_d g_m \qquad (10\text{-}42)$$

by setting $i_d = 0$ in Eq. (10-39).

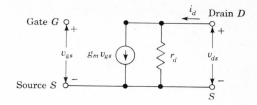

Fig. 10-15 The low-frequency small-signal FET model.

The FET Model A circuit which satisfies Eq. (10-39) is indicated in Fig. 10-15. This low-frequency small-signal model has a Norton's output circuit with a dependent current generator whose current is proportional to the gate-to-source voltage. The proportionality factor is the transconductance g_m, which is consistent with the definition of g_m in Eq. (10-40). The output resistance is r_d, which is consistent with the definition in Eq. (10-41). The input resistance between gate and source is infinite, since it is assumed that the reverse-biased gate in the FET, or the insulated gate in the MOSFET, takes no current. For the same reason the resistance between gate and drain is assumed to be infinite.

The order of magnitudes of the parameter values in the model for an FET is given in Table 10-4. Included in this table are also the values of interelectrode capacitances to be expected, but these are not included in the model of Fig. 10-15 since we are considering only low-frequency operation in this chapter.

The FET model of Fig. 10-15 should be compared with the h-parameter model of the bipolar junction transistor of Fig. 10-6. The latter has a current source in the output circuit, but the current generated depends upon the input *current*, whereas in the FET model the generator current depends upon the input *voltage*. Observe also that the very high input resistance of the FET (considered infinite in Fig. 10-15) is replaced by an input resistance h_{ie} of about 1 K for the bipolar transistor.

Small-Signal MOSFET Circuit Model If the small resistances of the source and drain regions are neglected, the small-signal equivalent circuit of

TABLE 10-4   Range of parameter values for an FET

Parameter	JFET	MOSFET†
g_m	0.1–10 mA/V	0.1–20 mA/V or more
r_d	0.1–1 M	1–50 K
C_{ds}	0.1–1 pF	0.1–1 pF
C_{gs}, C_{gd}	1–10 pF	1–10 pF
r_{gs}	$> 10^8$ Ω	$> 10^{10}$ Ω
r_{gd}	$> 10^8$ Ω	$> 10^{14}$ Ω

† Discussed in Sec. 9-2.

the MOSFET between terminals G $(=G_1)$, S, and D is identical with that given in Fig. 10-15 for the JFET. The transconductance g_m and the interelectrode capacitances have comparable values for the two types of devices. However, as noted in Table 10-4, the drain resistance r_d of the MOSFET is very much smaller than that of the JFET. It should also be noted in Table 10-4 that the input resistance r_{gs} and the feedback resistance r_{gd} are very much larger for the MOSFET than for the JFET.

10-10   THE LOW–FREQUENCY COMMON–SOURCE AND COMMON–DRAIN AMPLIFIERS

The common-source (CS) stage is indicated in Fig. 10-16a, and the common-drain (CD) configuration in Fig. 10-16b. The former is analogous to the bipolar transistor CE amplifier, and the latter to the CC stage. We shall analyze both of these circuits simultaneously by considering the generalized configuration in Fig. 10-17a. For the CS stage the output is v_{o1} taken at the drain and $R_s = 0$. For the CD stage the output is v_{o2} taken at the source and $R_d = 0$. The signal-source resistance is unimportant since it is in series with the gate, which draws negligible current. No biasing arrangements are indicated, but it is assumed that the stage is properly biased for linear operation (Sec. 10-15).

Replacing the FET by its low-frequency small-signal model of Fig. 10-15, the equivalent circuit of Fig. 10-17b is obtained. Applying KVL to the output circuit yields

$$i_d R_d + (i_d - g_m v_{gs})r_d + i_d R_s = 0 \tag{10-43}$$

From Fig. 10-17b the voltage from G to S is given by

$$v_{gs} = v_i - i_d R_s \tag{10-44}$$

Combining Eqs. (10-43) and (10-44) and remembering that $\mu = r_d g_m$ [Eq. (10-42)], we find

$$i_d = \frac{\mu v_i}{r_d + R_d + (\mu + 1)R_s} \tag{10-45}$$

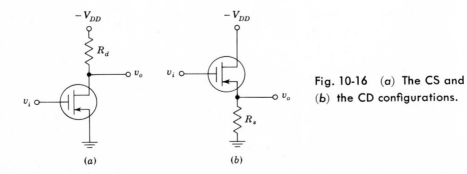

Fig. 10-16   (a) The CS and (b) the CD configurations.

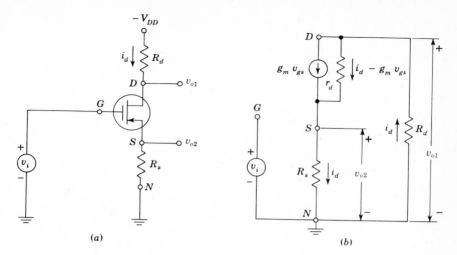

Fig. 10-17 (a) A generalized FET amplifier configuration. (b) The small-signal equivalent circuit.

The CS Amplifier with an Unbypassed Source Resistance Since $v_{o1} = -i_d R_d$, then

$$v_{o1} = \frac{-\mu v_i R_d}{r_d + R_d + (\mu + 1)R_s} \tag{10-46}$$

If R_s is bypassed with a large capacitance or if the source is grounded, the above equations are valid with $R_s = 0$. Under these circumstances,

$$A_V = \frac{v_{o1}}{v_i} = \frac{-\mu R_d}{r_d + R_d} = -g_m R_d' \tag{10-47}$$

where $\mu = r_d g_m$ [Eq. (10-42)] and $R_d' = R_d \| r_d$.

The CD Amplifier with a Drain Resistance Since $v_{o2} = i_d R_s$, then from Eq. (10-45)

$$v_{o2} = \frac{\mu v_i R_s}{r_d + R_d + (\mu + 1)R_s} \tag{10-48}$$

Note that there is no phase shift between input and output. If $R_d = 0$ and if $(\mu + 1)R_s \gg r_d$, then $A_V \approx \mu/(\mu + 1) \approx 1$ for $\mu \gg 1$. A voltage gain of unity means that the output (at the source) follows the input (at the gate). Hence the CD configuration is called a *source follower* (analogous to the *emitter follower* for a bipolar junction transistor).

10-11   THE OPERATING POINT

From our discussion of the bipolar junction transistor characteristics in Secs. 4-5 to 4-6, it is clear that the transistor functions most linearly when it is

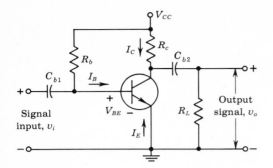

Fig. 10-18   The fixed-bias circuit.

constrained to operate in its active region.   To establish an operating point in this region it is necessary to provide appropriate direct potentials and currents, using external sources.   Once an operating point Q is established, such as the one shown in Fig. 10-2a, time-varying excursions of the input signal (base current, for example) should cause an output signal (collector voltage or collector current) of the same waveform.   If the output signal is not a faithful reproduction of the input signal, for example, if it is clipped on one side, the operating point is unsatisfactory and should be relocated on the collector characteristics.   The question now naturally arises as to how to choose the operating point.   In Fig. 10-18 we show a common-emitter circuit.   Figure 10-19 gives the output characteristics of the transistor used in Fig. 10-18.   Note that even if we are free to choose R_c, R_L, R_b, and V_{CC}, we may not operate the transistor everywhere in the active region because the various transistor ratings limit the range of useful operation.   These ratings (listed in the manufacturer's specification sheets) are maximum collector dissipation $P_{C,\max}$, maximum collector voltage $V_{C,\max}$, maximum collector current $I_{C,\max}$, and maximum emitter-to-base voltage $V_{EB,\max}$.   Figure 10-19 shows three of these bounds on typical collector characteristics.

Capacitive Coupling   Note that in the circuit of Fig. 10-1, neither side of the signal generator is grounded, and also that an auxiliary biasing supply V_{BB} is used.   Both of these difficulties are avoided by using a capacitor C_{b1} to couple the input signal to the transistor, as indicated in Fig. 10-18.   In this diagram one end of v_i is at ground potential, and the collector supply V_{CC} also provides the biasing base current I_B.   Under quiescent conditions (no input signal), C_{b1} (called a *blocking* capacitor) acts as an open circuit, because the reactance of a capacitor is infinite at zero frequency (dc).   The capacitances C_{b1} and C_{b2} are chosen large enough so that, at the lowest frequency of excitation, their reactances are small enough so that they can be considered to be short circuits.   These coupling capacitors block dc voltages but freely pass signal voltages.   For example, the quiescent collector voltage does not appear at the output, but v_o is an amplified replica of the input signal v_i.   The (ac or incremental) output signal voltage may be applied to the input of another amplifier without affecting its bias, because of the blocking capacitor C_{b2}.   The

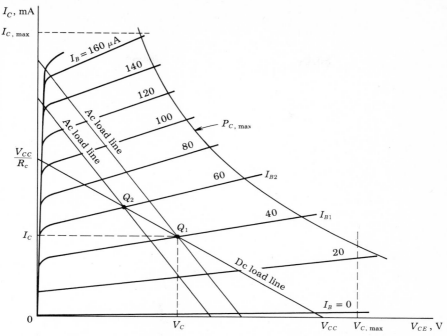

Fig. 10-19 Common-emitter collector characteristics; ac and dc load lines.

effect of the finite size of a blocking capacitor on the frequency response of an amplifier is considered in Sec. 11-7.

The Static and Dynamic Load Lines   We noted above that under dc conditions C_{b2} acts as an open circuit.   Hence the quiescent collector current and voltage are obtained by drawing a static (dc) load line corresponding to the resistance R_c through the point $i_C = 0$, $v_{CE} = V_{CC}$, as indicated in Fig. 10-19. If $R_L = \infty$ and if the input signal (base current) is large and symmetrical, we must locate the operating point Q_1 at the center of the load line.   In this way the collector voltage and current may vary approximately symmetrically around the quiescent values V_C and I_C, respectively.   If $R_L \neq \infty$, however, a *dynamic* (ac) load line must be drawn.   Since we have assumed that, at the signal frequency, C_{b2} acts as a short circuit, the effective load R_L' at the collector is R_c in parallel with R_L.   The dynamic load line must be drawn through the operating point Q_1 and must have a slope corresponding to $R_L' = R_c \| R_L$. This ac load line is indicated in Fig. 10-19, where we observe that the input signal may swing a maximum of approximately 40 μA, around Q_1 because, if the base current decreases by more than 40 μA, the transistor is driven off.

If a larger input swing is available, then in order to avoid cutoff during a part of the cycle, the quiescent point must be located at a higher current.   For example, by simple trial and error we locate Q_2 *on the dc load line* such that a line with a slope corresponding to the ac resistance R'_L and drawn through Q_2 gives as large an output as possible without too much distortion.   In Fig. 10-19 the choice of Q_2 allows an input peak current swing of about 60 μA.

The Fixed-Bias Circuit   The point Q_2 can be established by noting the required current I_{B2} in Fig. 10-19 and chcosing the resistance R_b in Fig. 10-18 so that the base current is equal to I_{B2}.   Therefore

$$I_B = \frac{V_{CC} - V_{BE}}{R_b} = I_{B2} \tag{10-49}$$

The voltage V_{BE} across the forward-biased emitter junction is (Table 4-1, page 85) approximately 0.2 V for a germanium transistor and 0.7 V for a silicon transistor in the active region.   Since V_{CC} is usually much larger than V_{BE}, we have

$$I_B \approx \frac{V_{CC}}{R_b} \tag{10-50}$$

The current I_B is constant, and the network of Fig. 10-18 is called the *fixed-bias circuit*.   In summary, we see that the selection of an operating point Q depends upon a number of factors.   Among these factors are the ac and dc loads of the stage, the available power supply, the maximum transistor ratings, the peak signal excursions to be handled by the stage, and the tolerable distortion.

10-12   BIAS STABILITY

In the preceding section we examined the problem of selecting an operating point Q on the load line of the transistor.   We now consider some of the problems of maintaining the operating point stable.

Let us refer to the biasing circuit of Fig. 10-18.   In this circuit the base current I_B is kept constant since $I_B \approx V_{CC}/R_b$.   Let us assume that the transistor of Fig. 10-18 is replaced by another of the same type.   In spite of the tremendous strides that have been made in the technology of the manufacture of semiconductor devices, transistors of a particular type still come out of production with a wide spread in the values of some parameters.   For example, Fig. 4-14 shows a range of $h_{FE} \approx \beta$ of about 3 to 1.   To provide information about this variability, a transistor data sheet, in tabulating parameter values, often provides columns headed minimum, typical, and maximum.

In Sec. 4-6 we see that the spacing of the output characteristics will increase or decrease (for equal changes in I_B) as β increases or decreases.   In Fig. 10-20 we have assumed that β is greater for the replacement transistor of Fig. 10-18, and since I_B is maintained constant at I_{B2} by the external biasing

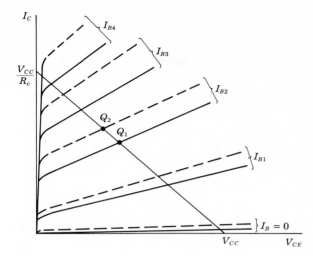

Fig. 10-20 Graphs showing the collector characteristics for two transistors of the same type. The dashed characteristics are for a transistor whose β is much larger than that of the transistor represented by the solid curves.

circuit, it follows that the operating point will move to Q_2. This new operating point may be completely unsatisfactory. Specifically, it is possible for the transistor to find itself in the saturation region. We now conclude that maintaining I_B constant will not provide operating-point stability as β changes. On the contrary, I_B should be allowed to change so as to maintain I_C and V_{CE} constant as β changes.

Thermal Instability   A second very important cause for bias instability is a variation in temperature. In Sec. 4-7 we note that the reverse saturation current I_{CO}† changes greatly with temperature. Specifically, I_{CO} doubles for every 10°C rise in temperature. This fact may cause considerable practical difficulty in using a transistor as a circuit element. For example, the collector current I_C causes the collector-junction temperature to rise, which in turn increases I_{CO}. As a result of this growth of I_{CO}, I_C will increase [Eq. (4-12)], which may further increase the junction temperature, and consequently I_{CO}. It is possible for this succession of events to become cumulative, so that the ratings of the transistor are exceeded and the device burns out.

Even if the drastic state of affairs described above does not take place, it is possible for a transistor which was biased in the active region to find itself in the saturation region as a result of this operating-point instability. To see how this may happen, we note that if $I_B = 0$, then, from Eq. (4-12), $I_C = I_{CO}(1 + \beta)$. As the temperature increases, I_{CO} increases, and even if we assume that β remains constant (actually, it also increases), it is clear that the $I_B = 0$ line in the CE output characteristics will move upward. The characteristics for other values of I_B will also move upward by the same amount (provided that β remains constant), and consequently the operating point will move if I_B is forced to remain constant. Hence the transistor could

† Throughout this section I_{CBO} is abbreviated I_{CO} (Sec. 4-7).

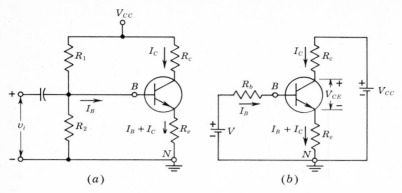

Fig. 10-21 (a) A self-biasing circuit. (b) Simplification of the base circuit in (a) by the use of Thévenin's theorem.

find itself almost in saturation at a temperature of $+100°C$, even though it would be biased in the middle of its active region at $+25°C$.

10-13   SELF–BIAS OR EMITTER BIAS

A circuit which is used to establish a stable operating point is the self-biasing configuration of Fig. 10-21a. The current in the resistance R_e in the emitter lead causes a voltage drop which is in the direction to reverse-bias the emitter junction. Since this junction must be forward-biased, the base voltage is obtained from the supply through the R_1R_2 network.

The physical reason for an improvement in stability with this circuit is the following: If I_C tends to increase, say, because I_{CO} has risen as a result of an elevated temperature, the current in R_e increases. As a consequence of the increase in voltage drop across R_e, the base current is decreased. Hence I_C will increase less than it would have, had there been no self-biasing resistor R_e.

Analysis of the Self-Bias Circuit If the circuit component values in Fig. 10-21a are specified, the quiescent point is found as follows:

If the circuit to the left between the base B and ground N terminals in Fig. 10-21a is replaced by its Thévenin equivalent, the two-mesh circuit of Fig. 10-21b is obtained, where

$$V \equiv \frac{R_2 V_{CC}}{R_2 + R_1} \qquad R_b \equiv \frac{R_2 R_1}{R_2 + R_1} \tag{10-51}$$

Obviously, R_b is the effective resistance seen looking back from the base terminal. Kirchhoff's voltage law around the base circuit yields

$$V = I_B R_b + V_{BE} + (I_B + I_C) R_e \tag{10-52}$$

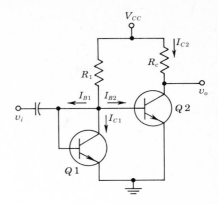

Fig. 10-22   Biasing technique for linear
integrated circuits.   The transistor $Q1$
with base connected to collector behaves
as a diode.

In the active region the collector current is given by Eq. (4-12), namely,

$$I_C = \beta I_B + (1 + \beta)I_{co} \tag{10-53}$$

Equations (10-52) and (10-53) can now be solved for I_B and I_C (since V_{BE} is known in the active region).   Note that with this method the currents (in the active region) are determined by the base circuit and the values of β and I_{co}. This method of solution is carried out in the illustrative example of Sec. 4-6.

10-14   BIASING TECHNIQUE FOR LINEAR INTEGRATED CIRCUITS

The self-bias circuit of Fig. 10-21a often requires a capacitor across R_e since otherwise the negative feedback, due to R_e, reduces the signal gain drastically (Sec. 10-7).   This bypass capacitance is much too large (of the order of micro-farads, Sec. 11-8) to be fabricated by integrated-circuit technology.   Hence the biasing technique shown in Fig. 10-22 has been developed for monolithic circuits.   In Fig. 10-22 the transistor $Q1$ is connected as a diode across the base-to-emitter junction of $Q2$ whose collector current is to be temperature-stabilized.   The collector current of $Q1$ is given by

$$I_{C1} = \frac{V_{CC} - V_{BE}}{R_1} - I_{B1} - I_{B2} \tag{10-54}$$

For $V_{BE} \ll V_{CC}$ and $I_{B1} + I_{B2} \ll I_{C1}$, Eq. (10-54) becomes

$$I_{C1} \approx \frac{V_{CC}}{R_1} = \text{constant} \tag{10-55}$$

If transistors $Q1$ and $Q2$ are identical and have the same V_{BE}, their collector currents will be equal.   Hence $I_{C2} = I_{C1} = \text{constant}$.   Even if the two transistors are not identical, experiments have shown that this biasing scheme gives collector-current matching between the biasing and operating transistors typically better than 5 percent and is stable over a wide temperature range.

If $R_c = \frac{1}{2}R_1$, then $V_{CE} = V_{CC} - I_{C2}R_c \approx V_{CC}/2$, which means that the amplifier will be biased at one-half the supply voltage V_{CC}, independent of the supply voltage as well as temperature, and dependent only on the matching of components within the integrated circuit.

10-15   BIASING THE FET

The selection of an appropriate operating point (I_D, V_{GS}, V_{DS}) for an FET amplifier stage is determined by considerations similar to those given to transistors, as discussed in Sec. 10-11. These considerations are output-voltage swing, distortion, power dissipation, voltage gain, and drift of drain current. In most cases it is not possible to satisfy all desired specifications simultaneously. In this section we examine several biasing circuits for field-effect devices.

The Transfer Characteristic   In amplifier applications the FET is almost always used in the region beyond pinch-off (also called the *constant-current, pentode*, or *current-saturation region*). Let the saturation drain current be designated by I_{DS}, and its value with the gate shorted to the source ($V_{GS} = 0$) by I_{DSS}. It has been found that the transfer characteristic, giving the relationship between I_{DS} and V_{GS}, can be approximated by the parabola

$$I_{DS} = I_{DSS}\left(1 - \frac{V_{GS}}{V_P}\right)^2 \tag{10-56}$$

This simple parabolic approximation gives an excellent fit, with the experimentally determined transfer characteristics for FETs made by the diffusion process.

Cutoff   Consider an FET operating at a fixed value of V_{DS} in the constant-current region. As V_{GS} is increased in the direction to reverse-bias the gate junction, the conducting channel will narrow. When $V_{GS} = V_P$, the channel width is reduced to zero, and from Eq. (10-56), $I_{DS} = 0$. With a physical device some leakage current $I_{D,\text{OFF}}$ still flows even under the cutoff condition $|V_{GS}| > |V_P|$. A manufacturer usually specifies a maximum value of $I_{D,\text{OFF}}$ at a given value of V_{GS} and V_{DS}. Typically, a value of a few nanoamperes may be expected for $I_{D,\text{OFF}}$ for a silicon FET.

The gate reverse current, also called *the gate cutoff current*, designated by I_{GSS}, gives the gate-to-source current, with the drain shorted to the source for $|V_{GS}| > |V_P|$. Typically, I_{GSS} is of the order of a few nanoamperes for a silicon device.

Source Self-Bias   The configuration shown in Fig. 10-23 can be used to bias junction FET devices or depletion-mode MOS transistors. For a specified

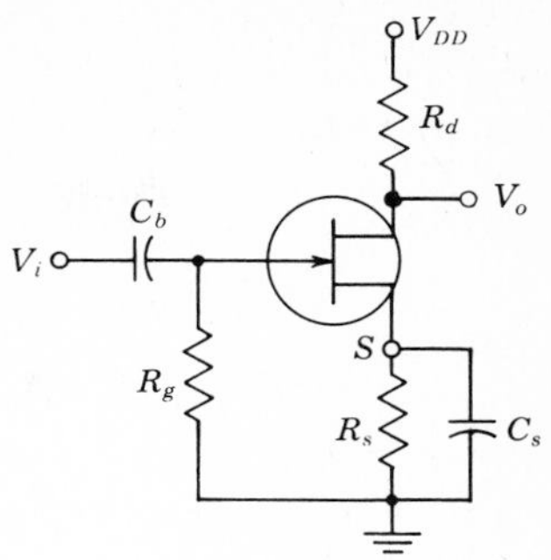

Fig. 10-23 Source self-bias circuit.

drain current I_D, the corresponding gate-to-source voltage V_{GS} can be obtained applying either Eq. (10-56) or the plotted drain or transfer characteristics. Since the gate current (and, hence, the voltage drop across R_g) is negligible, the source resistance R_s can be found as the ratio of V_{GS} to the desired I_D.

EXAMPLE The amplifier of Fig. 10-23 utilizes an n-channel FET for which $V_P = -2.0$ V and $I_{DSS} = 1.65$ mA. It is desired to bias the circuit at $I_D = 0.8$ mA, using $V_{DD} = 24$ V. Assume $r_d \gg R_d$ and $g_m = 1$ mA/V. Find (a) V_{GS}, (b) R_s, (c) R_d, such that the voltage gain is at least 10, with R_s bypassed with a very large capacitance C_s.

Solution a. Using Eq. (10-56), we have $0.8 = 1.65(1 + V_{GS}/2.0)^2$. Solving, $V_{GS} = -0.62$ V.

  b. $R_s = -\dfrac{V_{GS}}{I_D} = \dfrac{0.62}{0.8} = 0.77$ K $= 770\ \Omega$

  c. From Eq. (10-47), with $r_d \gg R_d$, $|A_V| = g_m R_d \geq 10$, or $R_d \geq 10/1 = 10$ K.

Biasing the Enhancement MOSFET The self-bias technique of Fig. 10-23 cannot be used to establish an operating point for the enhancement-type

Fig. 10-24 (a) Drain-to-gate bias circuit for enhancement-mode MOS transistors. (b) improved version of (a)

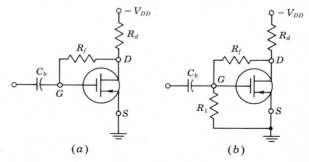

(a)   (b)

MOSFET because the voltage drop across R_s is in a direction to reverse-bias the gate, and a forward gate bias is required. The circuit of Fig. 10-24a can be used, and for this case we have $V_{GS} = V_{DS}$, since no current flows through R_f. If for reasons of linearity in device operation or maximum output voltage it is desired that $V_{GS} \neq V_{DS}$, then the circuit of Fig. 10-24b is suitable. We note that $V_{GS} = [R_1/(R_1 + R_f)]V_{DS}$. Both circuits discussed here offer the advantages of dc stabilization through the feedback introduced with R_f. However, the input impedance is reduced because, by Miller's theorem (Sec. C-4), R_f corresponds to an equivalent resistance $R_i = R_f/(1 - A_V)$ shunting the amplifier input.

REVIEW QUESTIONS

10-1 A transistor is excited by a large sinusoidal base current whose magnitude exceeds the quiescent value I_B for $0 < \omega t < \pi$ and is less than I_B for $\pi < \omega t < 2\pi$. Is the magnitude of the collector-current variation from the quiescent current greater at $\omega t = \pi/2$ or $3\pi/2$? Explain your answer with the aid of a graphical construction.

10-2 Is nonlinear distortion greater for a sinusoidal-input-base current or for a sinusoidal-input-base voltage? Explain with the aid of the input and output transistor characteristics.

10-3 Define in words (a) h_{ie}; (b) h_{fe}; (c) h_{re}; (d) h_{oe}. Indicate what variable is held constant and give the dimensions of each h parameter.

10-4 Draw the circuit of a CE transistor configuration and give its h-parameter model.

10-5 Repeat Rev. 10-4 for the CC configuration.

10-6 Repeat Rev. 10-4 for the CB configuration.

10-7 Which of the configurations (CB, CE, CC) has the (a) highest R_i; (b) lowest R_i; (c) highest R_o; (d) lowest R_o; (e) lowest A_V; (f) lowest A_I.

10-8 It is desired to have a high-gain amplifier with high input impedance and low output impedance. If a cascade of four stages is used, what configuration should be used for each stage?

10-9 Using the approximate h-parameter model, obtain the expression for a CE circuit for (a) A_I; (b) R_i; (c) A_V; (d) R_o.

10-10 Repeat Rev. 10-9 for the emitter-follower circuit.

10-11 Repeat Rev. 10-9 for the CE circuit with an emitter resistor.

10-12 Repeat Rev. 10-9 for the emitter follower with a collector resistor.

10-13 Define (a) *transconductance* g_m, (b) *drain resistance* r_d, and (c) *amplification factor* μ of an FET.

10-14 Give the order of magnitude of g_m, r_d, and μ for a MOSFET.

10-15 Show the small-signal model of an FET at low frequencies.

10-16 (a) Sketch the circuit of a CS amplifier. (b) Derive the expression for the voltage gain at low frequencies. (c) What is the maximum value of A_V?

10-17 What ratings limit the range of operation of a transistor?

10-18 Why is capacitive coupling used to connect a signal source to an amplifier?

10-19 For a capacitively coupled load, is the dc load larger or smaller than the ac load? Explain.

10-20 (a) Draw a fixed-bias circuit. (b) Explain why the circuit is unsatisfactory if the transistor is replaced by another of the same type.

10-21 (a) Draw a self-bias circuit. (b) Explain qualitatively why such a circuit is an improvement on the fixed-bias circuit, as far as stability is concerned.

10-22 (a) Draw a properly biased integrated-circuit linear amplifier. (b) How are the parameter values chosen so that the quiescent output voltage is $\frac{1}{2}V_{CC}$?

10-23 Draw a biasing circuit for a JFET or a depletion-type MOSFET.

10-24 Draw two biasing circuits for an enhancement-type MOSFET.

11 / FREQUENCY RESPONSE OF AMPLIFIERS

The frequency range of the amplifiers discussed in this chapter extends from a few cycles per second (hertz), or possibly from zero, up to some tens of megahertz. The original impetus for the study of such wideband amplifiers was supplied because they were needed to amplify the pulses occurring in a television signal. Therefore such amplifiers are often referred to as *video amplifiers*.

In this chapter, then, we consider the following problem: Given a low-level input waveform which is not necessarily sinusoidal but may contain frequency components from a few hertz to a few megahertz, how can this voltage signal be amplified with a minimum of distortion?

11-1   CLASSIFICATION OF AMPLIFIERS

Amplifiers are described in many ways, according to their frequency range, the method of operation, the ultimate use, the type of load, the method of interstage coupling, etc. The frequency classification includes dc (from zero frequency), audio (20 Hz to 20 kHz), video or pulse (up to a few megahertz), radio-frequency (a few kilohertz to hundreds of megahertz), and ultrahigh-frequency (hundreds or thousands of megahertz) amplifiers.

The position of the quiescent point and the extent of the characteristic that is being used determine the method of operation; whether the transistor or FET is operated as a linear or nonlinear amplifier. This chapter considers only the untuned audio or video voltage amplifier with a resistive load operated linearly.

11-2   DISTORTION IN AMPLIFIERS

The application of a sinusoidal signal to the input of an ideal linear amplifier will result in a sinusoidal output wave. Generally, the output waveform is not an exact replica of the input-signal waveform because of various types of distortion that may arise, either from the inherent nonlinearity in the characterisitcs of the transistors or FETs or from the influence of the associated circuit. The types of distortion that may exist either separately or simultaneously are called *nonlinear distortion, frequency distortion,* and *delay or phase-shift distortion.*

Nonlinear Distortion This type of distortion results from the production of new frequencies in the output which are not present in the input signal. These new frequencies, or harmonics, result from the existence of a nonlinear dynamic curve for the active device. This distortion is sometimes referred to as "amplitude distortion."

Frequency Distortion This type of distortion exists when the signa components of different frequencies are amplified differently. In either a transistor or an FET this distortion may be caused by the internal device capacitances, or it may arise because the associated circuit (for example, the coupling components or the load) is reactive. Under these circumstances, the gain A is a complex number whose magnitude and phase angle depend upon the frequency of the impressed signal. A plot of gain (magnitude) vs. frequency of an amplifier is called the *amplitude frequency-response characteristic.* If the frequency-response characteristic is not a horizontal straight line over the range of frequencies under consideration, the circuit is said to exhibit frequency distortion over this range.

Phase-Shift Distortion Phase-shift distortion results from unequal phase shifts of signals of different frequencies. This distortion is due to the fact that the phase angle of the complex gain A depends upon the frequency.

11-3   FREQUENCY RESPONSE OF AN AMPLIFIER

Consider a sinusoidal signal of angular frequency ω represented by $V_m \sin(\omega t + \phi)$. If the voltage gain of the amplifier has a magnitude A and if the signal suffers a phase change (lead angle) θ, then the output will be

$$A V_m \sin(\omega t + \phi + \theta) = A V_m \sin\left[\omega\left(t + \frac{\theta}{\omega}\right) + \phi\right]$$

Therefore, *if the amplification A is independent of frequency and if the phase shift θ is proportional to frequency (or is zero), then the amplifier will preserve the form of the input signal, although the signal will be shifted in time by*

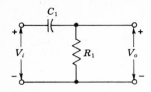

Fig. 11-1   A high-pass RC circuit used to calculate the low-frequency response of an amplifier.

an amount $D = \theta/\omega$. This discussion suggests that the extent to which an amplifier's amplitude response is not uniform, and its time delay is not constant with frequency, may serve as a measure of the lack of fidelity to be anticipated in it.

For an amplifier stage the frequency characteristics may be divided into three regions: There is a range, called the *midband frequencies*, over which the amplification is reasonably constant and equal to A_o and over which the delay is also quite constant.   For the present discussion we assume that the midband gain is normalized to unity, $A_o = 1$.   In the second (low-frequency) region, below midband, an amplifier stage may behave like the simple high-pass circuit of Fig. 11-1.   The response decreases with decreasing frequency, and the output usually approaches zero at dc ($f = 0$).   In the third (high-frequency) region, above midband, the circuit often behaves like the simple low-pass network of Fig. 11-2, and the response decreases with increasing frequency.   The total frequency characteristic, indicated in Fig. 11-3 for all three regions, will now be discussed.

Low-Frequency Response   In terms of the complex variable $s = j\omega = j2\pi f$, the reactance† of a capacitor is $1/sC$.   Hence, from the circuit of Fig. 11-1, we find

$$V_o = \frac{R_1}{R_1 + 1/sC_1} V_i = \frac{s}{s + 1/R_1C_1} V_i \tag{11-1}$$

The voltage transfer ratio, or voltage gain, $A_L = V_o/V_i$ becomes, in terms of the frequency $f = s/2\pi j$,

$$A_L(f) = \frac{1}{1 - j(f_L/f)} \tag{11-2}$$

where

$$f_L \equiv \frac{1}{2\pi R_1 C_1} \tag{11-3}$$

The magnitude $|A_L|$ and the phase lead θ_L of the gain are given by

$$|A_L(f)| = \frac{1}{\sqrt{1 + (f_L/f)^2}} \qquad \theta_L = \arctan \frac{f_L}{f} \tag{11-4}$$

† A discussion of reactive circuits and the complex frequency s is given in Sec. C-3.

Fig. 11-2 (a) A low-pass RC circuit used to calcu-late the high-frequency response of an amplifier. (b) The Norton's equiva-lent of the circuit in (a), where $I = V_i/R_2$.

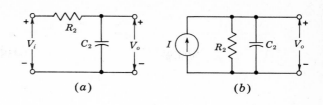

(a)   (b)

At the frequency $f = f_L$, $A_L = 1/\sqrt{2} = 0.707$, whereas in the midband region $(f \gg f_L)$, $A_L \rightarrow 1$. Hence f_L is that frequency at which the gain has fallen to 0.707 times its midband value A_o. This drop in signal level corre-sponds to a decibel† reduction of 20 log $(1/\sqrt{2})$, or 3 dB. Accordingly, f_L is referred to as the *low 3-dB frequency*. From Eq. (11-3) we see that f_L is that frequency for which the resistance R_1 equals the capacitive reactance $1/2\pi f_L C_1$.

High-Frequency Response In the high-frequency region, above the mid-band, the amplifier stage can often be approximated by the simple low-pass circuit of Fig. 11-2. In terms of the complex variable s, we find

$$V_o = \frac{1/sC_2}{R_2 + 1/sC_2} V_i = \frac{1}{1 + sR_2C_2} V_i \tag{11-5}$$

For real frequencies $(s = j\omega = j2\pi f)$ we obtain for the magnitude $|A_H(f)|$ and for the phase lead angle θ_H of the gain in the high-frequency region,

$$|A_H(f)| = \frac{1}{\sqrt{1 + (f/f_H)^2}} \qquad \theta_H = -\arctan \frac{f}{f_H} \tag{11-6}$$

where

$$f_H \equiv \frac{1}{2\pi R_2 C_2} \tag{11-7}$$

Since at $f = f_H$ the gain is reduced to $1/\sqrt{2}$ times its midband value, then f_H is called the *high 3-dB frequency*. It also represents that frequency at which the resistance R_2 equals the capacitive reactance $1/2\pi f_H C_2$. In the above expressions θ_L and θ_H represent the angle by which the output *leads* the input, neglecting the initial 180° phase shift through the amplifier if $A_o < 0$. The frequency dependence of the gains in the high- and low-frequency range is to be seen in Fig. 11-3. Such characteristics are called *Bode plots*.

Bandwidth The frequency range from f_L to f_H is called the *bandwidth* of the amplifier stage. We may anticipate in a general way that a signal, all of whose frequency components of appreciable amplitude lie well within the

† The voltage gain A expressed in decibels is given by 20 log A.

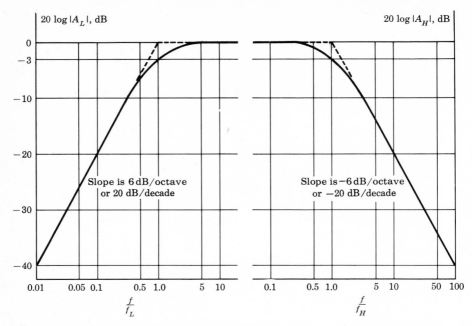

Fig. 11-3 A semi-log plot of the amplitude frequency-response (Bode) characteristic of an RC-coupled amplifier. The dashed curve is the idealized Bode
plot.

range f_L to f_H, will pass through the stage without excessive distortion.   This
criterion must be applied, however, with caution.

11-4   THE GENERALIZED VOLTAGE–GAIN FUNCTION

The high-frequency response of an amplifier is assumed in Fig. 11-2 to be
determined by a single time constant R_2C_2.   In reality, a multistage amplifier
may contain many energy-storage elements (capacitors), and the transfer function will be given by an equation of the form

$$A(s) = \frac{V_o(s)}{V_i(s)} = \frac{K(s - s_{z1})(s - s_{z2})(s - s_{z3})}{(s - s_{p1})(s - s_{p2})(s - s_{p3})(s - s_{p4})} \tag{11-8}$$

The values of s for which $A(s) = 0$ are called the *zeros* of the transfer function.
In Eq. (11-8) we have assumed three zeros, s_{z1}, s_{z2}, and s_{z3}.   The values of s
for which $A(s) = \infty$ are called the *poles* of the transfer function.   There are
four poles s_{p1}, . . . , s_{p4} in the amplifier represented by Eq. (11-8).   The simple
high-pass circuit of Fig. 11-1 has one zero $s_z = 0$ and one pole $s_p = -1/R_1C_1$,
as is evident from Eq. (11-1).   Equation (11-5) shows that the low-pass circuit

of Fig. 11-2 has no zeros and one pole at $s_p = -1/R_2C_2$.   Note that frequency f_p of a pole has a magnitude $s_p/2\pi$.

The Dominant Pole Consider a two-pole transfer function where f_{p1} is much smaller than f_{p2}.   Then a frequency-response plot such as that in Fig. 11-3 indicates that the upper 3-dB frequency is given approximately by f_{p1}. If $f_{p2} = 4f_{p1}$, an exact plot indicates (Prob. 11-5) that the 3-dB frequency is only 6 percent smaller than f_{p1}.   We conclude that *if a transfer function has several poles determining the high-frequency response, if the smallest of these is f_{p1} and if each other pole is at least two octaves† higher, then the amplifier behaves essentially as a single-time-constant circuit whose 3-dB frequency is f_{p1}.*   The frequency f_{p1} is called the *dominant pole*.

11-5   STEP RESPONSE OF AN AMPLIFIER

An alternative criterion of amplifier fidelity is the response of the amplifier to a particular input waveform.   Of all possible available waveforms, the most generally useful is the step voltage (Sec. C-5).   This waveform is one which permits small distortions to stand out clearly.   Additionally, from an experimental viewpoint, we note that excellent pulse (a short step) and square-wave (a repeated step) generators are available commercially.

As long as an amplifier can be represented by a dominant pole, the correlation between its frequency response and the output waveshape for a step input is that given below.   Quite generally, even for more complicated amplifier circuits, there continues to be an intimate relationship between the distortion of the leading edge of a step and the high-frequency response.   Similarly, there is a close relationship between the low-frequency response and the distortion of the flat portion of the step.   We should, of course, expect such a relationship, since the high-frequency response measures essentially the ability of the amplifier to respond faithfully to rapid variations in signal, whereas the low-frequency response measures the fidelity of the amplifier for slowly varying signals.   An important feature of a step is that it is a combination of the most abrupt voltage change possible and of the slowest possible voltage variation.

Rise Time The response of the low-pass circuit of Fig. 11-2 to a step input of amplitude V is exponential with a time constant R_2C_2.   Since the capacitor voltage cannot change instantaneously, the output starts from zero and rises toward the steady-state value V, as shown in Fig. 11-4.   The output is given by (Sec. C-5)

$$v_o = V(1 - \epsilon^{-t/R_2C_2}) \tag{11-9}$$

† Two frequencies are an octave apart if their ratio is 2.

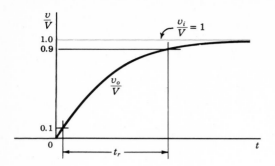

Fig. 11-4 Step-voltage response of the low-pass RC circuit. The rise time t_r is indicated.

The time required for v_o to reach one-tenth of its final value is readily found to be $0.1R_2C_2$, and the time to reach nine-tenths its final value is $2.3R_2C_2$. The difference between these two values is called the *rise time* t_r of the circuit and is shown in Fig. 11-4. The time t_r is an indication of how fast the amplifier can respond to a discontinuity in the input voltage. We have, using Eq. (11-7),

$$t_r = 2.2R_2C_2 = \frac{2.2}{2\pi f_H} = \frac{0.35}{f_H} \tag{11-10}$$

Note that the rise time is inversely proportional to the upper 3-dB frequency. For an amplifier with 1-MHz bandpass, $t_r = 0.35\ \mu s$.

Consider a pulse of width t_p. What must be the high 3-dB frequency f_H of an amplifier used to amplify this signal without excessive distortion? A reasonable answer to this question is: *Choose f_H equal to the reciprocal of the pulse width,* $f_H = 1/t_p$. From Eq. (11-10) we then have $t_r = 0.35t_p$. Using this relationship, the (shaded) pulse in Fig. 11-5 becomes distorted into the (solid) waveform, which is clearly recognized as a pulse.

Tilt or Sag If a step of amplitude V is impressed on the high-pass circuit of Fig. 11-1, the output is

$$v_o = V\epsilon^{-t/R_1C_1} \tag{11-11}$$

For times t which are small compared with the time constant R_1C_1, the response is given by

$$v_o \approx V\left(1 - \frac{t}{R_1C_1}\right) \tag{11-12}$$

From Fig. 11-6 we see that the output is tilted, and the percent tilt, or sag, in time t_1 is given by

$$P \equiv \frac{V - V'}{V} \times 100 = \frac{t_1}{R_1C_1} \times 100\% \tag{11-13}$$

It is found that this same expression is valid for the tilt of each half cycle of a symmetrical square wave of peak-to-peak value V and period T provided

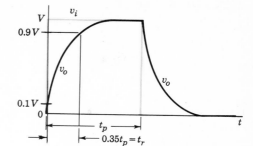

Fig. 11-5   Pulse response for the case
$f_H = 1/t_p$.

that we set $t_1 = T/2$.   If $f = 1/T$ is the frequency of the square wave, then, using Eq. (11-3), we may express P in the form

$$P = \frac{T}{2R_1C_1} \times 100 = \frac{1}{2fR_1C_1} \times 100 = \frac{\pi f_L}{f} \times 100\% \qquad (11\text{-}14)$$

Note that the tilt is directly proportional to the lower 3-dB frequency.   If we wish to pass a 50-Hz square wave with less than 10 percent sag, then f_L must not exceed 1.6 Hz.

Square-Wave Testing   An important experimental procedure (called *square-wave testing*) is to observe with an oscilloscope the output of an amplifier excited by a square-wave generator.   It is possible to improve the response of an amplifier by adding to it certain circuit elements, which then must be adjusted with precision.   It is a great convenience to be able to adjust these elements and to see simultaneously the effect of such an adjustment on the amplifier output waveform.   The alternative is to take data, after each successive adjustment, from which to plot the amplitude and phase responses. Aside from the extra time consumed in this latter procedure, we have the problem that it is usually not obvious which of the attainable amplitude and phase responses corresponds to optimum fidelity.   On the other hand, the step response gives immediately useful information.

It is possible, by judicious selection of two square-wave frequencies, to examine individually the high-frequency and low-frequency distortion.   For example, consider an amplifier which has a high-frequency time constant of

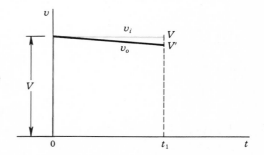

Fig. 11-6   When a step v_i is applied to a high-pass RC circuit the response v_o exhibits a tilt.

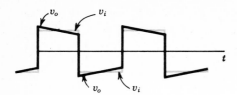

Fig. 11-7 A square-wave (shaded) input signal is distorted by an amplifier with a lower 3-dB frequency f_L. The output (solid) waveform shows a tilt where the input is horiozntal.

1 μs and a low-frequency time constant of 0.1 s. A square wave of half period equal to several microseconds, on an appropriately fast oscilloscope sweep, will display the rounding of the leading edge of the waveform and will not display the tilt. At the other extreme, a square wave of half period approximately 0.01 s on an appropriately slow sweep will display the tilt, and not the distortion of the leading edge. Such a waveform is indicated in Fig. 11-7.

11-6   BANDPASS OF CASCADED STAGES

The high 3-dB frequency for n cascaded stages is f_H^* and equals the frequency for which the overall voltage gain falls 3 dB to $1/\sqrt{2}$ of its midband value. To obtain the overall transfer function of *noninteracting* stages, the transfer gains of the individual stages are multiplied together. Hence, if there are identical stages each with a dominant pole f_H, then f_H^* can be calculated from

$$\left[\frac{1}{\sqrt{1 + (f_H^*/f_H)^2}} \right]^n = \frac{1}{\sqrt{2}}$$

to be

$$\frac{f_H^*}{f_H} = \sqrt{2^{1/n} - 1} \tag{11-15}$$

For example, for $n = 2$, $f_H^*/f_H = 0.64$. Hence two cascaded stages, each with a bandwidth $f_H = 10$ kHz, have an overall bandwidth of 6.4 kHz. Similarly, three cascaded 10-kHz stages give a resultant upper 3-dB frequency of 5.1 kHz, etc.

If the low 3-dB frequency for n identical noninteracting cascaded stages is f_L^*, then, corresponding to Eq. (11-15), we find

$$\frac{f_L^*}{f_L} = \frac{1}{\sqrt{2^{1/n} - 1}} \tag{11-16}$$

We see that a cascade of stages has a lower f_H and a higher f_L than a single stage, resulting in a shrinkage in bandwidth.

If the amplitude response for a single stage is plotted on log-log paper, the resulting graph will approach a straight line whose slope is 6 dB per octave both at the low and at the high frequencies, as indicated in Fig. 11-3. For an n-stage amplifier it follows that the amplitude response falls $6n$ dB per octave, or equivalently, $20n$ dB per decade.

Interacting Stages   If in a cascade of stages the input impedance of one stage is low enough to act as an appreciable shunt on the output impedance of the preceding stage, then it is no longer possible to isolate stages.   Under these circumstances individual 3-dB frequencies for each stage cannot be defined.   When the overall transfer function of the cascade is obtained, it is found to contain n poles (in addition to k zeros).   If the cascade has a dominant pole f_D which is much smaller than all other poles, it follows that $f_H^* \approx f_D$, or the high 3-dB frequency equals the dominant-pole frequency.   (From here on we shall drop the asterisk on f_H^*.)

Consider now the situation discussed in Sec. 11-4, where there is a dominant frequency f_D, a second pole whose frequency is only two octaves away, and all other poles are at very much higher frequencies.   Then the upper 3-dB frequency f_H is given by

$$\frac{1}{\sqrt{1 + (f_H/f_D)^2}} \frac{1}{\sqrt{1 + (f_H/4f_D)^2}} = \frac{1}{\sqrt{2}} \tag{11-17}$$

Since we expect the 3-dB frequency to be approximately equal to the dominant frequency, substitute $f_H = f_D$ into the second term in Eq. (11-17) to obtain

$$1 + \left(\frac{f_H}{f_D}\right)^2 = \frac{2}{1 + (\frac{1}{4})^2} \tag{11-18}$$

or

$$f_H = 0.94 f_D \tag{11-19}$$

This calculation verifies that the high 3-dB frequency is less than 6 percent smaller than the dominant frequency provided that the next higher pole frequency is at least two octaves away.

If the pole frequencies are not widely separated, it is found (Prob. 11-11) that f_H (with $f_1 = f_{p1}$, $f_2 = f_{p2}$, etc.) is given (to within 10 percent) by

$$\frac{1}{f_H} = 1.1 \sqrt{\frac{1}{f_1^2} + \frac{1}{f_2^2} + \cdots + \frac{1}{f_n^2}} \tag{11-20}$$

If this equation is applied to the case considered above, $f_1 = f_D$ and $f_2 = 4f_D$ and all other poles much higher, the result is $f_H = 0.89f_D$, in close agreement with Eq. (11-19).   If Eq. (11-20) is applied to the case where $f_1 = f_2$ and all other poles are much higher, then $f_H = 0.65f_1$ (instead of the exact value of $0.64f_1$).   For three equal poles, Eq. (11-20) yields $f_H = 0.53f_1$ (instead of the exact value of $0.51f_1$).

Step Response   If the rise time of isolated individual cascaded stages is $t_{r1}, t_{r2}, \ldots, t_{rn}$ and if the input waveform rise time is t_{ro}, then, corresponding to Eq. (11-20) for the resultant upper 3-dB frequency, we have that the output-signal rise time t_r is given (to within 10 percent) by

$$t_r = 1.1 \sqrt{t_{ro}^2 + t_{r1}^2 + t_{r2}^2 + \cdots + t_{rn}^2} \tag{11-21}$$

If, upon application of a voltage step, one circuit produces a tilt of P_1 percent and if a second stage gives a tilt of P_2 percent, the effect of cascading these two noninteracting circuits is to produce a tilt of $P_1 + P_2$ percent. This result applies only if the individual tilts and the combined tilt are small enough so that in each case the waveform falls approximately linearly with time.

11-7   THE RC–COUPLED AMPLIFIER

A cascaded arrangement of common-emitter (CE) transistor stages is shown in Fig. 11-8. The output Y_1 of one stage is coupled to the input X_2 of the next stage via a blocking capacitor C_b which is used to keep the dc component of the output voltage at Y_1 from reaching the input X_2. The collector circuit resistor is R_c, the emitter resistor R_e, and the resistors R_1 and R_2 are used to establish the bias. The bypass capacitor, used to prevent loss of amplification due to negative feedback (Chap. 10), is C_z in the emitter circuit. Also present are junction capacitances, to be taken into account when we consider the high-frequency response, which is limited by their presence. In any practical mechanical arrangement of the amplifier components, there are also capacitances associated with device sockets and the proximity to the chassis of components (for example, the body of C_b) and signal leads. These stray capacitances are also considered later. We assume that the active device operates linearly, so that small-signal models are used throughout this chapter.

Low-Frequency Response of an RC-coupled Stage The effect of the bypass capacitor C_z on the low-frequency response is discussed in the next section. For the present we assume that this capacitance is arbitrarily large and acts as an ac (incremental) short circuit across R_e. The low-frequency equivalent circuit is obtained by neglecting all shunt and all junction capacitances. A single intermediate stage in the cascade of Fig. 11-8 is represented

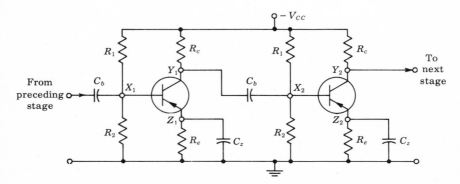

Fig. 11-8   A cascade of common-emitter (CE) transistor stages.

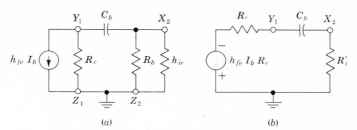

Fig. 11-9 (a) The low-frequency model of an RC-coupled bipolar transistor amplifier ($R_b \equiv R_1 \| R_2$). (b) An equivalent representation ($R'_i \equiv R_b \| h_{ie}$).

schematically in Fig. 11-9. Each transistor is replaced by its approximate h-parameter model of Fig. 10-5. The resistance R_b represents R_1 in parallel with R_2. The input resistance of a stage is $R_i = h_{ie}$. Replacing the current generator $h_{fe}I_b$ shunted by R_c by its Thévenin's equivalent of a voltage generator $h_{fe}I_bR_c$ in series with R_c, we obtain the circuit of Fig. 11-9b. In this figure R'_i represents h_{ie} in parallel with R_b. Note that Fig. 11-9b is a single-time-constant high-pass circuit. Hence, from Eq. (11-3) the low 3-dB frequency is

$$f_L = \frac{1}{2\pi(R_c + R'_i)C_b} \tag{11-22}$$

This result is easy to remember since the time constant equals C_b multiplied by the sum of the effective resistances R_c to the left of the blocking capacitor and R'_i to the right of C_b.

For an FET amplifier an analysis similar to that given above yields a low 3-dB frequency given by Eq. (11-22), with the collector resistor replaced by the FET output resistance R'_o (the drain resistor R_d in parallel with r_d) and with R'_i equal to the gate-to-ground resistance R_g.

$$f_L = \frac{1}{2\pi(R'_o + R_g)C_b} \tag{11-23}$$

EXAMPLE (a) It is desired to have a low 3-dB frequency of not more that 10 Hz for an RC-coupled FET amplifier for which $R'_o = 1$ K and $R_g = 1$ M. What minimum value of coupling capacitance is required? (b) Repeat part (a) if bipolar junction transistors are used with $R_c = 1$ K and $h_{ie} = 1$ K. Since R_1 and R_2 are much larger than h_{ie}, $R'_i \approx h_{ie} = 1$ K.

Solution   *a.* From Eq. (11-23) we have

$$f_L = \frac{1}{2\pi(10^3 + 10^6)C_b} \leq 10$$

or

$$C_b \leq \frac{1}{62.8 \times 10^6} \text{ F} = 0.016 \ \mu\text{F}$$

b. From Eq. (11-22)

$$C_b \geq \frac{1}{62.8(10^3 + 10^3)} \quad F = 8.0 \ \mu F$$

Note that, because the input resistance of a transistor is much smaller than that of an FET, a coupling capacitor is required with the transistor which is 500 times larger than that required with the FET. Fortunately, it is possible to obtain physically small electrolytic capacitors having such high capacitance values at the low voltages at which transistors operate. Since the coupling capacitances required for good low-frequency response are far larger than those obtainable in integrated form, *cascaded integrated stages must be direct-coupled.*

11-8 EFFECT OF AN EMITTER BYPASS CAPACITOR ON LOW–FREQUENCY RESPONSE

If an emitter resistor R_e is used for self bias in an amplifier and if it is desired to avoid the degeneration, and hence the loss of gain due to R_e, we might attempt to bypass this resistor with a very large capacitance C_z. The circuit is indicated in Fig. 11-8. It is shown below that the effect of this capacitor is to affect adversely the low-frequency response.

At high frequencies the reactance of C_z is very small and hence acts as a virtual short circuit across R_e. Under these circumstances the gain A_V is high and equal to the value for a CE amplifier given in Eq. (10-16). At zero frequency C_z is effectively out of the circuit and the gain is low, given by Eq. (10-31) for a CE amplifier with an emitter resistor R_e. Hence, A_V drops as f falls from its midband region (high frequency) toward zero frequency. If the drop in voltage exceeds 3 dB, then a low 3-dB frequency exists.

Consider the single stage of Fig. 11-10a. To simplify the analysis we assume that $R_1 \| R_2 \gg R_s$ and that the load R_c is small enough so that the simplified hybrid model of Fig. 10-5 is valid. The equivalent circuit subject to these assumptions is shown in Fig. 11-10b. The blocking capacitor C_b is omitted from Fig. 11-10b; its effect is considered in Sec. 11-7.

The output voltage V_o is given by

$$V_o = -I_b h_{fe} R_c = -\frac{V_s h_{fe} R_c}{R_s + h_{ie} + Z_e'} \tag{11-24}$$

where

$$Z_e' \equiv (1 + h_{fe}) \frac{R_e}{1 + j\omega C_z R_e} \tag{11-25}$$

Substituting Eq. (11-25) in Eq. (11-24) and solving for the voltage gain A_V, we find that $A_V = V_o/V_s$ has one pole and one zero whose frequencies f_o and f_p,

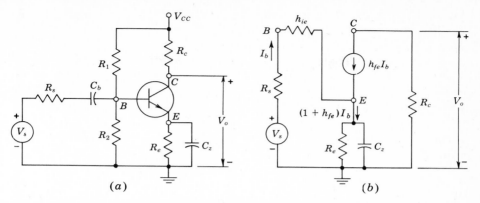

Fig. 11-10 (a) An amplifier with a bypassed emitter resistor; (b) the low-frequency simplified h-parameter model of the circuit in (a).

respectively, are given (Prob. 11-19) by

$$f_o \equiv \frac{1}{2\pi C_z R_e} \qquad f_p \equiv \frac{1 + R'/R}{2\pi C_z R_e} \qquad (11\text{-}26)$$

where

$$R \equiv R_s + h_{ie} \qquad \text{and} \qquad R' \equiv (1 + h_{fe})R_e \qquad (11\text{-}27)$$

Note that f_o determines the zero and f_p the pole of the gain A_V. Since, usually, $R'/R \gg 1$, then $f_p \gg f_o$, so that the pole and zero are widely separated. For example, assuming $R_s = 0$, $R_e = 1$ K, $C_z = 100$ μF, $h_{fe} = 50$, $h_{ie} = 1.1$ K, and $R_c = 2$ K, we find $f_o = 1.6$ Hz and $f_p = 76$ Hz.

Square-Wave Response If $f_p \gg f_o$, the network in Fig. 11-10 behaves like a single-time-constant circuit whose low 3-dB frequency is f_p. The percentage tilt to a square wave is given by Eq. (11-14), or

$$P = \frac{\pi f_p}{f} \times 100 = \frac{1 + R'/R}{2fC_z R_e} \times 100\% \qquad (11\text{-}28)$$

Since $R'/R \gg 1$,

$$P \approx \frac{R' \times 100}{2fC_z R R_e} = \frac{1 + h_{fe}}{2fC_z(R_s + h_{ie})} \times 100\% \qquad (11\text{-}29)$$

Practical Considerations Let us calculate the size of C_z so that we may reproduce a 50-Hz square wave with a tilt of less than 10 percent. Using the parameters given above, we obtain

$$C_z = \frac{(51)(100)}{(2)(50)(1,100)(10)} \text{ F} = 4,600 \text{ }\mu\text{F}$$

Such a large value of capacitance is impractical, and it must be concluded that if very small tilts are to be obtained for very low frequency signals, the emitter resistor must be left unbypassed. The flatness will then be obtained at the sacrifice of gain because of the degeneration caused by R_e. If the loss in amplification cannot be tolerated, R_e cannot be used. For an IC amplifier the emitter is grounded ($R_e = 0$), and bias is obtained by placing a diode between base and emitter, as indicated in Fig. 10-22.

Response Due to Both Emitter and Coupling Capacitors If in a given stage both C_z and the coupling capacitor C_b are present, we can assume, first, C_z to be infinite and compute the low 3-dB frequency due to C_b alone. We then calculate f_L due to C_z by assuming C_b to be infinite. If the two cutoff frequencies are significantly different (by two octaves or more), the higher of the two is approximately the low 3-dB frequency for the stage.

If a dominant pole does not exist at low frequencies, the response must be obtained by writing the network equations for a sinusoidal excitation ($s = j2\pi f$) and with both C_z and C_b finite. The transfer function V_o/V_s is plotted as a function of frequency, and the low 3-dB frequency is read from this plot.

The low-frequency analysis of an FET amplifier with a source resistor R_s bypassed with a capacitor C_s is considered in Prob. 11-22.

In the remainder of this chapter we consider the transistor at high frequencies. First a model for the transistor is found and the high-frequency response of a single stage is obtained. Then the high-frequency behavior of cascaded stages is studied in Sec. 11-13.

11-9 THE HYBRID–π COMMON–EMITTER TRANSISTOR MODEL

In Chap. 10 it is emphasized that the common-emitter circuit is the most important practical configuration. Hence we now seek a CE model which will be valid at high frequencies. Such a circuit, called *the hybrid-π, or Giacoletto, model*, is indicated in Fig. 11-11. Analyses of circuits using this model are not too difficult and give results which are in excellent agreement with experiment at all frequencies for which the transistor gives reasonable amplification. Furthermore, the resistive components in this circuit can be obtained from the low-frequency h parameters. All parameters (resistances and capacitances) in the model are assumed to be independent of frequency. They may vary with the quiescent operating point, but under given bias conditions are reasonably constant for small-signal swings.

Discussion of Circuit Components The internal node B' is not physically accessible. The ohmic base-spreading resistance $r_{bb'}$ is represented as a lumped parameter between the external base terminal and B' (Sec. 4-8).

For small changes in the voltage $V_{b'e}$ across the emitter junction, the

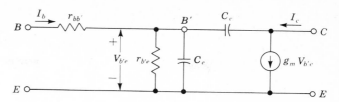

Fig. 11-11 The hybrid-π model for a transistor in the
CE configuration.

excess-minority-carrier concentration injected into the base is proportional to
$V_{b'e}$, and therefore the resulting small-signal collector current, with the collector shorted to the emitter, is proportional to $V_{b'e}$. This effect accounts for
the current generator $g_m V_{b'e}$ in Fig. 11-11 where g_m is the transconductance.

The increase in minority carriers in the base results in increased recombination base current, and this effect is taken into account by inserting a
resistance $r_{b'e}$ between B' and E. The excess-minority-carrier storage in
the base is accounted for by the diffusion capacitance C_e connected between
B' and E. The collector-junction barrier capacitance C_c is included between
B' and C.

In Fig. 11-11 we have omitted the feedback resistance $r_{b'c}$ because its
value ($\sim$4 M) is much larger than the other resistors in the circuit. An output resistance r_{ce} should also be included between the output terminals to
account for the fact the transistor output characteristics (Fig. 4-9) are not
horizontal. However, r_{ce} ($\sim$80 K) is in parallel with the load R_L across the
output. Since usually $r_{ce} \gg R_L$, very little error is introduced by omitting r_{ce}
from Fig. 11-11.

The Hybrid-π Capacitances The hybrid-π model for a transistor as
shown in Fig. 11-11 includes two capacitances. The collector-junction capacitance $C_c = C_{b'c}$ is the measured CB output capacitance with the input open
($I_E = 0$), and is usually specified by manufacturers as C_{ob}. Since in the active
region the collector junction is reverse-biased, then C_c is the transition capaci-
tance and varies as $V_{CB}{}^{-n}$, where n is $\frac{1}{2}$ or $\frac{1}{3}$ for an abrupt or graded junction,
respectively (Sec. 2-6).

The capacitance C_e represents the sum of the emitter-diffusion capacitance
C_{De} and the emitter-junction capacitance C_{Te}. For a forward-biased emitter
junction, C_{De} is usually much larger than C_{Te}, and hence

$$C_e = C_{De} + C_{Te} \approx C_{Dc} \tag{11-30}$$

Using the minority-carrier distribution of base charge of Fig. 4-3c, the following equation for C_{De} is obtained:

$$C_{De} = \frac{W_B{}^2 I_E}{2 D_B V_T} = g_m \frac{W_B{}^2}{2 D_B} \approx C_e \tag{11-31}$$

where W_B is the base width, D_B is the diffusion constant for the minority carriers in the base, and g_m is given by Eq. (11-37). Note that Eq. (11-31) indicates that the *diffusion capacitance is proportional to the emitter bias current* I_E.

Experimentally, C_e is determined from a measurement of f_T, *the frequency at which the* CE *short-circuit current gain drops to unity.* We verify in Sec. 11-11 that

$$C_e \approx \frac{g_m}{2\pi f_T} \tag{11-32}$$

The emitter-diffusion capacitance is much larger than the collector-transition capacitance, $C_e \gg C_c$. For example, a transistor with $C_e \approx 100\ \text{pF}$ would have $C_c \approx 3\ \text{pF}$.

The hybrid-π model is found experimentally to be valid for frequencies up to approximately $f_T/3$.

The high-frequency response of a transistor amplifier depends upon the parameters in the hybrid-π model, namely, g_m, $r_{b'e}$, $r_{bb'}$, C_e, and C_c. From the equations in this and the next section it follows that the high-frequency response can be determined from the four parameters h_{fe}, h_{ie}, f_T, and C_c.

11-10   HYBRID–π PARAMETERS

We now demonstrate that the resistive components in the hybrid-π model can be obtained from the h parameters in the CE configuration. At low frequencies C_e and C_c can be neglected, and the hybrid-π model valid at low frequencies is indicated in Fig. 11-12a. Comparing this network with the h-parameter model of Fig. 11-12b, we obtain from the output circuit

$$g_m V_{b'e} = h_{fe} I_b \tag{11-33}$$

Since $V_{b'e} = I_b r_{b'e}$,

$$h_{fe} = g_m r_{b'e} \qquad \text{or} \qquad r_{b'e} = \frac{h_{fe}}{g_m} \tag{11-34}$$

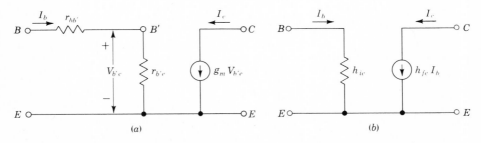

Fig. 11-12   (a) The hybrid-π model at low frequencies. (b) The h-parameter model at low frequencies.

Comparing the input circuits in Fig. 11-12a and b we find that

$$h_{ie} = r_{bb'} + r_{b'e} \qquad \text{or} \qquad r_{bb'} = h_{ie} - r_{b'e} \tag{11-35}$$

Since $I_c = g_m V_{b'e}$ and $|I_c| \approx |I_e|$

$$|g_m| \approx \frac{|I_e|}{|V_{b'e}|} = \frac{1}{r_e} \tag{11-36}$$

where the incremental emitter-junction diode resistance is $r_e = V_T/I_E$ from Eq. (2-7).† Hence,

$$|g_m| \approx \frac{|I_E|}{V_T} \approx \frac{|I_C|}{V_T} \tag{11-37}$$

where $|I_E|$ is the quiescent emitter current and where, from Eq. (2-3), $V_T = T/11,600$. We see that g_m *is directly proportional to current and inversely proportional to temperature.* Also note that *the transconductance is independent of the geometry* (the method of fabrication or the dimensions) *of the transistor.* At room temperature

$$g_m = 38.5 \, |I_C| \tag{11-38}$$

For $I_C = 1.3$ mA, $g_m = 38.5 \times 1.3 = 50$ mA/V. For $I_C = 10$ mA, $g_m = 385$ mA/V. These values are much larger than the transconductances obtained with an FET (Table 10-4).

Assume that the CE h parameters h_{ie} and h_{fe} at low frequencies are known. First g_m is calculated at the quiescent collector current I_C from Eq. (11-37). Then $r_{b'e}$ is found from Eq. (11-34) and $r_{bb'}$ from Eq. (11-35). For $I_C = 1.3$ mA, $h_{fe} = 50$, and $h_{ie} = 1,100 \, \Omega$, we find

$$g_m = 50 \text{ mA/V} \qquad r_{b'e} = 1,000 \, \Omega \qquad r_{bb'} = 100 \, \Omega$$

11-11   THE CE SHORT–CIRCUIT CURRENT GAIN

Consider a single-stage CE transistor amplifier, or the last stage of a cascade. The load R_L on this stage is the collector-circuit resistor, so that $R_c = R_L$. In this section we assume that $R_L = 0$, whereas the circuit with $R_L \neq 0$ is analyzed in the next section. To obtain the *frequency response* of the transistor amplifier, we use the hybrid-π model of Fig. 11-11.

The approximate equivalent circuit from which to calculate the short-circuit current gain is shown in Fig. 11-13. A current source furnishes a sinusoidal input current of magnitude I_i, and the load current is I_L. An approximation is involved, in that we have neglected the current delivered directly to the output through C_c. We see shortly that this approximation is justified.

---

† Since the recombination current in the emitter space-charge region does not reach the collector, the factor η in Eq. (2-7) is taken as unity in the calculation of g_m.

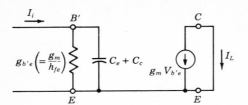

Fig. 11-13 Approximate equivalent circuit for the calculation of the short-circuit CE current gain. Note that $g_{b'e} = 1/r_{b'e}$.

The load current is $I_L = -g_m V_{b'e}$, where

$$V_{b'e} = \frac{I_i}{g_{b'e} + j\omega(C_e + C_c)} \tag{11-39}$$

The current amplification under short-circuited conditions is

$$A_i = \frac{I_L}{I_i} = \frac{-g_m}{g_{b'e} + j\omega(C_e + C_c)} \tag{11-40}$$

Using the results given in Eq. (11-34), we have

$$A_i = \frac{-h_{fe}}{1 + j(f/f_\beta)} \qquad |A_i| = \frac{h_{fe}}{[1 + (f/f_\beta)^2]^{\frac{1}{2}}} \tag{11-41}$$

where

$$f_\beta = \frac{g_{b'e}}{2\pi(C_e + C_c)} = \frac{1}{h_{fe}} \frac{g_m}{2\pi(C_e + C_c)} \tag{11-42}$$

At $f = f_\beta$, $|A_i|$ is equal to $1/\sqrt{2} = 0.707$ of its low-frequency value h_{fe}. The frequency range up to f_β is referred to as the *bandwidth* of the circuit. Note that the value of A_i at $\omega = 0$ is $-h_{fe}$, in agreement with the definition of $-h_{fe}$ as the low-frequency short-circuit CE current gain.

The Parameter f_T   We now introduce f_T, which is defined as the *frequency at which the short-circuit common-emitter current gain attains unit magnitude*. Since $h_{fe} \gg 1$, we have, from Eqs. (11-41) and (11-42), that f_T is given by

$$f_T \approx h_{fe}f_\beta = \frac{g_m}{2\pi(C_e + C_c)} \approx \frac{g_m}{2\pi C_e} \tag{11-43}$$

since $C_e \gg C_c$.   Hence, from Eq. (11-41),

$$A_i \approx \frac{-h_{fe}}{1 + jh_{fe}(f/f_T)} \tag{11-44}$$

The parameter f_T is an important high-frequency characteristic of a transistor. Like other transistor parameters, its value depends on the operating conditions of the device.

Since $f_T \approx h_{fe}f_\beta$, this parameter may be given a second interpretation. It represents the *short-circuit current-gain–bandwidth product;* that is, for the CE configuration with the output shorted, f_T is the product of the low-frequency

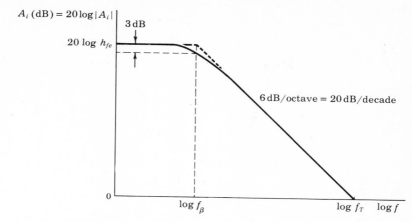

Fig. 11-14 The short-circuit CE current gain vs. frequency (plotted on a log-log scale).

current gain and the upper 3-dB frequency. For our typical transistor (Sec. 11-10), $f_T = 80$ MHz and $f_\beta = 1.6$ MHz. It is to be noted from Eq. (11-43) that there is a sense in which gain may be sacrificed for bandwidth, and vice versa. Thus, if two transistors are available with equal f_T, the transistor with lower h_{fe} will have a correspondingly larger bandwidth.

In Fig. 11-14, A_i expressed in decibels (i.e., $20 \log |A_i|$) is plotted against frequency on a logarithmic frequency scale. When $f \ll f_\beta$, $A_i \approx -h_{fe}$, and A_i (dB) approaches asymptotically the horizontal line A_i (dB) $= 20 \log h_{fe}$. When $f \gg f_\beta$, $|A_i| \approx h_{fe}f_\beta/f = f_T/f$, so that A_i (dB) $= 20 \log f_T - 20 \log f$. Accordingly, A_i (dB) $= 0$ dB at $f = f_T$. And for $f \gg f_\beta$, the plot approaches as an asymptote a straight line passing through the point $(f_T, 0)$ and having a slope which causes a decrease in A_i (dB) of 6 dB per octave (f is multiplied by a factor of 2, and $20 \log 2 = 6$ dB), or 20 dB per decade. The intersection of the two asymptotes occurs at the "corner" frequency $f = f_\beta$, where A_i is down by 3 dB. Hence f_β is also called the 3-dB frequency.

Earlier, we neglected the current delivered directly to the output through C_c. Now we may see that this approximation is justified. The magnitude of this current is $\omega C_c V_{b'e}$, whereas the current due to the controlled generator is $g_m V_{b'e}$. The ratio of currents is $\omega C_c/g_m$. At the highest frequency of interest f_T, we have, from Eq. (11-43),

$$\frac{\omega C_c}{g_m} = \frac{2\pi f_T C_c}{g_m} = \frac{C_c}{C_e + C_c} \ll 1$$

Measurement of f_T The frequency f_T is often inconveniently high to allow a direct experimental determination of f_T. However, a procedure is available which allows a measurement of f_T at an appreciably lower frequency. We note from Eq. (11-41) that, for $f \gg f_\beta$, we may neglect the unity in the

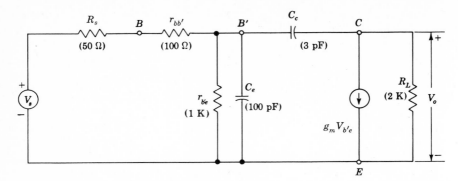

Fig. 11-15 Equivalent circuit for the frequency analysis of a CE amplifier stage driven by a voltage source.

denominator and write $|A_i|f \approx h_{fe}f_\beta = f_T$ from Eq. (11-43). Accordingly, at some particular frequency f_1 (say, f_1 is five or ten times f_β), we measure the gain $|A_{i1}|$. The parameter f_T may be calculated now from $f_T = f_1|A_{i1}|$. In the case of our typical transistor, for which $f_T = 80$ MHz and $f_\beta = 1.6$ MHz, the frequency f_1 may be $f_1 = 5 \times 1.6 = 8.0$ MHz, a much more convenient frequency than 80 MHz.

The experimentally determined value of f_T is used to calculate the value of C_e in the hybrid-π circuit [Eq. (11-43)].

11-12   SINGLE–STAGE CE TRANSISTOR AMPLIFIER RESPONSE

In the preceding section we assume that the transistor is driven from an ideal current source, that is, a source of infinite resistance. We now remove this restriction and consider that the source V_s has a finite resistance R_s. The equivalent circuit from which to obtain the response is shown in Fig. 11-15. This circuit is obtained from Fig. 11-11 by adding V_s in series with R_s between B and E and by placing R_L between C and E.

The Transfer Function   We wish to calculate the transfer function V_o/V_s as a function of the complex-frequency variable s. Introducing $R'_s \equiv R_s + r_{bb'} = 1/G'_s$ and noting that the admittance of a capacitor C is sC, we obtain in Sec. C-4 the following KCL equations at nodes B' and C, respectively (the so-called nodal equations):

$$G'_s V_s = [G'_s + g_{b'e} + s(C_e + C_c)]V_{b'e} - sC_c V_o \qquad (11\text{-}45)$$

$$0 = (g_m - sC_c)V_{b'e} + \left(\frac{1}{R_L} + sC_c\right)V_o \qquad (11\text{-}46)$$

Solving Eqs. (11-45) and (11-46), we find that the transfer function V_o/V_s has one zero s_0 and two poles s_1 and s_2. For the numerical values indicated in

Fig. 11-15 and with $g_m = 50$ mA/V, we find (expressed in radians per second)

$$s_0 = 1.67 \times 10^9 \qquad s_1 = -1.75 \times 10^7 \qquad s_2 = -7.30 \times 10^8$$

From a plot of the magnitude of the transfer function versus frequency we find $f_H = 2.8$ MHz.

Since $s_2/s_1 = 73.0/1.75 = 41.7 \gg 4$, s_1 is a dominant pole (Sec. 11-4). The dominant-pole frequency is

$$f_D = \frac{|s_1|}{2\pi} = \frac{1.75 \times 10^7}{2\pi} \ \text{Hz} = 2.78 \ \text{MHz}$$

in agreement with the value of f_H found graphically.

Approximate Analysis We can obtain a very simple approximate expression for the transfer function by applying Miller's theorem to the circuit of Fig. 11-15. Proceeding as in Sec. C-4, we obtain the circuit of Fig. 11-16, with $K \equiv V_{ce}/V_{b'e}$. Since $|K| \gg 1$, the output capacitance is C_c and the output time constant is $C_c R_L = 6$ ns. Neglecting C_c, it follows that $K = -g_m R_L$ and the input capacitance is

$$C \equiv C_e + C_c(1 + g_m R_L) = 100 + (3)(101) = 403 \ \text{pF} \qquad (11\text{-}47)$$

The input-loop resistance R which is in parallel with C is $R_s' \| r_{b'e}$, so that

$$R = \frac{150 \times 1,000}{150 + 1,000} = 130 \ \Omega$$

and the input time constant is $RC = (130)(403)$ ps $= 53$ ns. Since this time constant is almost nine times as large as the output time constant $C_c R_L = 6$ ns, the amplifier has a dominant pole the frequency of which is determined by the larger (input) time constant. Hence, the high 3-dB frequency is approximately

$$f_H = \frac{1}{2\pi RC} = \frac{1}{2\pi \times 53 \times 10^{-9}} \ \text{Hz} = 3.00 \ \text{MHz}$$

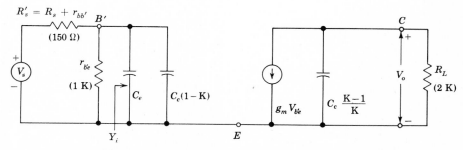

Fig. 11-16 The equivalent circuit of the CE amplifier stage, using the Miller effect.

Note that the use of Miller's theorem greatly simplifies the analysis and yields a value of f_H which is in error by only about 7 percent.

11-13   HIGH–FREQUENCY RESPONSE OF TWO CASCADED CE TRANSISTOR STAGES

Since there is interaction between CE cascaded transistor stages, the analysis of a multistage amplifier is complicated and tedious. Fortunately, it is possible to make certain approximations in the analysis, and thus reduce the complexity of bandwidth calculations while keeping the error under approximately 20 percent.

Figure 11-8 shows two CE transistors in cascade. For high-frequency calculations each transistor is replaced by its small-signal hybrid-π model, as indicated in Fig. 11-17. We have included a voltage source V_s with $R_s = 50\ \Omega$ and have assumed that capacitors C_b and C_z represent short circuits for high frequencies. The base biasing resistors R_1 and R_2 in Fig. 11-8 are assumed to be large compared with R_s. The symbol R_{L1} represents the parallel combination of R_1, R_2, and collector circuit resistance R_c of the first stage. The symbol R_{L2} is the total load resistance of the second stage. A complete stage is included in each shaded block.

To describe completely, this network requires four (nodal) equations {instead of only two [Eqs. (11-45) and (11-46)] for a single stage}. The transfer function V_4/V_s has four poles and two zeros. So much effort is required to find these poles and zeros that it is advisable to use a computer. Several computer programs, such as CORNAP, are available for obtaining the frequency response of multistage amplifiers. For the numerical values given in Fig. 11-17 the frequency-response curve yields a high 3-dB frequency $f_H = 540$ kHz. For the single-stage amplifier in Sec. 11-12, having the same parameters, $f_H = 2.8$ MHz. Note that two cascaded (interacting) stages result in a bandwidth only about 20 percent that of a single stage.

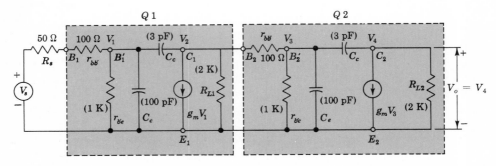

Fig. 11-17   Two-stage interacting CE amplifier ($g_m = 50$ mA/V).

Two-Stage-Cascade Simplified Analysis   If a computer is not available to help with the computations, we must make simplified assumptions in order to proceed with the analysis. We follow the method outlined in Sec. 11-12.

The effect of C_c is approximated using Miller's theorem and the midband value of the stage gain. Thus C_c of Q_2 is replaced by a capacitance

$$C_c(1 + g_m R_{L2}) = 3(1 + 50 \times 2) = 303 \text{ pF}$$

across the input of $Q2$. Similarly, C_c of $Q1$ is replaced by

$$C_c[1 + g_m R_{L1} \| (r_{b'e} + r_{bb'})] = 3(1 + 50 \times 0.709) = 109 \text{ pF}$$

across the input of $Q1$. The circuit is now considerably simplified since, there are only two independent capacitors in the network. Hence, the simplified network has only two poles, which can be readily calculated. Furthermore, it turns out that one of these is a dominant pole, the value of what is **585** kHz, which is higher than **540** kHz obtained by using the exact method, by 8 percent.

11-14   THE COMMON–SOURCE AMPLIFIER AT HIGH FREQUENCIES

The small-signal high-frequency model of an FET is obtained from the low-frequency model of Fig. 10-15 by adding three capacitors: C_{gs} between gate and source, C_{gd} between gate and drain, and C_{ds} between drain and source. Hence, the CS amplifier of Fig. 11-18a has the small-signal high-frequency equivalent circuit indicated in Fig. 11-18b. The output voltage V_o between D and S is easily found with the aid of the theorem in Sec. C-2, namely, $V_o = IZ$, where I is the short-circuit current and Z is the impedance seen

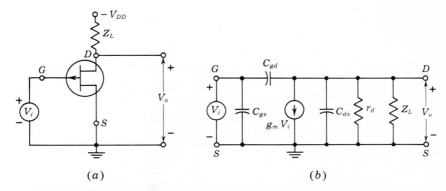

Fig. 11-18   (a) The common-source amplifier circuit; (b) small-signal equivalent circuit at high frequencies. (The biasing network is not indicated.)

between the terminals.   To find Z, the independent generator V_i is (imagined) short-circuited, so that $V_i = 0$, and hence there is no current in the dependent generator $g_m V_i$.   We then note that Z is the parallel combination of the impedances corresponding to Z_L, r_d, C_{ds}, and C_{gd}.   Hence

$$Y = \frac{1}{Z} = Y_L + g_d + Y_{ds} + Y_{gd} \tag{11-48}$$

where $Y_L = 1/Z_L = $ admittance corresponding to Z_L
$\quad\ g_d = 1/r_d = $ conductance corresponding to r_d
$\quad Y_{ds} = j\omega C_{ds} = $ admittance corresponding to C_{ds}
$\quad Y_{gd} = j\omega C_{gd} = $ admittance corresponding to C_{gd}

The current in the direction from D to S in a zero-resistance wire connecting the output terminals is

$$I = -g_m V_i + V_i Y_{gd} \tag{11-49}$$

The amplification A_V with the load Z_L in place is given by

$$A_V = \frac{V_o}{V_i} = \frac{IZ}{V_i} = \frac{I}{V_i Y} \tag{11-50}$$

or from Eqs. (11-48) and (11-49)

$$A_V = \frac{-g_m + Y_{gd}}{Y_L + g_d + Y_{ds} + Y_{gd}} \tag{11-51}$$

At low frequencies the FET capacitances can be neglected and hence

$$Y_{ds} = Y_{gd} = 0$$

Under these conditions Eq. (11-51) reduces to

$$A_V = \frac{-g_m}{Y_L + g_d} = \frac{-g_m r_d Z_L}{r_d + Z_L} = -g_m Z_L' \tag{11-52}$$

where $Z_L' \equiv Z_L \| r_d$.   This equation agrees with Eq. (10-47), with Z_L replaced by R_d.

Input Capacitance (Miller Effect)   Consider an FET with a drain-circuit resistance R_d.   From the previous discussion it follows that within the audio-frequency range, the gain is given by the expression $A_V = -g_m R_d' \equiv K$ where R_d' is $R_d \| r_d$.   From Sec. C-4, C_i is increased by $(1 - K)C_{gd}$, or

$$C_i = C_{gs} + (1 + g_m R_d')C_{gd} \tag{11-53}$$

This increase in input capacitance C_i over the capacitance from gate to source is called the *Miller effect*.

The input capacitance is important in the operation of cascaded amplifiers.   In such a system the output from one stage is used as the input to a second amplifier. Hence the input impedance of the second stage acts as a shunt across the output of the first stage and R_d is shunted by the capacitance C_i. Since the reactance of a capacitor decreases with increasing frequencies, the resultant output impedance of the first stage will be correspondingly low for the high frequencies. This will result in a decreasing gain at the higher frequencies.

REVIEW QUESTIONS

11-1 Define the following types of distortion: (a) *nonlinear;* (b) *frequency;* (c) *phase-shift distortion.*

11-2 Under what conditions does an amplifier preserve the form of the input signal?

11-3 Define the *frequency-response magnitude characteristic* of an amplifier.

11-4 Define *dominant pole.*

11-5 (a) Sketch the high-frequency *step response* of a low-pass single-pole amplifier. (b) Define the *rise time t_r*. (c) What is the relationship between t_r and the high 3-dB frequency f_H?

11-6 (a) The input to a low-pass amplifier is a pulse of width t_p. Sketch the output waveshape. (b) What must be the relationship between t_p and f_H in order to amplify the pulse without excessive distortion?

11-7 (a) Sketch the response of an amplifier to a low-frequency square wave. (b) Define *tilt.* (c) How is the tilt related to the low 3-dB frequency f_L?

11-8 Derive the expression for the high 3-dB frequency f_H^* of n identical noninteracting stages in terms of f_H for one stage.

11-9 (a) Is f_H^* for two stages greater or smaller than f_H for a single stage? Explain. (b) Repeat for f_L^* versus f_L.

11-10 (a) Give an approximate expression relating f_H^* and the 3-dB frequencies of n nonidentical stages. (b) For two identical stages, what is f_H^*/f_H? Repeat for three stages.

11-11 Give an approximate relationship between the output rise time t_r, the rise time t_{ro} of an input signal, and the rise times of n nonidentical stages.

11-12 (a) Sketch two *RC*-coupled CE transistor stages. (b) Show the low-frequency model for one stage. (c) What is the expression for f_L?

11-13 Repeat Rev. 11-12 for CS JFET stages.

11-14 Draw the small-signal high-frequency CE model of a transistor.

11-15 (a) What is the physical origin of the two capacitors in the hybrid-π model? (b) What is the order of magnitude of each capacitance?

11-16 What is the order of magnitude of each resistance in the hybrid-π model?

11-17 How does g_m vary with (a) $|I_C|$; (b) T?

11-18 Prove that $h_{fe} = g_m r_{b'e}$.

11-19 (a) Prove that $h_{ie} = r_{bb} + r_{b'e}$. (b) Assuming $r_{bb'} \ll r_{b'e}$, how does h_{ie} vary with $|I_C|$?

11-20 Derive the expression for the CE short-circuit current gain A_i as a function of frequency.

11-21 (a) Define f_β. (b) Define f_T. (c) What is the relationship between f_β and f_T?

11-22   Consider a CE stage with a resistive load R_L.   (a) Using Miller's theorem, what is the midband input capacitance?   (b) Assuming the output time constant is small compared with the input time constant, what is the high 3-dB frequency f_H for the current gain?

11-23   Explain why the 3-dB frequency for current gain is not the same as f_H for voltage gain.

11-24   (a) Outline the general method for obtaining the high-frequency response of two interacting transistor amplifier stages.   (b) Outline an approximate method of solution.

11-25   Draw the high-frequency small-signal model of an FET.

11-26   Obtain the effective input capacitance of a CS amplifier.

12 / FEEDBACK AMPLIFIERS

In this chapter we introduce the concept of feedback and show how to modify the characteristics of an amplifier by combining a portion of the output signal with the external input signal. Many advantages are to be gained from the application of negative (degenerative) feedback, and these are studied. Examples of feedback amplifier circuits at low frequencies are given.

Instability and compensation of feedback amplifiers are discussed. Sinusoidal oscillators are also considered.

12-1   THE FEEDBACK CONCEPT

In any amplifier circuit we may sample the output voltage or current by means of a suitable sampling network and apply this signal to the input through a feedback two-port network, as shown in Fig. 12-1. At the input the feedback signal is combined with the external (source) signal through a mixer network and is fed into the amplifier proper.

Signal Source   This block in Fig. 12-1 is either a signal voltage V_s in series with a resistor R_s (a Thévenin's representation as in Fig. C-1c) or a signal current I_s in parallel with a resistor R_s (a Norton's representation as in Fig. C-1d).

Feedback Network   This block in Fig. 12-1 is usually a passive two-port network which may contain resistors, capacitors, and inductors. Most often it is simply a resistive configuration.

Sampling Network   Two sampling blocks are shown in Fig. 12-2. In Fig. 12-2a the output voltage is sampled by connecting the feedback network *in shunt* across the output. This type of connection is referred to as *voltage*, or *node, sampling*. Another feedback connection

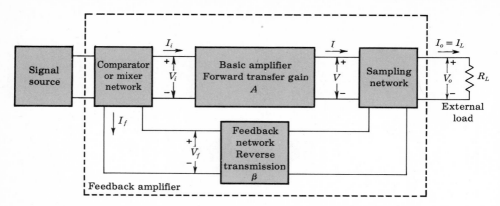

Fig. 12-1 Representation of any single-loop feedback connection around a basic amplifier. The transfer gain A may represent A_V, A_I, G_M, or R_M.

which samples the output current is shown in Fig. 12-2b, where the feedback network is connected *in series* with the output. This type of connection is referred to as *current*, or *loop, sampling*. Other sampling networks are possible.

Comparator, or Mixer, Network Two mixing blocks are shown in Fig. 12-3. Figure 12-3a and b shows the simple and very common *series* (loop) *input* and *shunt* (*node*) *input* connections, respectively. A differential amplifier (Sec. 13-3) is often also used as the mixer. Such an amplifier has two inputs and gives an output proportional to the difference between the signals at the two inputs.

Transfer Ratio, or Gain The symbol A in Fig. 12-1 represents the ratio of the output signal to the input signal of the basic amplifier. The transfer

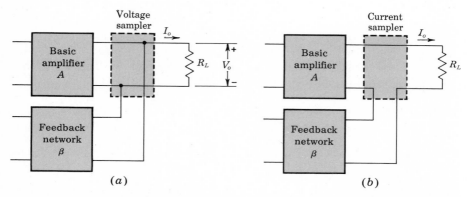

Fig. 12-2 Feedback connections at the output of a basic amplifier, sampling the output (a) voltage and (b) current.

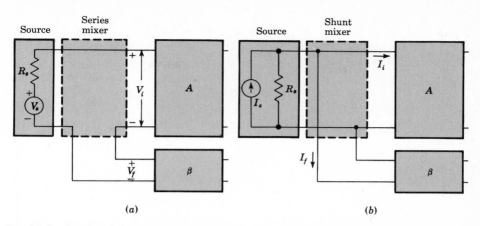

Fig. 12-3 Feedback connections at the input of a basic amplifier. (*a*) Series comparison. (*b*) Shunt mixing.

ratio V/V_i is the voltage amplification, or the *voltage gain*, A_V. Similarly, the transfer ratio I/I_i is the current amplification, or *current gain*, A_I for the amplifier. The ratio I/V_i of the basic amplifier is the transconductance G_M, and V/I_i is the transresistance R_M. Although G_M and R_M are defined as the ratio of two signals, one of these is a current and the other is a voltage waveform. Hence the symbol G_M or R_M does not represent an amplification in the usual sense of the word. Nevertheless, it is convenient to refer to each of the four quantities A_V, A_I, G_M, and R_M as a *transfer gain of the basic amplifier without feedback* and to use the symbol A to represent any one of these quantities.

The symbol A_f is defined as the ratio of the output signal to the input signal of the amplifier configuration of Fig. 12-1 and is called the *transfer gain of the amplifier with feedback*. Hence A_f is used to represent any one of the four ratios $V_o/V_s \equiv A_{Vf}$, $I_o/I_s \equiv A_{If}$, $I_o/V_s \equiv G_{Mf}$, and $V_o/I_s \equiv R_{Mf}$. The relationship between the transfer gain A_f with feedback and the gain A of the ideal amplifier without feedback is derived below [Eq. (12-4)].

Advantages of Negative Feedback When any increase in the output signal results in a feedback signal into the input in such a way as to cause a decrease in the output signal, the amplifier is said to have *negative feedback*. The usefulness of negative feedback lies in the fact that, in general, the characteristics of an amplifier may be improved by the proper use of negative feedback. For example, the normally high input resistance of a voltage amplifier can be made higher, and its normally low output resistance can be lowered. Also, the transfer gain A_f of the amplifier with feedback can be stabilized against variations of the h or hybrid-π parameters of the transistors or the parameters of the other active devices used in the amplifier. Another impor-

tant advantage of the proper use of negative feedback is the significant improvement in the frequency response and in the linearity of operation of the
feedback amplifier compared with that of the amplifier without feedback.

It should be pointed out that all the advantages mentioned above are obtained at the expense of the gain A_f with feedback, which is lowered in comparison with the transfer gain A of an amplifier without feedback. Also, under
certain circumstances, discussed in Sec. 12-15, a negative-feedback amplifier
may become unstable and break into oscillations.

12-2   THE TRANSFER GAIN WITH FEEDBACK

Any one of the output connections of Fig. 12-2 may be combined with any
of the input connections of Fig. 12-3 to form the feedback amplifier of Fig.
12-1. The analysis of the feedback amplifier can then be carried out by
replacing each active element (transistor or FET) by its small-signal model
and by writing Kirchhoff's loop, or nodal, equations. That approach, however,
does not place in evidence the main characteristics of feedback.

As a first step toward a method of analysis which emphasizes the benefits
of feedback, consider Fig. 12-4, which represents a generalized feedback amplifier. The four possible combinations of voltage or current sampling and
mixing are shown in Fig. 12-5.

These four topologies are referred to as (1) *voltage-series feedback*, (2)
current-series feedback, (3) *current-shunt feedback*, and (4) *voltage-shunt
feedback*. In Fig. 12-4, the source resistance R_s is considered to be part of
the amplifier, and the transfer gain $A(A_V, G_M, A_I, R_M)$ includes the effect of
the loading of the β network (as well as R_L) upon the amplifier. The input
signal X_s, the output signal X_o, the feedback signal X_f, and the difference
signal X_d, each represents either a voltage or a current. These signals and also
the ratios A and β are summarized in Table 12-1. The symbol indicated by
the circle in Fig. 12-4 represents a mixing, or comparison, network, whose
output is the sum of the inputs, taking the sign shown at each input into
account. Thus

$$X_d = X_s - X_f = X_i \qquad (12\text{-}1)$$

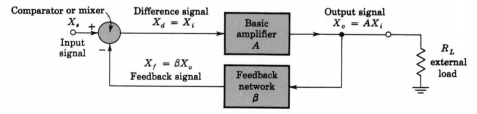

Fig. 12-4   Schematic representation of a single-loop feedback amplifier.

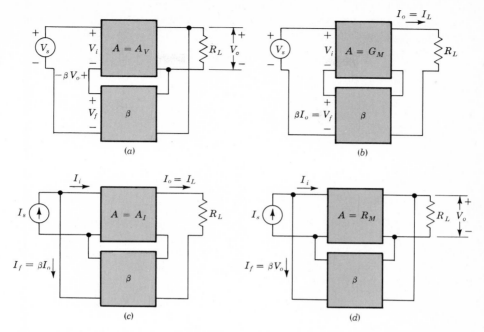

Fig. 12-5 Feedback-amplifier topologies. The source resistance is considered to be part of the amplifier. (*a*) Voltage-series feedback. (*b*) Current-series feedback. (*c*) Current-shunt feedback. (*d*) Voltage-shunt feedback.

Since X_d represents the difference between the applied signal and that fed back to the input, X_d is called the *difference, error,* or *comparison, signal.*

The reverse transmission factor β is defined by

$$\beta \equiv \frac{X_f}{X_o} \tag{12-2}$$

TABLE 12-1   Voltage and current signals in feedback amplifiers

Signal or ratio	Type of feedback			
	Voltage-series Fig. 12-5*a*	Current-series Fig. 12-5*b*	Current-shunt Fig. 12-5*c*	Voltage-shunt Fig. 12-5*d*]
X_0	Voltage	Current	Current	Voltage
X_s, X_f, X_d	Voltage	Voltage	Current	Current
A	A_V	G_M	A_I	R_M
β	V_f/V_o	V_f/I_o	I_f/I_o	I_f/V_o

The factor β is often a positive or a negative real number, but in general, β is a complex function of the signal frequency. (This symbol should not be confused with the symbol β used previously for the CE short-circuit current gain.) The symbol X_o is the output voltage, or the output (load) current.

The transfer gain A is defined by

$$A \equiv \frac{X_o}{X_i} \tag{12-3}$$

By substituting Eqs. (12-1) and (12-2) into (12-3), we obtain for A_f the gain with feedback,

$$A_f \equiv \frac{X_o}{X_s} = \frac{A}{1 + \beta A} \tag{12-4}$$

The quantity A in Eqs. (12-3) and (12-4) represents the transfer gain of the corresponding amplifier without feedback, but *including the loading of the β network, R_L and R_s.* In the following section many of the desirable features of feedback are deduced, starting with the fundamental relationship given in Eq. (12-4).

If $|A_f| < |A|$, the feedback is termed *negative*, or *degenerative*. If $|A_f| > |A|$, the feedback is termed *positive*, or *regenerative*. From Eq. (12-4) we see that, in the case of negative feedback, the gain of the basic ideal amplifier with feedback is divided by the factor $|1 + \beta A|$, which exceeds unity.

Loop Gain   The signal X_d in Fig. 12-4 is multiplied by A in passing through the amplifier, is multiplied by β in transmission through the feedback network, and is multiplied by -1 in the mixing or differencing network. Such a path takes us from the input terminals around the loop consisting of the amplifier and feedback network back to the input; the product $-A\beta$ is called the *loop gain*, or *return ratio*. The difference between unity and the loop gain is called the *return difference* $D = 1 + A\beta$. Also, the amount of feedback introduced into an amplifier is often expressed in decibels by the definition

$$N = \text{dB of feedback} \equiv 20 \log \left| \frac{A_f}{A} \right| = 20 \log \left| \frac{1}{1 + A\beta} \right| \tag{12-5}$$

If negative feedback is under consideration, N will be a negative number.

Fundamental Assumptions   Three conditions must be satisfied for the feedback network of Fig. 12-4 in order that Eq. (12-4) be true.

1. The input signal is transmitted to the output through the amplifier A and *not* through the β network.
2. The feedback signal is transmitted from the output to the input through the β block, and *not* through the amplifier.

3. The reverse transmission factor β of the feedback network is independent of the load and the source resistances R_L and R_s.

These conditions are valid (to a reasonable approximation) for practical feedback amplifier connections.

12-3   GENERAL CHARACTERISTICS OF NEGATIVE–FEEDBACK AMPLIFIERS

Since negative feedback reduces the transfer gain, why is it used?   The answer is that many desirable characteristics are obtained for the price of gain reduction.   We now examine some of the advantages of negative feedback.

Desensitivity of Transfer Amplification   The variation due to aging, temperature, replacement, etc., of the circuit components and transistor or FET characteristics is reflected in a corresponding lack of stability of the amplifier transfer gain.   The fractional change in amplification with feedback divided by the fractional change without feedback is called the *sensitivity* of the transfer gain.   If Eq. (12-4) is differentiated with respect to A, the absolute value of the resulting equation is

$$\left| \frac{dA_f}{A_f} \right| = \frac{1}{|1 + \beta A|} \left| \frac{dA}{A} \right| \tag{12-6}$$

Hence the sensitivity is $1/|1 + \beta A|$.   If, for example, the sensitivity is 0.1, the percentage change in gain with feedback is one-tenth the percentage variation in amplification if no feedback is present.   The reciprocal of the sensitivity is called the *desensitivity* D, or

$$D \equiv 1 + \beta A \tag{12-7}$$

The fractional change in gain without feedback is divided by the desensitivity D when feedback is added.   [In passing, note that the desensitivity is another name for the *return difference*, and that the amount of feedback is $-20 \log D$ (Eq. 12-5).]   For an amplifier with 20 dB of negative feedback, $D = 10$, and hence, for example, a 5 percent change in gain without feedback is reduced to a 0.5 percent variation after feedback is introduced.

Note from Eq. (12-4) that the transfer gain is divided by the desensitivity after feedback is added.   Thus

$$A_f = \frac{A}{D} \tag{12-8}$$

In particular, if $|\beta A| \gg 1$, then

$$A_f = \frac{A}{1 + \beta A} \approx \frac{A}{\beta A} = \frac{1}{\beta} \tag{12-9}$$

and the gain may be made to depend entirely on the feedback network. The worst offenders with respect to stability are usually the active devices (transistors) involved. If the feedback network contains only stable passive elements, the improvement in stability may indeed be pronounced.

Since A represents either A_V, G_M, A_I, or R_M, then A_f represents the corresponding transfer gains with feedback: either A_{Vf}, G_{Mf}, A_{If}, or R_{Mf}. The topology determines which transfer ratio (Table 12-1) is stabilized. For example, for voltage-series feedback, Eq. (12-9) signifies that $A_{Vf} \approx 1/\beta$, and it is the voltage gain which is stabilized. For current-series feedback, Eq. (12-9) is $G_{Mf} \approx 1/\beta$, and hence, for this topology, it is the transconductance gain which is desensitized. Similarly, it follows from Eq. (12-9) that the current gain is stabilized for current-shunt feedback ($A_{If} \approx 1/\beta$) and the transresistance gain is desensitized for voltage-shunt feedback ($R_{Mf} \approx 1/\beta$).

Feedback is used to improve stability in the following way: Suppose an amplifier of gain A_1 is required. We start by building an amplifier of gain $A_2 = DA_1$, in which D is a large number. Feedback is now introduced to divide the gain by the factor D. The stability will be improved by the same factor D, since both gain and instability are divided by the desensitivity D. If now the instability of the amplifier of gain A_2 is not appreciably larger than the instability of an amplifier of gain without feedback equal to A_1, this procedure will have been useful. It often happens as a matter of practice that amplifier gain may be increased appreciably without a corresponding loss of stability. For example, the voltage gain of a transistor may be increased by increasing the collector resistance R_c.

Frequency Distortion It follows from Eq. (12-9) that if the feedback network does not contain reactive elements, the overall gain is not a function of frequency. Under these circumstances a substantial reduction in frequency and phase distortion is obtained. The frequency response of feedback amplifiers is discussed in Sec. 12-6.

Nonlinear Distortion Suppose that a large amplitude signal is applied to a stage of an amplifier so that the operation of the device extends slightly beyond its range of linear operation, and as a consequence the output signal is slightly distorted. Negative feedback is now introduced, and the input signal is increased by the same amount by which the gain is reduced, so that the output-signal amplitude remains the same. For simplicity, let us consider that the input signal is sinusoidal and that the distortion consists, simply, of a second-harmonic signal generated within the active device. We assume that the second-harmonic component, in the absence of feedback, is equal to B_2. Because of the effects of feedback, a component B_{2f} actually appears in the output. The relationship that exists between B_{2f} and B_2 is

$$B_{2f} = \frac{B_2}{1 + \beta A} = \frac{B_2}{D} \tag{12-10}$$

Fig. 12-6 Voltage series feedback circuit increases
the input resistance.

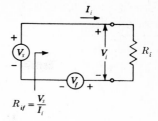

$$R_{if} = \frac{V_s}{I_i}$$

Since A and β are generally functions of the frequency, they must be evaluated at the second-harmonic frequency.

The signal X_s to the feedback amplifier may be the actual signal externally available, or it may be the output of an amplifier preceding the feedback stage or stages under consideration. To multiply the input to the feedback amplifier by the factor $|1 + A\beta|$, it is necessary either to increase the nominal gain of the preamplifying stages or to add a new stage. If the full benefit of the feedback amplifier in reducing nonlinear distortion is to be obtained, these preamplifying stages must not introduce additional distortion, because of the increased output demanded of them. Since, however, appreciable harmonics are introduced only when the output swing is large, most of the distortion arises in the last stage. The preamplifying stages are of smaller importance in considerations of harmonic generation.

12-4   INPUT RESISTANCE

We now discuss qualitatively the effect of the topology of a feedback amplifier upon the input resistance. If the feedback signal is returned to the input in *series* with the applied voltage (regardless of whether the feedback is obtained by sampling the output current or voltage), it *increases the input resistance.* Since the feedback voltage V_f opposes V_s, the input current I_i is less than it would be if V_f were absent. Hence the input resistance with feedback $R_{if} \equiv V_s/I_i$ (Fig. 12-6) is greater than the input resistance without feedback R_i.

Negative feedback in which the feedback signal is returned to the input in *shunt* with the applied signal (regardless of whether the feedback is obtained by sampling the output current or voltage) *decreases the input resistance.* Since $I_s = I_i + I_f$ (Fig. 12-7), then the current I_s drawn from the signal source is increased over what it would be if there were no feedback current. Hence $R_{if} \equiv V_i/I_s$ (Fig. 12-7) is decreased because of this type of feedback.

Fig. 12-7 Current-shunt feedback circuit decreases the input resistance.

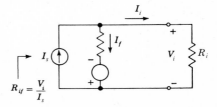

$$R_{if} = \frac{V_i}{I_s}$$

TABLE 12-2   Effect of negative feedback on amplifier characteristics

	Type of feedback			
	Voltage-series	Current-series	Current-shunt	Voltage-shunt
Reference	Fig. 12-5a	Fig. 12-5b	Fig. 12-5c	Fig. 12-5d
R_{of}	Decreases	Increases	Increases	Decreases
R_{if}	Increases	Increases	Decreases	Decreases
Desensitizes	A_{Vf}	G_{Mf}	A_{If}	R_{Mf}
Bandwidth	Increases	Increases	Increases	Increases
Nonlinear distortion	Decreases	Decreases	Decreases	Decreases

Table 12-2 summarizes the characteristics of the four types of negative-feedback configurations: *For series comparison, $R_{if} \gg R_i$, whereas for shunt mixing, $R_{if} < R_i$.*

12-5   OUTPUT RESISTANCE

We now discuss qualitatively the effect of the topology of a feedback amplifier upon the output resistance. Negative feedback which samples the output *voltage*, regardless of how this output signal is returned to the input, tends to *decrease the output resistance.* For example if R_L increases so that V_o increases, the effect of feeding this voltage back to the input in a degenerative manner (negative feedback) is to cause V_o to increase less than it would if there were no feedback. Hence the output voltage tends to remain constant as R_L changes, which means that $R_{of} \ll R_L$. This argument leads to the conclusion that this type of feedback (sampling the output voltage) reduces the output resistance.

By reasoning similar to that given above negative feedback which samples the output *current* will tend to hold this current constant. Hence an output-current source is created ($R_{of} \gg R_L$), and we conclude that this type of sampling connection increases the output resistance.

In summary (Table 12-2): *For voltage sampling, $R_{of} < R_o$, whereas for current sampling, $R_{of} > R_o$.*

12-6   EFFECT OF FEEDBACK ON AMPLIFIER BANDWIDTH

To study the effect of feedback on bandwidth we shall assume in this section that the transfer gain A without feedback is given by a single-pole transfer function.

$$A = \frac{A_o}{1 + j(f/f_H)} \tag{12-11}$$

where A_o (real and negative) is the midband gain without feedback, and f_H is the high 3-dB frequency.   The gain with feedback is given by Eq. (12-4), or using Eq. (12-11),

$$A_f = \frac{A_o/[1 + j(f/f_H)]}{1 + \beta A_o/[1 + j(f/f_H)]} = \frac{A_o}{1 + \beta A_o + j(f/f_H)}$$

By dividing numerator and denominator by $1 + \beta A_o$, this equation may be put in the form

$$A_f = \frac{A_{of}}{1 + j(f/f_{Hf})} \tag{12-12}$$

where

$$A_{of} \equiv \frac{A_o}{1 + \beta A_o} \qquad \text{and} \qquad f_{Hf} \equiv f_H(1 + \beta A_o) \tag{12-13}$$

We see that the *midband amplification with feedback* A_{of} equals the midband amplification without feedback A_o divided by $1 + \beta A_o$.   Also, the *high* 3-dB *frequency with feedback* f_{Hf} equals the corresponding 3-dB frequency without feedback f_H multiplied by the same factor, $1 + \beta A_o$.   The gain-frequency product has not been changed by feedback because, from Eqs. (12-13),

$$A_{of}f_{Hf} = A_o f_H \tag{12-14}$$

By starting with Eq. (11-2) for the low-frequency gain of a single *RC*-coupled stage and proceeding as above, we can show that the *low* 3-dB *frequency with feedback* f_{Lf} is decreased by the same factor as is the gain, or

$$f_{Lf} = \frac{f_L}{1 + \beta A_o} \tag{12-15}$$

For an audio or video amplifier, $f_H \gg f_L$, and hence the bandwidth is $f_H - f_L \approx f_H$.   Under these circumstances, Eq. (12-14) may be interpreted to mean that the gain-bandwidth product is the same with or without feedback.

The frequency response of a multipole feedback amplifier is much more complicated than that for the single-pole transfer function discussed above.   If the amplifier without feedback has more than two poles, then it may become unstable (for example, it may break out into steady oscillation) if too much feedback is added.   A two-pole feedback amplifier is always stable, but the step response may be unsatisfactory because the output step may exhibit ringing (damped oscillation).†

12-7   EXAMPLES OF FEEDBACK AMPLIFIERS

In Chap. 10 two practical feedback amplifiers are considered; the emitter follower and the CE amplifier with an unbypassed emitter resistor.   These will

† "Integrated Electronics," chap. 14.

now be reexamined in the light of the feedback theory discussed in this chapter. Other feedback circuits are considered in subsequent sections.

Voltage-Series Feedback   The emitter-follower circuit is given in Fig. 12-8. The feedback signal is the voltage V_f across R_L and this voltage is in series with the applied signal V_s. The sampled signal V_o is also across R_L. Hence, this is a case of voltage-series feedback and the feedback factor is given by Eq. (12-2)

$$\beta = \frac{V_f}{V_o} = 1 \tag{12-16}$$

If the amplifier gain A_V without feedback is large so that $\beta A_V \gg 1$ then, from Eq. (12-9), the voltage gain after feedback is added is

$$A_{Vf} \approx \frac{1}{\beta} = 1 \tag{12-17}$$

Note that the effect of the feedback has been to reduce the gain to unity and, thus, to stabilize the amplifier because A_{Vf} is independent of the transistor parameters, the other circuit components, the temperature, etc.

From the first column in Table 12-2 the output resistance of the emitter follower should be low and the input resistance should be high. These conclusions are verified by referring to Table 10-2 for the CC configuration, since $h_{fe} \gg 1$.

The above considerations are equally valid for the source follower of Fig. 10-16b, which is another example of the voltage-series configuration.

Current-Series Feedback   The CE circuit with an emitter resistor R_e is indicated in Fig. 12-9. The feedback signal is the voltage V_f across R_e and this voltage is in series with the input signal V_s. The sampled signal is the output or load current I_o. Hence this is a case of current-series feedback and the feedback factor is, from Eq. (12-2), $\beta = V_f/I_o$. Since the base current

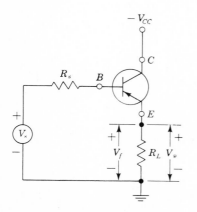

Fig. 12-8   The emitter-follower is an example of voltage-series feedback.

can be neglected compared with the collector current, then I_o is the emitter current in R_e. From Fig. 12-9, $V_f = -I_oR_e$ and, therefore,

$$\beta = \frac{V_f}{I_o} = -R_e \tag{12-18}$$

In passing, note that although I_o is proportional to V_o, it is *not* correct to conclude that this is voltage-series feedback. Thus, if the output signal is taken as the voltage V_o, then

$$\beta = \frac{V_f}{V_o} = \frac{-I_oR_e}{I_oR_L} = -\frac{R_e}{R_L}$$

Since β is now a function of the load R_L, the third basic assumption given in Sec. 12-2 is violated.

From Table 12-1, it is the transconductance gain which is stabilized and, from Eq. (12-9),

$$A_f = G_{Mf} \approx \frac{1}{\beta} = -\frac{1}{R_e} \tag{12-19}$$

By definition [Eq. (12-4)] $A_f = X_o/X_s$ and, hence, for the current-series configuration,

$$G_{Mf} = \frac{I_o}{V_s} \approx -\frac{1}{R_e} \tag{12-20}$$

where use is made of Eq. (12-19). Since $I_o = -V_s/R_e$, *the load current is directly proportional to the input voltage, and this current depends only upon R_e, and not upon any other circuit or transistor parameter.* As an example, consider that this circuit is used as the driver for the deflection current I_o in a magnetic cathode-ray oscilloscope. The load is then the deflection-yoke impedance, which is essentially an inductance whose reactance is proportional to frequency. Yet, from Eq. (12-20) the load current is independent of the

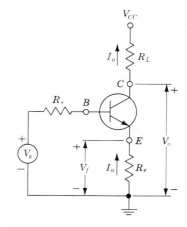

Fig. 12-9 Amplifier with an unbypassed emitter resistance as an example of current series feedback.

characteristics of the yoke. If a deflection which varies linearly with time is desired, it is only necessary to generate a voltage waveform V_s which increases linearly with time (we are assuming that the deflection of the spot on the tube face is proportional to the yoke current).

The voltage gain is given by

$$A_{Vf} = \frac{I_o R_L}{V_s} = G_{Mf} R_L \approx - \frac{R_L}{R_e} \tag{12-21}$$

Note that if R_L and R_e are stable resistors, the voltage gain (as well as the transconductance gain) is stable.

From the second column of Table 12-2, the input and output resistances of the amplifier in Fig. 12-9 should both be high because of the feedback. These conclusions [as well as the result in Eq. (12-21)] are verified by referring to the second column of Table 10-2.

12-8   CURRENT–SHUNT FEEDBACK

Figure 12-10 shows two transistors in cascade with feedback from the second emitter to the first base through the resistor R'. We now verify that this connection produces negative feedback. The voltage V_{i2} is much larger than V_{i1} because of the voltage gain of $Q1$. Also, V_{i2} is 180° out of phase with V_{i1}. Because of emitter-follower action, V_{e2} is only slightly smaller than V_{i2}, and these voltages are in phase. Hence V_{e2} is larger in magnitude than V_{i1} and is 180° out of phase with V_{i1}. If the input signal increases so that I'_s increases, I_f also increases, and $I_i = I'_s - I_f$ is smaller than it would be if there were no feedback. This action is characteristic of *negative* feedback.

We now show that the configuration of Fig. 12-10 approximates a current-shunt feedback pair. Since $V_{e2} \gg V_{i1}$, and neglecting the base current of $Q2$ compared with the collector current,

$$I_f = \frac{V_{i1} - V_{e2}}{R'} \approx - \frac{V_{e2}}{R'} = \frac{(I_o - I_f)R_e}{R'} \tag{12-22}$$

or

$$I_f = \frac{R_e I_o}{R' + R_e} = \beta I_o \tag{12-23}$$

where $\beta = R_e/(R' + R_e)$. Since the feedback current is proportional to the output current, this circuit is an example of a current-shunt feedback amplifier. From Table 12-2 we expect the transfer (current) gain A_{If} to be stabilized.

$$A_{If} = \frac{I_o}{I_s} \approx \frac{1}{\beta} = \frac{R' + R_e}{R_e} \tag{12-24}$$

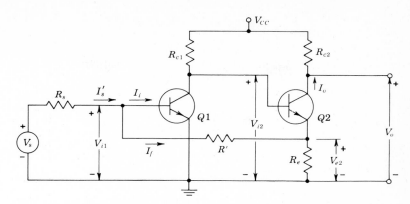

Fig 12-10   Second-emitter to first-base feedback pair.   (The input blocking capacitor and the biasing resistors are not indicated.)

and hence we have verified that A_{If} is desensitized provided that R' and R_e are stable resistances.   Note that $I_s \equiv V_s/R_s$ is the Norton's equivalent current.

From Table 12-2 we expect the input resistance to be low and the output resistance to be high.   The voltage gain with feedback is, using Eq. (12-24)

$$A_{Vf} = \frac{V_o}{V_s} = \frac{I_o R_{c2}}{I_s R_s} \approx \frac{R' + R_e}{R_e} \frac{R_{c2}}{R_s} \tag{12-25}$$

Note that if R_e, R', R_{c2}, and R_s are stable resistors, then A_{Vf} is stable (independent of the transistor parameters, the temperature, or supply-voltage variations).

12-9   VOLTAGE–SHUNT FEEDBACK

Figure 12-11 shows a common-emitter stage with a resistor R' connected from the output to the input.   We first show that this configuration conforms to voltage-shunt topology, and then obtain approximate expressions for trans-resistance and the voltage gain with feedback.

In the circuit of Fig. 12-11, the output voltage V_o is much greater than the input voltage V_i and is 180° out of phase with V_i.   Hence

$$I_f = \frac{V_i - V_o}{R'} \approx -\frac{V_o}{R'} = \beta V_o \tag{12-26}$$

where $\beta = -1/R'$.   Since the feedback current is proportional to the output voltage, this circuit is an example of a *voltage-shunt feedback amplifier*.   From Table 12-2 we expect the transfer gain (the transresistance) R_{Mf} to be desensitized.   From Eq. (12-9)

$$R_{Mf} \equiv \frac{V_o}{I_s} \approx \frac{1}{\beta} = -R' \tag{12-27}$$

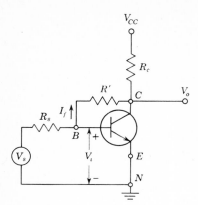

Fig. 12-11   Voltage-shunt feedback.

Note that the transresistance equals the negative of the feedback resistance from output to input of the transistor and is stable if R' is a stable resistance.

From Table 12-2 we expect both the input and output resistance to be low because of the voltage-shunt feedback. For this important circuit the voltage gain with feedback is

$$A_{Vf} = \frac{V_o}{V_s} = \frac{V_o}{I_s R_s} \approx \frac{1}{\beta R_s} = -\frac{R'}{R_s} \tag{12-28}$$

where use is made of Eq. (12-27). Note that if R' and R_s are stable resistors, then A_{Vf} is stable (independent of the transistor parameters, the temperature, and supply-voltage variations).

The approximate formula, Eq. (12-28), leads to the erroneous conclusion that A_{Vf} increases without limit as $R_s \to 0$. The difficulty arises because Eq. (12-27) is valid only if $\beta R_M \gg 1$, and this inequality is not satisfied if $R_s = 0$. (It turns out that $R_M = 0$ if $R_s = 0$.) The correct result for A_{Vf} with $R_s = 0$ is obtained by inspection of Fig. 12-11. This circuit is now that of a CE amplifier *without feedback* and with a load resistor $R_L = R_c \| R'$. From Table 10-2, the correct value of the voltage gain is $-h_{fe} R_L / h_{ie}$ if $R_s = 0$.

12-10   METHOD OF ANALYSIS OF A FEEDBACK AMPLIFIER

The circuits considered in the previous section were analyzed as feedback amplifiers by assuming that the magnitude of the loop gain βA is much greater than unity. Very often the assumption $|\beta A| \gg 1$ is not justified, and it becomes necessary to separate the feedback amplifier into two blocks—the basic amplifier A and the feedback network β. With a knowledge of A and β we can calculate the transfer function A_f of the feedback system from Eq. (12-4). The basic amplifier configuration *without feedback but taking the loading of the β network into account* is obtained by applying the following rules:

To find the input circuit:

1. Set $V_o = 0$ for voltage sampling. In other words, short-circuit the output node to ground.
2. Set $I_o = 0$ for current sampling. In other words, open-circuit the output loop.

To find the output circuit:

1. Set $V_i = 0$ for shunt comparison. In other words, short-circuit the input node to ground.
2. Set $I_i = 0$ for series comparison. In other words open-circuit the input loop.

These procedures ensure that the feedback is reduced to zero without altering the loading on the basic amplifier.

The complete analysis of a feedback amplifier is obtained by carrying out the following steps:

1. Identify the topology. (*a*) Is the feedback signal X_f a voltage or a current? In other words, is X_f applied in series or in shunt with the external excitation? (*b*) Is the sampled signal X_o a voltage or a current? In other words, is the sampled signal taken at the output node or from the output loop?
2. Draw the basic amplifier circuit without feedback, following the rules listed above.
3. Use a Thévenin's source if X_f is a voltage and a Norton's source if X_f is a current.
4. Replace each active device by the proper model (for example, the hybrid-π model for a transistor at high frequencies or the h-parameter model at low frequencies).
5. Indicate X_f and X_o on the circuit obtained by carrying out steps **2, 3,** and **4.** Evaluate $\beta = X_f/X_o$.
6. Evaluate A by applying KVL and KCL to the equivalent circuit obtained after step 4.
7. From A and β, find D and A_f.

We now carry out the analysis for an emitter follower. In subsequent sections other feedback amplifier configurations are considered in detail.

The Emitter Follower The circuit is given in Fig. 12-8 and repeated in Fig. 12-12*a* for convenience. The feedback signal is the voltage V_f across R_e, and the sampled signal is V_o across R_e. Hence this is a case of voltage-series feedback.

We now draw the basic amplifier without feedback. To find the input circuit, set $V_o = 0$, and hence V_s in series with R_s appears between B and E.

To find the output circuit, set $I_i = I_b = 0$ (the input loop is opened), and hence R_e appears only in the output loop. Following these rules, we obtain the circuit of Fig. 12-12b. If the transistor is replaced by its low-frequency approximate model of Fig. 10-5, the result is Fig. 12-12c. From this figure $V_o = V_f$ and $\beta = V_o/V_f = 1$.

This topology stabilizes the voltage gain. A_V is calculated by inspection of Fig. 12-12c. Since R_s is considered as part of the amplifier, then $V_i = V_s$, and

$$A_V = \frac{V_o}{V_i} = \frac{h_{fe}I_bR_e}{V_s} = \frac{h_{fe}R_e}{R_s + h_{ie}} \qquad (12\text{-}29)$$

$$D = 1 + \beta A_V = 1 + \frac{h_{fe}R_e}{R_s + h_{ie}} = \frac{R_s + h_{ie} + h_{fe}R_e}{R_s + h_{ie}} \qquad (12\text{-}30)$$

$$A_{Vf} = \frac{A_V}{D} = \frac{h_{fe}R_e}{R_s + h_{ie} + h_{fe}R_e} \qquad (12\text{-}31)$$

For $h_{fe}R_e \gg R_s + h_{ie}$, $A_{Vf} \approx 1$, as it should be for an emitter follower.

Note that the feedback desensitizes voltage gain with respect to changes in h_{fe}. Also note that A_{Vf} is independent of R_c.

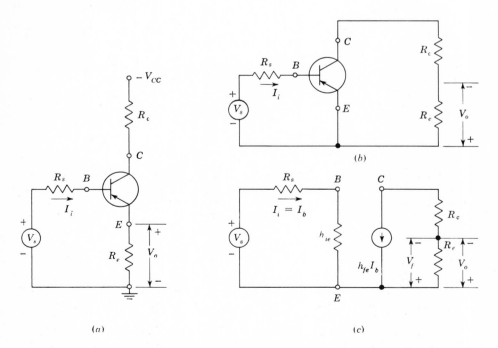

Fig. 12-12 (a) An emitter follower. (b) The amplifier without feedback and (c) the transistor replaced by its approximate low-frequency model.

12-11   A VOLTAGE–SERIES FEEDBACK PAIR

Figure 12-13 shows two cascaded stages whose voltage gains are A_{V1} and A_{V2}, respectively. The output of the second stage is returned through the feedback network R_1R_2 in opposition to the input signal V_s. Clearly, then, this is a case of voltage-series negative feedback. We expect the input resistance to increase, the output resistance to decrease, and the voltage gain to be stabilized (desensitized).

The first basic assumption listed in Sec. 12-2 is not strictly satisfied for the circuit of Fig. 12-13a because I' represents transmission through the feedback network from input to the output. We shall neglect I' compared with I on the realistic assumption that the current gain of the second stage is much larger than unity. Under these circumstances very little error is made in using the feedback formulas developed in this chapter.

The input of the basic circuit without feedback is found by setting $V_o = 0$, and hence R_2 appears in parallel with R_1. The output of the basic amplifier without feedback is found by opening the input loop (set $I' = 0$), and hence R_1 is placed in series with R_2. Following these rules results in Fig. 12-13b, to which has been added the series feedback voltage V_f across R_1 in the output circuit. Clearly,

$$\beta = \frac{V_f}{V_o} = \frac{R_1}{R_1 + R_2} \tag{12-32}$$

Second-Collector to First-Emitter Feedback Pair   The circuit of Fig. 12-14 shows a two-stage amplifier which makes use of voltage-series feedback by connecting the second collector to the first emitter through the voltage divider R_1R_2. Capacitors C_1, C_2, C_5, and C_6 are dc blocking capacitors, and capaci-

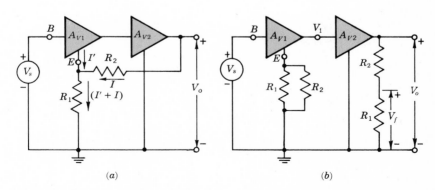

(a)   (b)

Fig. 12-13   (a) Voltage-series feedback pair.   (b) Equivalent circuit, without external feedback, but including the loading of R_2.

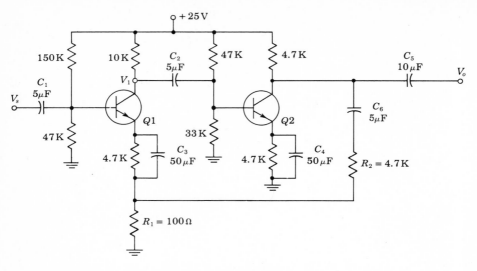

Fig. 12-14　Second-collector to first-emitter feedback pair.

tors C_3 and C_4 are bypass capacitors for the emitter bias resistors. All these capacitances represent negligible reactances at the frequencies of operation of this circuit. For this amplifier the voltage gain A_{Vf} is given approximately by $1/\beta$, and is thus stabilized against temperature changes and transistor replacement. A more accurate determination of A_{Vf}, is given in the following illustrative problem.

EXAMPLE　Calculate A_{Vf} for the amplifier of Fig. 12-14. Assume $R_s = 0, h_{fe} = 50,$ $h_{ie} = 1.1$ K, $h_{re} = h_{oe} = 0$, and identical transistors.

Solution　We first calculate the overall voltage gain without feedback from $A_V = A_{V1}A_{V2}$. The effective load R'_{L1} of transistor $Q1$ is

$$R_{L1} = 10\|47\|33\|1.1 \text{ K} = 942 \text{ }\Omega$$

From Fig. 12-13b we see that the effective load R'_{L2} of transistor $Q2$ is the collector resistance $R_{c2} = 4.7$ K in parallel with $R_1 + R_2 = 4.8$ K,

$$R'_{L2} = 4.7\|4.8 = 2.37 \text{ K}$$

From Fig. 12-13b we see that the effective emitter resistance R_e of $Q1$ is $R_1\|R_2$ or

$$R_e = R_1\|R_2 = 0.1\|4.7 \text{ K} = 0.098 \text{ K} = 98 \text{ }\Omega$$

The voltage gain A_{V1} of $Q1$ is, from Eq. (10-30) and Fig. 12-13b with $V_i = V_s,$

$$A_{V1} \equiv \frac{V_1}{V_i} = \frac{-h_{fe}R'_{L1}}{h_{ie} + (1 + h_{fe})R_e} = \frac{-50 \times 0.942}{1.1 + 51 \times 0.098} = -7.72$$

The voltage gain A_{V2} of $Q2$ is, from Eq. (10-16),

$$A_{V2} \equiv \frac{V_o}{V_1} = -h_{fe}\frac{R'_{L2}}{h_{ie}} = -50 \times \frac{2.37}{1.1} = -108$$

Hence the voltage gain A_V of the two stages in cascade without feedback is

$$A_V \equiv \frac{V_o}{V_i} = A_{V1}A_{V2} = 7.72 \times 108 = 834$$

$$\beta = \frac{R_1}{R_1 + R_2} = \frac{100}{4,800} = \frac{1}{48} \quad \text{and} \quad A_V\beta = \frac{834}{48} = 17.4$$

$$D = 1 + \beta A_V = 18.4$$

$$A_{Vf} = \frac{A_V}{D} = \frac{834}{18.4} = 45.4$$

This value is to be compared with the approximate solution (based upon $A_V \rightarrow \infty$) given by $A_{Vf} = 1/\beta = 48$.

It is interesting to note that there is internal (local) feedback in the first stage of Fig. 12-14 because the R_1R_2 parallel combination acts as an emitter resistor. This first stage is an example of current-series feedback, which is analyzed in the next section.

12-12   CURRENT–SERIES FEEDBACK TRANSISTOR STAGE

The CE amplifier with an unbypassed emitter resistor R_e was considered briefly in Sec. 12-7. In this section we examine the circuit by separating it into the A and β circuit blocks.

The Transistor Configuration   The circuit is given in Fig. 12-15a. The feedback signal is the voltage V_f across R_e and the sampled signal is the load current I_o. [For the present argument, we neglect base current compared with collector (load) current.] Hence this is a case of current-series feedback.

The input circuit of the amplifier without feedback is obtained by opening the output loop. Hence R_e must appear in the input side. Similarly, the output circuit is obtained by opening the input loop, and this places R_e also in the output side. The resulting equivalent circuit is given in Fig. 12-15b. No ground can be indicated in this circuit because to do so would again couple the input to the output via R_e; that is, it would reintroduce feedback. And the circuit of Fig. 12-15b represents the basic amplifier without feedback, but taking the loading of the β network into account.

This topology stabilizes the transconductance G_M. In Fig. 12-15c the transistor is replaced by its low-frequency approximate h-parameter model. Since the feedback voltage V_f appears across R_e in the output circuit, then,

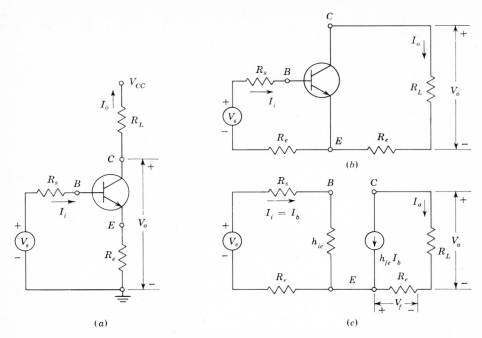

Fig. 12-15 (a) Amplifier with an unbypassed emitter resistance as an example of current-series feedback. (b) The amplifier without feedback, but including the loading of R_e. (c) The h-parameter model used for the transistor in (b).

from Fig. 12-15c

$$\beta = \frac{V_f}{I_o} = \frac{-I_o R_e}{I_o} = -R_e \tag{12-33}$$

Since the input signal V_i without feedback is the V_s of Fig. 12-15c, then

$$G_M = \frac{I_o}{V_i} = \frac{-h_{fe}I_b}{V_s} = \frac{-h_{fe}}{R_s + h_{ie} + R_e} \tag{12-34}$$

$$D = 1 + \beta G_M = 1 + \frac{h_{fe}R_e}{R_s + h_{ie} + R_e} = \frac{R_s + h_{ie} + (1 + h_{fe})R_e}{R_s + h_{ie} + R_e} \tag{12-35}$$

$$G_{Mf} = \frac{G_M}{D} = \frac{-h_{fe}}{R_s + h_{ie} + (1 + h_{fe})R_e} \tag{12-36}$$

Note that if $(1 + h_{fe})R_e \gg R_s + h_{ie}$, and since $h_{fe} \gg 1$, then $G_{Mf} \approx -1/R_e$, in agreement with $G_{Mf} \approx 1/\beta$. If R_e is a stable resistor, the transconductance gain with feedback is stabilized (desensitized). The load current is given by

$$I_o = G_{Mf}V_s = \frac{-h_{fe}V_s}{R_s + h_{ie} + (1 + h_{fe})R_e} \approx -\frac{V_s}{R_e} \tag{12-37}$$

The voltage gain is given by

$$A_{Vf} = \frac{I_o R_L}{V_s} = G_{Mf} R_L = \frac{-h_{fe} R_L}{R_s + h_{ie} + (1 + h_{fe}) R_e} \tag{12-38}$$

Subject to the approximations made above, $A_{Vf} \approx -R_L/R_e$ [Eq. (12-21)] and the voltage gain is stable if R_L and R_e are stable resistors. Note that the above results agree exactly with those derived in Sec. 10-7. Compare Eq. (12-38) and Eq. (10-30).

EXAMPLE The circuit of Fig. 12-15a is to have an overall transconductance gain of -1 mA/V, a voltage gain of -4, and a desensitivity of 50. If $R_s = 1$ K, $h_{fe} = 150$, and $r_{bb'}$ is negligible, find (a) R_e, (b) R_L, and (c) the quiescent collector current I_C at room temperature.

Solution   a. $G_{Mf} = \dfrac{G_M}{D} = \dfrac{G_M}{50} = -1$ mA/V

or

$\qquad G_M = -50$ mA/V

Since $\beta = -R_e$, then

$\qquad D = 1 + \beta G_M = 1 + 50 R_e = 50$

or

$\qquad R_e = 0.98$ K ≈ 1 K

b. $A_{Vf} = G_{Mf} R_L$

or

$\qquad R_L = \dfrac{A_{Vf}}{G_{Mf}} = \dfrac{-4}{-1} = 4$ K

c. From Eqs. (11-34) and (11-35)

$\qquad h_{ie} = r_{bb'} + r_{b'e} \approx \dfrac{h_{fe}}{g_m} = \dfrac{h_{fe} V_T}{I_C}$

or

$\qquad I_C = \dfrac{h_{fe} V_T}{h_{ie}}$

To calculate h_{ie}, note from Eq. (12-34) that

$\qquad G_M = -50 = \dfrac{-h_{fe}}{R_s + h_{ie} + R_e} = \dfrac{-150}{1 + h_{ie} + 1}$

or

$\qquad h_{ie} = 1$ K

Hence,

$\qquad I_C = \dfrac{(150)(0.026)}{1} = 3.9$ mA

12-13   CURRENT–SHUNT FEEDBACK PAIR

The feedback amplifier of Fig. 12-10 is studied by separating it into the A and β circuit blocks.   In Sec. 12-8 it is demonstrated that this configuration represents current-shunt feedback.

The Amplifier without Feedback   The input circuit of the amplifier without feedback is obtained by opening the output loop at the emitter of $Q2$. This places R' in series with R_e from base to emitter of $Q1$.   The output circuit is found by shorting the input node (the base of $Q1$).   This places R' in parallel with R_e.   The resultant equivalent circuit is given in Fig. 12-16.   Since the feedback signal is a current, the source is represented by a Norton's equivalent circuit with $I_s = V_s/R_s$.

The feedback signal is the current I_f in the resistor R', which is in the output circuit.   From Fig. 12-16, with $I_{b2} < I_{c2} = |I_o|$

$$\beta = \frac{I_f}{I_o} = \frac{R_e}{R' + R_e} \qquad\qquad (12\text{-}39)$$

in agreement with Eq. (12-23).

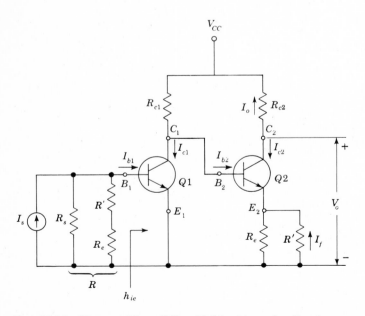

Fig. 12-16   The amplifier of Fig. 12-10 without feedback, but including the loading of R'.

EXAMPLE The circuit of Fig. 12-10 has the following parameters: $R_{c1} = 3$ K, $R_{c2} = 500$ Ω, $R_e = 50$ Ω, $R' = R_s = 1.2$ K, $h_{fe} = 50$, $h_{ie} = 1.1$ K, and $h_{re} = h_{oe} = 0$. Find A_{Vf}

Solution Since the current gain is stabilized, we first calculate A_{If} from A_I. We can then obtain A_{Vf} from A_{If}. Referring to Fig. 12-16,

$$A_I = \frac{-I_{c2}}{I_s} = \frac{-I_{c2}}{I_{b2}} \frac{I_{b2}}{I_{c1}} \frac{I_{c1}}{I_{b1}} \frac{I_{b1}}{I_s} \tag{12-40}$$

Using the low-frequency approximate h-parameter models for $Q1$ and $Q2$,

$$\frac{-I_{c2}}{I_{b2}} = -h_{fe} = -50 \qquad \frac{I_{c1}}{I_{b1}} = +h_{fe} = +50 \tag{12-41}$$

$$\frac{I_{b2}}{I_{c1}} = \frac{-R_{c1}}{R_{c1} + R_{i2}} = \frac{-3}{3 + 3.55} = -0.458 \tag{12-42}$$

because, from Eq. (10-29),

$$R_{i2} = h_{ie} + (1 + h_{fe})(R_e \| R') = 1.1 + (51)\left(\frac{0.05 \times 1.20}{1.25}\right) = 3.55 \text{ K}$$

If R is defined by

$$R \equiv R_s \| (R' + R_e) = \frac{(1.2)(1.25)}{1.2 + 1.25} = 0.612 \text{ K} \tag{12-43}$$

then from Fig. 12-16

$$\frac{I_{b1}}{I_s} = \frac{R}{R + h_{ie}} = \frac{0.61}{0.61 + 1.1} = 0.358 \tag{12-44}$$

Substituting the numerical values in Eqs. (12-41), (12-42), and (12-44) into Eq. (12-40) yields

$$A_I = (-50)(-0.458)(50)(0.358) = +410$$

$$\beta = \frac{R_e}{R' + R_e} = \frac{50}{1,250} = 0.040$$

$$D = 1 + \beta A_I = 1 + (0.040)(410) = 17.4$$

$$A_{If} = \frac{A_I}{D} = \frac{410}{17.4} = 23.6$$

$$A_{Vf} = \frac{V_o}{V_s} = \frac{-I_{c2}R_{c2}}{I_s R_s} = \frac{A_{If}R_{c2}}{R_s} = \frac{(23.6)(0.5)}{1.2} = 9.8$$

The approximate expression of Eq. (12-25) yields

$$A_{Vf} \approx \frac{R_{c2}}{\beta R_s} = \frac{0.5}{(0.040)(1.2)} = 10.4$$

which is in error by 6 percent.

12-14   VOLTAGE–SHUNT FEEDBACK TRANSISTOR STAGE

The amplifier stage of Fig. 12-11 (and repeated in Fig. 12-17) will be analyzed without the assumption made in Sec. 12-9 that $|\beta A| = |\beta R_M| \gg 1$ from which we obtained the result of Eq. (12-28) for the voltage gain A_{Vf}.

The Amplifier without Feedback   The input circuit of the amplifier without feedback is obtained by shorting the output node ($V_o = 0$). This places R' from base to emitter of the transistor. The output circuit is found by shorting the input node ($V_i = 0$), thus connecting R' from collector to emitter. The resultant equivalent circuit is given in Fig. 12-17b. Since the feedback signal is a current, the source is represented by a Norton's equivalent with $I_s = V_s/R_s$.

The feedback signal is the current I_f in the resistor R' which is in the output circuit. From Fig. 12-17b

$$\beta = \frac{I_f}{V_o} = -\frac{1}{R'} \tag{12-45}$$

in agreement with Eq. (12-27).

The first assumption in Sec. 12-2 is not satisfied exactly. If the amplifier is deactivated by reducing h_{fe} to zero, a current I_f passes through the β network (the resistor R') from input to output. This current is given by

$$I_f = \frac{V_s}{R_s + R' + R_c}$$

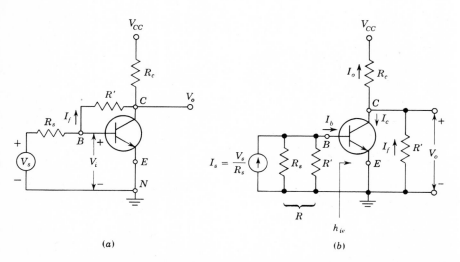

(a)

(b)

Fig. 12-17 (a) Voltage-shunt feedback. (b) The amplifier without feedback, but including the loading of R'.

The output current I_o with the amplifier activated is

$$I_o = \frac{V_o}{R_c} = \frac{A_{Vf}V_s}{R_c}$$

Hence the condition that the forward transmission through the feedback network can be neglected is $|I_o| \gg |I_f|$, or

$$|A_{Vf}| \gg \frac{R_c}{R_s + R' + R_c} \qquad\qquad (12\text{-}46)$$

Since the voltage gain is at least unity, this inequality is easily satisfied by selecting $R_s + R' \gg R_c$.

EXAMPLE   The circuit of Fig. 12-17 has the following parameters: $R_c = 4$ K, $R' = 40$ K, $R_s = 10$ K, $h_{ie} = 1.1$ K, $h_{fe} = 50$, and $h_{re} = h_{oe} = 0$.   Find A_{Vf}.

Solution   Since the transresistance is stabilized, we first calculate R_{Mf} from R_M. Define R_c' and R by

$$R_c' \equiv R_c \| R' = \frac{(4)(40)}{44} = 3.64 \text{ K}$$

and

$$R \equiv R_s \| R' = \frac{(10)(40)}{50} = 8 \text{ K}$$

From Fig. 12-17

$$R_M = \frac{V_o}{I_s} = \frac{-I_c R_c'}{I_s} = \frac{-h_{fe}I_b R_c'}{I_s} = \frac{-h_{fe}R_c'R}{R + h_{ie}}$$

$$R_M = \frac{(-50)(3.64)(8)}{8 + 1.1} = -160 \text{ K}$$

$$\beta = -\frac{1}{R'} = -\frac{1}{40} = -0.025 \text{ mA/V}$$

$$D = 1 + \beta R_M = 1 + 0.025 \times 160 = 5.00$$

$$R_{Mf} = \frac{R_M}{D} = \frac{-160}{5.00} = -32.0 \text{ K}$$

$$A_{Vf} = \frac{V_o}{V_s} = \frac{V_o}{I_s R_s} = \frac{R_{Mf}}{R_s}$$

or

$$A_{Vf} = \frac{-32.0}{10} = -3.20$$

It is instructive to examine the approximate expression for the voltage gain given in Eq. (12-28).

$$A_{Vf} \approx \frac{-R'}{R_s} = -\frac{40}{10} = -4.00$$

which differs from the exact value -3.20 by about 25 percent.

12-15 INSTABILITY IN FEEDBACK AMPLIFIERS

Negative feedback for which $|1 + A\beta| > 1$ has been considered in some detail in the foregoing sections. If $|1 + A\beta| < 1$, the feedback is termed *positive*, or *regenerative*. Under these circumstances, the resultant transfer gain A_f will be greater than A, the nominal gain without feedback, since $|A_f| = |A|/|1 + A\beta| > |A|$. Regeneration as an effective means of increasing the amplification of an amplifier was first suggested by Armstrong. Because of the reduced stability of an amplifier with positive feedback, this method is seldom used.

To illustrate the instability in an amplifier with positive feedback, consider the following situation: No signal is applied, but because of some transient disturbance, a signal X_o appears at the output terminals. A portion of this signal, $-\beta X_o$, will be fed back to the input circuit, and will appear in the output as an increased signal $-A\beta X_o$. If this term just equals X_o, then the spurious output has regenerated itself. In other words, if $-A\beta X_o = X_o$ (that is, if $-A\beta = 1$), the amplifier will oscillate. Hence, if an attempt is made to obtain large gain by making $|A\beta|$ almost equal to unity, there is the possibility that the amplifier may break out into spontaneous oscillation. This would occur if, because of variation in supply voltages, aging of transistors, etc., $-A\beta$ becomes equal to unity. There is little point in attempting to achieve amplification at the expense of stability. In fact, because of all the advantages enumerated in Sec. 12-1, feedback in amplifiers is almost always negative. However, combinations of positive and negative feedback are used.

The Condition for Stability If an amplifier is designed to have negative feedback in a particular frequency range but breaks out into oscillation at some high or low frequency, it is useless as an amplifier. Hence, in the design of a feedback amplifier, it must be ascertained that the circuit is stable at *all* frequencies, and not merely over the frequency range of interest. In the sense used here, the system is stable if a transient disturbance results in a response which dies out. A system is unstable if a transient disturbance persists indefinitely or increases until it is limited only by some nonlinearity in the circuit. The question of stability involves a study of the poles of the transfer function since these determine the transient behavior of the network. If a pole exists with a positive real part, this will result in a disturbance increasing exponentially with time. Hence the condition which must be satisfied, if a system is to be stable, is that the poles of the transfer function must all lie in the left-hand half of the complex-frequency plane. If the system without feedback is stable, the poles of A do lie in the left-hand half plane. It follows from Eq. (12-4), therefore, that *the stability condition requires that the zeros of* $1 + A\beta$ *all lie in the left-hand half of the complex-frequency plane.*

It is difficult to apply the above condition to a practical amplifier. An alternative but equivalent criterion for stability is contained in the following statement: *No oscillations are possible if the magnitude of the loop gain* $|A\beta|$ *is less than unity when the phase angle of* $A\beta$ *is* 180°. This condition is sought for in practice to ensure that the amplifier will be stable.

Consider, for example, a three-pole transfer function. To be specific, we can assume that this gain represents three cascaded stages, each with a dominant pole due to shunt capacitance. To simplify the discussion, consider that the amplifier stages are noninteracting, and that the poles are equal; $\omega_1 = \omega_2 = \omega_3$. There is a definite maximum value of the feedback fraction $|\beta|$ allowable for stable operation. To see this, note that at low frequencies there is a 180° phase shift in each stage and 540°, or equivalently 180°, for the three stages. In other words, the midband gain A_o is negative. Since we are considering negative feedback, then $1 + \beta A_o > 1$ and β must be negative (it is assumed to be real). At high frequencies there is a phase shift due to the shunting capacitances, and at the frequency for which the phase shift per stage is 60°, the total phase shift of A is zero, and of $A\beta$ is 180°. If the magnitude of the gain at this frequency is called A_{60}, then β must be chosen so that $A_{60}|\beta|$ is less than unity, if the possibility of oscillations is to be avoided. Similarly, because of the phase shift introduced by the blocking capacitors between stages, there is a low frequency for which the phase shift per stage is also 60°, and hence there is the possibility of oscillation at this low frequency also, unless the maximum value of β is restricted as outlined above.

It is convenient to plot the magnitude, usually in decibels, and also the phase of $A\beta$ as a function of frequency. If we can show that $|A\beta|$ is less than unity when the phase angle of $A\beta$ is 180°, the closed-loop amplifier will be stable.

Gain Margin   The gain margin is defined as the value of $|A\beta|$ in decibels at the frequency at which the phase angle of $A\beta$ is 180°. If the gain margin is negative, this gives the decibel rise in open-loop gain, which is theoretically permissible without oscillation. If the gain margin is positive, the amplifier is potentially unstable. This definition is illustrated in Fig. 12-18.

Phase Margin   The phase margin is 180° minus the magnitude of the angle of $A\beta$ at the frequency at which $|A\beta|$ is unity (zero decibels). The magnitudes of these quantities give an indication of how stable an amplifier is. For example, a linear amplifier of good stability requires gain and phase mar-

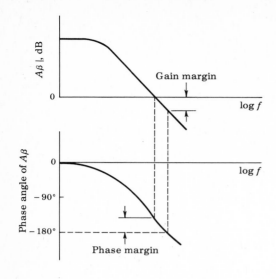

Fig. 12-18   Bode plots relating to the definitions of gain and phase margins.

gins of at least 10 dB and 50°, respectively.   This definition is illustrated in Fig. 12-18.

Compensation   From the above discussion we know that a multistage amplifier will oscillate if the open-loop gain $|A\beta|$ is unity when the phase shift of $A\beta$ is 180°.   In order to compensate an amplifier so as to prevent this instability, it is necessary to reshape the magnitude and phase plots of βA so that $|\beta A| < 1$ when the angle of βA is 180°.   There are three general methods of accomplishing this goal:

1. *Dominant-pole, or lag, compensation.*   This method inserts an extra pole into the transfer function at a lower frequency than the existing poles. Such a circuit introduces a phase lag into the amplifier.   This dominant pole also reduces the high 3-dB frequency and, hence, the bandwidth.

2. *Lead compensation.*   The amplifier or the feedback network is modified so as to add a zero to the transfer function, thereby increasing the phase.

3. *Pole-zero, or lag-lead, compensation.*   This technique adds both a pole (a lag) and a zero (a lead) to the transfer gain.   This zero is chosen so as to cancel the lowest pole.

A manufacturer usual indicates the best method for adding external elements (usually capacitors) in order to compensate a feedback amplifier and thereby prevent oscillations.

12-16   SINUSOIDAL OSCILLATORS

From the considerations in the preceding section it is clear that a feedback amplifier will function as an oscillator provided that

$$-A\beta = 1 \tag{12-47}$$

This condition is called the *Barkhausen criterion*. Since $-A\beta$ is the loop gain (Sec. 12-2), Eq. (12-47) states that *the frequency of a sinusoidal oscillator is determined by the condition that the loop-gain phase shift is zero*. This condition determines the frequency, provided that the circuit will oscillate at all. A second condition to be met is the following: *if oscillations are to be sustained, then the magnitude of the loop gain must equal unity*.

The above conditions imply both that $|A\beta| = 1$ and that the phase of $-A\beta$ is zero. The above principles are consistent with the feedback formula $A_f = A/(1 + \beta A)$. For if $-\beta A = 1$, then $A_f \to \infty$, which may be interpreted to mean that there exists an output voltage even in the absence of an externally applied signal voltage.

Practical Considerations Referring to Fig. 12-4, it appears that if $|\beta A|$ at the oscillator frequency is precisely unity, then, with the feedback signal connected to the input terminals, the removal of the external generator will make no difference. If $|\beta A|$ is less than unity, the removal of the external generator will result in a cessation of oscillations. But now suppose that $|\beta A|$ is greater than unity. Then, for example, a 1-V signal appearing initially at the input terminals will, after a trip around the loop and back to the input terminals, appear there with an amplitude larger than 1 V. This larger voltage will then reappear as a still larger voltage, and so on. It seems, then, that if $|\beta A|$ is larger than unity, the amplitude of the oscillations will continue to increase without limit. But of course, such an increase in the amplitude can continue only as long as it is not limited by the onset of nonlinearity of operation in the active devices associated with the amplifier. Such a non-linearity becomes more marked as the amplitude of oscillation increases. This onset of nonlinearity to limit the amplitude of oscillation is an essential feature of the operation of all practical oscillators, as the following considerations will show: The condition $|\beta A| = 1$ does not give a range of acceptable values of $|\beta A|$, but rather a single and precise value. Now suppose that initially it were even possible to satisfy this condition. Then, because circuit components and, more importantly, transistors change characteristics (drift) with age, temperature, voltage, etc., it is clear that if the entire oscillator is left to itself, in a very short time $|\beta A|$ will become either less or larger than unity. In the former case the oscillation simply stops, and in the latter case we are back to the point of requiring nonlinearity to limit the amplitude. An oscillator in which the loop gain is exactly unity is an abstraction completely unrealizable in practice. It is accordingly necessary, in the adjustment of a practical oscillator, always to arrange to have $|\beta A|$ somewhat larger (say 5 percent) than unity in order to ensure that, with incidental variations in transistor and circuit parameters, $|\beta A|$ shall not fall below unity.

If $|\beta A| \gg 1$, then the active devices will be driven excessively in a non-linear fashion. Each transistor will be taken from cutoff into saturation, back into cutoff, then into saturation again, and so forth. The oscillation waveform will then approach that of a square wave rather than that of a sinusoid (Sec. 14-4).

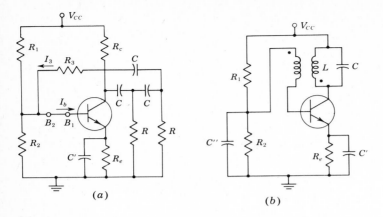

Fig. 12-19 (a) Transistor phase-shift oscillator. (b) A resonant-circuit oscillator.

Examples of Sinusoidal Oscillators   Many different topologies exist for sinusoidal oscillators depending upon the β feedback network, which now is a function of frequency.   If the output (the collector) of a transistor amplifier is connected back to the input (the base) by means of a three-section RC filter, then the so-called *phase-shift* oscillator of Fig. 12-19a is obtained.   Note that the transistor is biased in its active region (where the gain is greatest)

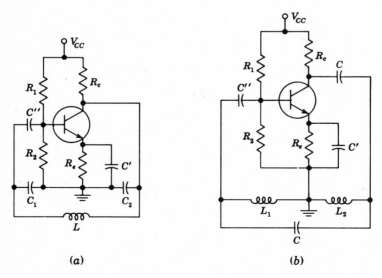

Fig. 12-20 (a) A transistor Colpitts oscillator. (b) A transistor Hartley oscillator.

by means of the resistors R_1, R_2, and R_e (Sec. 10-13). The bypass capacitor C' is used to avoid the reduction in gain due to the degeneration caused by R_e. This same self-biasing arrangement is used in all the oscillators in Figs. 12-19 and 12-20.

In Fig. 12-19b feedback is obtained by means of the transformer, which must introduce 180° phase shift. In Fig. 12-20a the β network consists of C_1, C_2, and L, whereas in Fig. 12-20b the feedback is supplied through L_1, L_2, and C. In each case the circuit components must be properly chosen to satisfy the Barkhausen conditions in order to sustain sinusoidal oscillations. Alternative forms of the feedback network consist of a Wien bridge, a piezo-electric crystal, and many other configurations. Also, FETs may be used for the active elements instead of bipolar transistors.

REVIEW QUESTIONS

12-1 Draw a feedback amplifier in block-diagram form. Identify each block, and state its function.

12-2 (a) What are the four possible topologies of a feedback amplifier? (b) Identify the output signal X_o and the feedback signal X_f for each topology (either as a current or voltage). (c) Identify the transfer gain A for each topology (for example, give its dimensions). (d) Define the feedback factor β.

12-3 (a) What is the relationship between the transfer gain with feedback A_f and that without feedback A? (b) Define *negative feedback*. (c) Define *positive feedback*. (d) Define the amount of feedback in decibels.

12-4 State the three fundamental assumptions which are made in order that the expression $A_f = A/(1 + A\beta)$ be satisfied exactly.

12-5 (a) Define *desensitivity D*. (b) For large values of D, what is A_f? What is the significance of this result?

12-6 List five characteristics of an amplifier which are modified by negative feedback.

12-7 State whether the input resistance R_{if} is increased or decreased for each of the four possible topologies.

12-8 Repeat Rev. 12-7 for the output resistance R_{of}.

12-9 Consider a feedback amplifier with a single-pole transfer function. (a) What is the relationship between the high 3-db frequency with and without feedback? (b) Repeat part a for the low 3-dB frequency. (c) Repeat part a for the gain-bandwidth product.

12-10 Draw the circuit of a single-stage voltage-series feedback amplifier.

12-11 Find A_f for a source follower using the feedback method of analysis.

12-12 Find A_f for an emitter follower.

12-13 Draw a circuit of a current-series feedback amplifier.

12-14 Find A_f for a CE stage with an unbypassed emitter resistor.

12-15 Find A_f for a CS stage with an unbypassed resistance in the source lead.

12-16 Draw the circuit of a feedback pair with current-shunt topology.

12-17 Draw the circuit of a voltage-shunt feedback amplifier.

12-18 (a) Define *positive* feedback. (b) What is the relationship between A_f and A for positive feedback?

12-19 State the condition on $1 + A\beta$ which a feedback amplifier must satisfy in order to be stable.

12-20 Define with the aid of graphs (a) gain margin and (b) phase margin.

12-21 (a) Describe *dominant-pole compensation*. (b) What is a disadvantage of this technique?

12-22 Describe *pole-zero compensation*.

12-23 Describe *lead compensation*.

12-24 Give the two Barkhausen conditions required in order for sinusoidal oscillations to be sustained.

12-25 Sketch the circuit of a phase-shift oscillator using a bipolar junction transistor.

12-26 Sketch the circuit of a resonant-circuit oscillator.

13 / OPERATIONAL AMPLIFIER LINEAR ANALOG SYSTEMS

The operational amplifier (abbreviated OP AMP) is a direct-coupled high-gain amplifier to which feedback is added to control its overall response characteristic. It is used to perform a wide variety of linear functions (and also many nonlinear operations) and is often referred to as the *basic linear* (or more accurately, *analog*) *integrated circuit*. In this chapter the following linear systems are discussed: analog computers, voltage-to-current and current-to-voltage converters, amplifiers of various types (for example, dc, instrumentation, tuned, and video amplifiers), voltage followers, and active filters.

Operational amplifier nonlinear analog systems are considered in the following chapter.

13-1 THE BASIC OPERATIONAL AMPLIFIER

The schematic diagram of the OP AMP is shown in Fig. 13-1a, and the equivalent circuit in Fig. 13-1b. A large number of operational amplifiers have a differential input, with voltages V_2 and V_1 applied to the inverting and noninverting terminals, respectively. A single-ended amplifier may be considered as a special case where one of the input terminals is grounded. Nearly all OP AMPS have only one output terminal.

Ideal Operational Amplifier The ideal OP AMP has the following characteristics:

1. Input resistance $R_i = \infty$.
2. Output resistance $R_o = 0$.
3. Voltage gain $A_V = -\infty$.
4. Bandwidth $= \infty$.
5. $V_o = 0$ when $V_1 = V_2$ independent of the magnitude of V_1.
6. Characteristics do not drift with temperature.

311

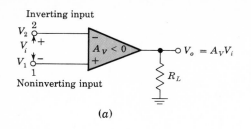

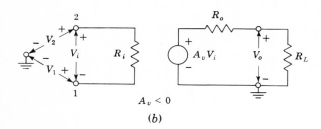

Fig. 13-1 (a) Basic operational amplifier. (b) Low-frequency equivalent circuit of operational amplifier ($V_i = V_2 - V_1$). The open-circuit voltage gain is A_v and the gain under load is A_V.

In Fig. 13-2 we show the ideal operational amplifier with feedback impedances (Z and Z') and the + terminal grounded. This is the basic inverting circuit. The voltage gain A_{Vf} with feedback is obtained as follows: Since $R_i \to \infty$, the current I through Z also passes through Z', as indicated in Fig. 13-2a. In addition, we note that $V_i = V_o/A_V \to 0$ as $|A_V| \to \infty$, so that the input is effectively shorted. Hence

$$A_{Vf} = \frac{V_o}{V_s} = \frac{-IZ'}{IZ} = -\frac{Z'}{Z} \tag{13-1}$$

This topology represents voltage-shunt feedback, discussed in Sec. 12-9. The circuit of Fig. 13-2a is a generalization of Fig. 12-11a, with the single transistor replaced by the multistage OP AMP and the resistors R_s and R' replaced by impedances Z and Z', respectively. From Eq. (12-28) the voltage gain A_{Vf} with feedback is given by Eq. (13-1).

The operation of the circuit may now be described in the following terms: At the input to the amplifier proper there exists a *virtual ground*, or *short circuit*. The term "virtual" is used to imply that, although the feedback from output to input through Z' serves to keep the voltage V_i at zero, no current actually flows into this short. The situation is depicted in Fig. 13-2b where the virtual ground is represented by the heavy double-headed arrow. This figure does not represent a physical circuit, but it is a convenient mnemonic aid from which to calculate the output voltage for a given input signal. This symbolism is very useful in connection with analog computations discussed in Sec. 13-7.

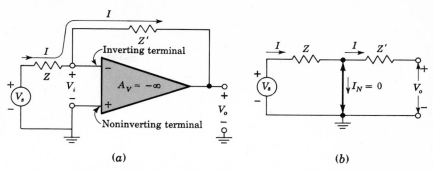

(a)   (b)

Fig. 13-2   (a) Inverting operational amplifier with added voltage-shunt feedback.   (b) Virtual ground in the operational amplifier.

Noninverting Operational Amplifier   Very often there is a need for an amplifier whose output is equal to, and in phase with, the input, and in addition $R_i = \infty$ and $R_o = 0$, so that the source and load are in effect isolated. An emitter follower approximates these specifications. More ideal characteristics can be obtained by using an operational amplifier having a noninverting terminal for signals and an inverting terminal for the feedback voltage, as shown in Fig. 13-3.

If we assume again that $R_i = \infty$, we have

$$V_2 = \frac{R}{R + R'} V_o$$

Since $V_o = A_V(V_2 - V_s)$, then for a finite V_o and $-A_V = \infty$ it follows that $V_s = V_2$ (there is a virtual short at the input terminals) and

$$A_{Vf} = \frac{V_o}{V_s} = \frac{V_o}{V_2} = \frac{R + R'}{R} \tag{13-2}$$

Hence the closed-loop gain is always greater than unity. If $R = \infty$ and/or $R' = 0$, then $A_{Vf} = +1$ and the amplifier acts as a *voltage follower*.

Fig. 13-3   Noninverting operational amplifier with resistive feedback. If $R \gg R'$, the output follows the input; $V_o \approx V_s$.

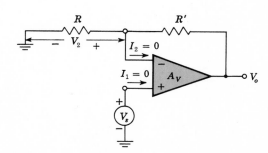

In the analysis of noninverting OP AMP circuits we shall use the facts that (1) *no current flows into either input* and (2) *the potentials of the two inputs are equal.*

13-2 THE DIFFERENTIAL AMPLIFIER

The function of a differential amplifier (abbreviated DIFF AMP) is, in general, to amplify the difference between two signals. The need for DIFF AMPS arises in many physical measurements where response from dc to many megahertz is required. It is also the basic stage of an integrated operational amplifier with differential input.

Figure 13-4 represents a linear active device with two input signals v_1, v_2 and one output signal v_o, each measured with respect to ground. In an ideal DIFF AMP the output signal v_o should be given by

$$v_o = A_d(v_1 - v_2) \tag{13-3}$$

where A_d is the gain of the differential amplifier. Thus it is seen that any signal which is common to both inputs will have no effect on the output voltage. However, a practical DIFF AMP cannot be described by Eq. (13-3), because, in general, the output depends not only upon the *difference signal v_d* of the two signals, but also upon the average level, called the *common-mode signal v_c*, where

$$v_d \equiv v_1 - v_2 \qquad \text{and} \qquad v_c \equiv \tfrac{1}{2}(v_1 + v_2) \tag{13-4}$$

For example, if one signal is $+50\ \mu V$ and the second is $-50\ \mu V$, the output will not be exactly the same as if $v_1 = 1{,}050\ \mu V$ and $v_2 = 950\ \mu V$, even though the difference $v_d = 100\ \mu V$ is the same in the two cases.

The Common-Mode Rejection Ratio The foregoing statements are now clarified, and a figure of merit for a difference amplifier is introduced. The output of Fig. 13-4 can be expressed as a linear combination of the two input voltages v_1 and v_2. From the linear relationships in Eq. (13-4) it is clear that v_o must be a linear combination of v_d and v_c. Thus,

$$v_o = A_d v_d + A_c v_c \tag{13-5}$$

where the voltage gain for the difference signal is A_d, and that for the common-mode signal is A_c. We can measure A_d directly by setting $v_1 = -v_2 = 0.5$ V, so that $v_d = 1$ V and $v_c = 0$. Under these conditions the measured output

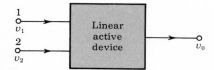

Fig. 13-4 The output is a linear function of v_1 and v_2 for an ideal differential amplifier.
$v_o = A_d(v_1 - v_2).$

voltage v_o gives the gain A_d for the difference signal [Eq. (13-4)]. Similarly, if we set $v_1 = v_2 = 1$ V, then $v_d = 0$, $v_c = 1$ V, and $v_o = A_c$. The output voltage now is a direct measurement of the common-mode gain A_c.

Clearly, we should like to have A_d large, whereas ideally, A_c should equal zero. A quantity called the *common-mode rejection ratio*, which serves as a figure of merit for a DIFF AMP, is defined by

$$\text{CMRR} \equiv \rho \equiv \left| \frac{A_d}{A_c} \right| \tag{13-6}$$

From Eqs. (13-5) and (13-6) we obtain an expression for the output in the following form:

$$v_o = A_d v_d \left(1 + \frac{1}{\rho} \frac{v_c}{v_d} \right) \tag{13-7}$$

From this equation we see that the amplifier should be designed so that ρ is large compared with the ratio of the common-mode signal to the difference signal. For example, if $\rho = 1,000$, $v_c = 1$ mV, and $v_d = 1$ μV, the second term in Eq. (13-7) is equal to the first term. Hence, for an amplifier with a common-mode rejection ratio of 1,000, a 1-μV difference of potential between the two inputs gives the same output as a 1-mV signal applied with the same polarity to both inputs.

13-3   THE EMITTER–COUPLED DIFFERENTIAL AMPLIFIER

The circuit of Fig. 13-5 is an excellent DIFF AMP if the emitter resistance R_e is large. This statement can be justified as follows: If $V_{s1} = V_{s2} = V_s$, then from Eqs. (13-4) and (13-5) we have $V_d = V_{s1} - V_{s2} = 0$ and $V_o = A_c V_s$. However, if $R_e = \infty$, then, because of the symmetry of Fig. 13-5, we obtain $I_{e1} = I_{e2} = 0$. If $I_{b2} \ll I_{c2}$, then $I_{c2} \approx I_{e2}$, and it follows that $V_o = 0$. Hence the common-mode gain A_c becomes very small, and the common-mode rejec-

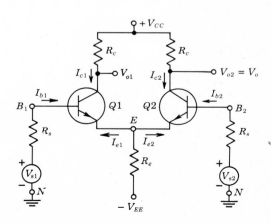

Fig. 13-5 Symmetrical emitter-coupled difference amplifier.

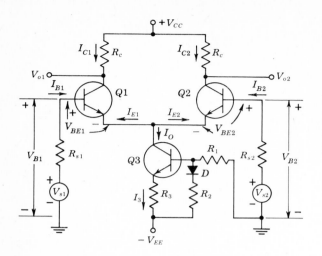

Fig. 13-6 Differential amplifier with constant-current stage in the emitter circuit. Nominally, $R_{s1} = R_{s2}$.

tion ratio is very large for a very large value of R_e and a symmetrical circuit. There are, however, practical limitations on the magnitude of R_e because of the quiescent dc voltage drop across it; the emitter supply V_{EE} must become larger as R_e is increased, in order to maintain the quiescent current at its proper value. If the operating currents of the transistors are allowed to decrease, this will lead to higher h_{ie} values and lower values of h_{fe}. Both of these effects will tend to decrease the common-mode rejection ratio.

Differential Amplifier Supplied with a Constant Current   Frequently, in practice, R_e is replaced by a transistor circuit, as in Fig. 13-6, in which R_1, R_2, and R_3 can be adjusted to give the same quiescent conditions for $Q1$ and $Q2$ as the original circuit of Fig. 13-5.   This modified circuit of Fig. 13-6 presents a very high effective emitter resistance R_e for the two transistors $Q1$ and $Q2$. This statement follows from the fact that R_e represents the output resistance of a CE amplifier with an emitter resistor R_3.   It can be verified that R_e is hundreds of kilohms even if R_3 is as small as 1 K.   Since R_e represents a very high value of dynamic resistance, the incremental current through R_e must be very small.   In other words, the current I_o in R_e must remain essentially constant.

We now verify that $Q3$ does indeed act as an approximately constant-current source, subject to the condition that the base current of $Q3$ is negligible. Applying KVL to the base circuit of $Q3$, we have

$$I_3 R_3 + V_{BE3} = V_D + (V_{EE} - V_D) \frac{R_2}{R_1 + R_2} \tag{13-8}$$

where V_D is the diode voltage.   Hence

$$I_o \approx I_3 = \frac{1}{R_3} \left(\frac{V_{EE} R_2}{R_1 + R_2} + \frac{V_D R_1}{R_1 + R_2} - V_{BE3} \right) \tag{13-9}$$

If the circuit parameters are chosen so that

$$\frac{V_D R_1}{R_1 + R_2} = V_{BE3} \tag{13-10}$$

then

$$I_O = \frac{V_{EE} R_2}{R_3(R_1 + R_2)} \tag{13-11}$$

Since this current is independent of the signal voltages V_{s1} and V_{s2}, then $Q3$ acts to supply the DIFF AMP consisting of $Q1$ and $Q2$ with the constant current I_O.

The above result for I_O has been rendered independent of temperature because of the added diode D. Without D the current would vary with temperature because V_{BE3} decreases approximately 2.5 mV/°C (Sec. 2-4). The diode has this same temperature dependence, and hence the two variations cancel each other and I_O does not vary appreciably with temperature. Since the cutin voltage V_D of a diode has approximately the same value as the base-to-emitter voltage V_{BE3} of a transistor, then Eq. (13-10) cannot be satisfied with a single diode. Hence two diodes in series are used for V_D.

The above discussion assumes that the resistances do not vary with temperature T. Since these have a negative temperature coefficient of resistance then, in practice, Eq. (13-10) is not satisfied, but $R_1/(R_1 + R_2)$ is chosen experimentally so that I_O in Eq. (13-9) is almost independent of T.

Consider that $Q1$ and $Q2$ are identical and that $Q3$ is a true constant-current source. Under these circumstances we can demonstrate that the common-mode gain is zero. Assume that $V_{s1} = V_{s2} = V_s$, so that from the symmetry of the circuit, the collector current I_{c1} (the increase over the quiescent value for $V_s = 0$) in $Q1$ equals the current I_{c2} in $Q2$. However, since the total current increase $I_{c1} + I_{c2} = 0$ if $I_O = $ constant, then $I_{c1} = I_{c2} = 0$ and $A_c = V_{o2}/V_s = -I_{c2}R_c/V_s = 0$.

Practical Considerations The differential amplifier is often used in dc applications. It is difficult to design dc amplifiers using transistors because of drift due to variations of h_{FE}, V_{BE}, and I_{CBO} with temperature. A shift in any of these quantities changes the output voltage and cannot be distinguished from a change in input-signal voltage. Using the techniques of integrated circuits (Chap. 5), it is possible to construct a DIFF AMP with $Q1$ and $Q2$ having almost identical characteristics. Under these conditions any parameter changes due to temperature will cancel and the output will not vary.

Differential amplifiers may be cascaded to obtain larger amplifications for the difference signal. Outputs V_{o1} and V_{o2} are taken from each collector (Fig. 13-6) and are coupled directly to the two bases, respectively, of the next stage.

Finally, the differential amplifier may be used as an emitter-coupled phase inverter. For this application the signal is applied to one base, whereas the second base is not excited (but is, of course, properly biased). The output voltages taken from the collectors are equal in magnitude and 180° out of phase.

13-4   TRANSFER CHARACTERISTICS OF A DIFFERENTIAL AMPLIFIER

It is important to examine the transfer characteristic (I_C versus $V_{B1} - V_{B2}$) of the DIFF AMP of Fig. 13-6 to understand its advantages and limitations. We first consider this circuit qualitatively. When V_{B1} is below the cutoff point of $Q1$, all the current I_O flows through $Q2$ (assume for this discussion that V_{B2} is constant). As V_{B1} carries $Q1$ above cutoff, the current in $Q1$ increases, while the current in $Q2$ decreases, and the sum of the currents in the two transistors remain constant and equal to I_O. The total range ΔV_O over which the output can vary is $R_C I_O$ and is therefore adjustable through an adjustment of I_O.

If we assume that $Q1$ and $Q2$ are matched, it can be shown that

$$I_{C1} = \frac{I_O}{1 + \exp\left[-(V_{B1} - V_{B2})/V_T\right]} \tag{13-12}$$

and I_{C2} is given by the same expression with V_{B1} and V_{B2} interchanged. The transfer characteristics described by Eq. (13-12) for the normalized collector currents I_{C1}/I_O (and I_{C2}/I_O) are shown in Fig. 13-7, where the abscissa is the normalized differential input $(V_{B1} - V_{B2})/V_T$.

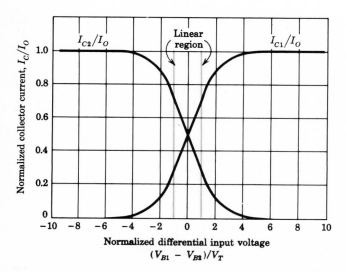

Fig. 13-7 Transfer characteristics of the basic differential amplifier circuit.

If Eq. (13-12) is differentiated with respect to $V_{B1} - V_{B2}$, we have the transconductance g_{md} of the DIFF AMP with respect to the differential input voltage, or

$$\frac{dI_{C1}}{d(V_{B1} - V_{B2})} = g_{md} = \frac{I_O}{4V_T} \tag{13-13}$$

where g_{md} is evaluated at $V_{B1} = V_{B2}$. This equation indicates that, for the same value of I_O, the effective transconductance of the differential amplifier is one-fourth that of a single transistor [Eq. (11-37)].

The following conclusions can be drawn from the transfer curves of Fig. 13-7:

1. The differential amplifier is a very good limiter, since when the input $(V_{B1} - V_{B2})$ exceeds $\pm 4V_T$ ($\approx \pm 100$ mV at room temperature), very little further increase in the output is possible.

2. The slope of these curves defines the transconductance, and it is clear that g_{md} starts from zero, reaches a maximum of $I_O/4V_T$ when $I_{C1} = I_{C2} = \frac{1}{2}I_O$, and again approaches zero.

3. The value of g_{md} is proportional to I_O [Eq. (13-13)]. Since the output voltage change V_{o2} is given by [with $\Delta V_B = v_b$, Eq. (10-1)]

$$V_{o2} = g_{md}R_c\Delta(V_{B1} - V_{B2}) = g_{md}R_c(V_{b1} - V_{b2}) \tag{13-14}$$

it is possible to change the differential gain by varying the value of the current I_O. This means that automatic gain control (AGC) is possible with the DIFF AMP.

4. The transfer characteristics are linear in a small region around the operating point $V_{B1} = V_{B2}$ where the input $V_{B1} - V_{B2}$ varies approximately $\pm V_T$ (± 26 mV at room temperature).

13-5   AN EXAMPLE OF AN IC OPERATIONAL AMPLIFIER

An integrated OP AMP usually consists of a cascade of four stages. As indicated in Fig. 13-8, the first stage is a DIFF AMP with a double-ended output, the second stage is a DIFF AMP with a single-ended output, the third stage is an emitter follower, and the last stage is a dc level translator and output driver. The values of the circuit components and the topology of the section called the *level translator* are such that if the dc difference voltage to the input DIFF AMP is zero, then the output voltage V_o is approximately zero. The output stage terminates in a totem-pole driver similar to that used with the TTL logic gate of Fig. 6-22.

Note that the fourth stage in Fig. 13-8 is itself connected as a voltage-shunt feedback amplifier and, therefore, $A_{V4} = -R_{10}/R_9$. The overall OP AMP voltage gain is $-A_{V1}A_{V2}R_{10}/R_9$, since the emitter-follower gain is approximately unity. The gains of the DIFF AMP stages A_{V1} and A_{V2} are stable

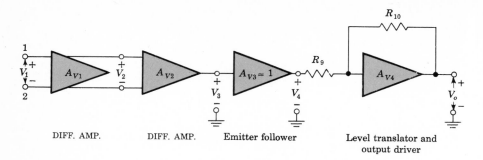

DIFF. AMP.   DIFF. AMP.   Emitter follower   Level translator and
                                                                  output driver

Fig. 13-8 The Motorola MC1530 as a four-stage cascaded amplifier.

for the reasons given in Sec. 13-4, and, hence, the overall amplification of the IC op amp is large and stable.

The integrated operational amplifier has gained wide acceptance as a versatile, predictable, and economic system building block. It offers all the advantages of monolithic integrated circuits: small size, high reliability, reduced cost, temperature tracking, and low offset voltage and current (terms which are defined in the next section).

13-6 OFFSET ERROR VOLTAGES AND CURRENTS

In Sec. 13-1 we observe that the ideal operational amplifier shown in Fig. 13-1a is perfectly balanced, that is, $V_o = 0$ when $V_1 = V_2$. A real operational amplifier exhibits an unbalance caused by a mismatch of the input transistors. This mismatch results in unequal bias currents flowing through the input terminals, and also requires that an input offset voltage be applied between the two input terminals to balance the amplifier output.

In this section we are concerned with the dc error voltages and currents that can be measured at the input and output terminals.

Input Bias Current The input bias current is one-half the sum of the separate currents entering the two input terminals of a balanced amplifier, as shown in Fig. 13-9a. Since the input stage is that shown in Fig. 13-6, the input bias current is $I_B = (I_{B1} + I_{B2})/2$ when $V_o = 0$.

Input Offset Current The input offset current I_{io} is the difference between the separate currents entering the input terminals of a balanced amplifier. As shown in Fig. 13-9a, we have $I_{io} = I_{B1} - I_{B2}$ when $V_o = 0$.

Input Offset Current Drift The input offset current drift $\Delta I_{io}/\Delta T$ is the ratio of the change of input offset current to the change of temperature.

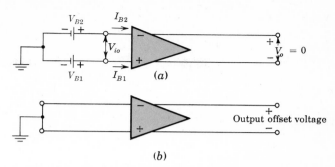

Fig. 13-9   (a) Input offset voltage.   (b) Output offset voltage.

Input Offset Voltage   The input offset voltage V_{io} is that voltage which must be applied between the input terminals to balance the amplifier, as shown in Fig. 13-9a.

Input Offset Voltage Drift   The input offset voltage drift $\Delta V_{io}/\Delta T$ is the ratio of the change of input offset voltage to the change in temperature.

Output Offset Voltage   The output offset voltage is the difference between the dc voltages present at the two output terminals (or at the output terminal and ground for an amplifier with one output) when the two input terminals are grounded (Fig. 13-9b).

Power Supply Rejection Ratio   The power supply rejection ratio (PSRR) is the ratio of the change in input offset voltage to the corresponding change in one power supply voltage, with all remaining power supply voltages held constant.

Slew Rate   The slew rate is the time rate of change of the closed-loop amplifier output voltage under large-signal conditions.

The various parameters of a typical monolithic operational amplifier are given in Table 13-1.

TABLE 13-1   Typical parameters of monolithic operational amplifier

Open-loop gain A_d	50,000
Input offset voltage V_{io}	1 mV
Input offset current I_{io}	10 nA
Input bias current I_B	100 nA
Common-mode rejection ratio ρ	100 dB
PSRR	20 μV/V
I_{io} drift	0.1 nA/°C
V_{io} drift	1.0 μV/°C
Slew rate	10 V/μs

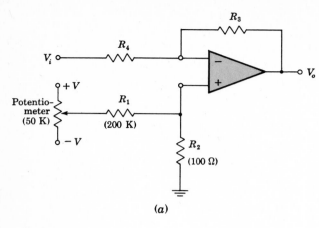

(a)

Fig. 13-10 Universal offset-voltage balancing circuits for (a) inverting and (b) noninverting operational amplifiers.

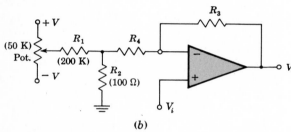

(b)

Universal Balancing Techniques   When we use an operational amplifier, it is often necessary to balance the offset voltage.   This means that we must apply a small dc voltage in the input so as to cause the dc output voltage to become zero.   The techniques shown here allow offset-voltage balancing without regard to the internal circuitry of the amplifier.   The circuit shown in Fig. 13-10a supplies a small voltage effectively in series with the noninverting input terminal in the range $\pm V[R_2/(R_1 + R_2)] = \pm 7.5$ mV if ± 15-V supplies are used and $R_1 = 200$ K, $R_2 = 100$ Ω.   Thus this circuit is useful for balancing inverting amplifiers even when the feedback element R_3 is a capacitor or a nonlinear element.   If the operational amplifier is used as a noninverting amplifier, the circuit of Fig. 13-10b is used for balancing the offset voltage.

13-7   BASIC OPERATIONAL AMPLIFIER APPLICATIONS

An OP AMP may be used to perform many mathematical operations.   This feature accounts for the name *operational amplifier*.   Some of the basic applications are given in this section.   Consider the ideal OP AMP of Fig. 13-11a.

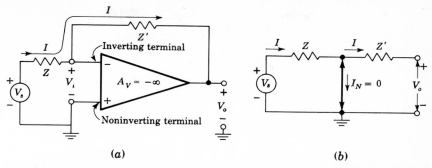

Fig. 13-11 (a) Inverting operational amplifier with voltage-shunt feedback. (b) Virtual ground in the OP AMP. (This is Fig 13-2 repeated.)

Recalling (Sec. 13-1) that the equivalent circuit of Fig. 13-11b has a virtual ground (which takes no current), it follows that the voltage gain is given by

$$A_{Vf} = \frac{V_o}{V_s} = -\frac{Z'}{Z} \tag{13-15}$$

Based upon this equation we can readily obtain an *analog inverter*, a *scale changer*, a *phase shifter*, and an *adder*.

Sign Changer, or Inverter   If $Z = Z'$ in Fig. 13-11, then $A_{Vf} = -1$, and the sign of the input signal has been changed. Hence such a circuit acts as a phase inverter. If two such amplifiers are connected in cascade, the output from the second stage equals the signal input without change of sign. Hence the outputs from the two stages are equal in magnitude but opposite in phase, and such a system is an excellent *paraphase amplifier*.

Scale Changer   If the ratio $Z'/Z = k$, a real constant, then $A_{Vf} = -k$, and the scale has been multiplied by a factor $-k$. Usually, in such a case of multiplication by a constant, -1 or $-k$, Z and Z' are selected as precision resistors.

Phase Shifter   Assume that Z and Z' are equal in magnitude but differ in angle. Then the operational amplifier shifts the phase of a sinusoidal input voltage while at the same time preserving its amplitude. Any phase shift from 0 to 360° (or ±180°) may be obtained.

Adder, or Summing Amplifier   The arrangement of Fig. 13-12 may be used to obtain an output which is a linear combination of a number of input signals. Since a virtual ground exists at the OP AMP input, then

$$i = \frac{v_1}{R_1} + \frac{v_2}{R_2} + \cdots + \frac{v_n}{R_n}$$

and

$$v_o = -R'i = -\left(\frac{R'}{R_1}v_1 + \frac{R'}{R_2}v_2 + \cdots + \frac{R'}{R_n}v_n\right) \tag{13-16a}$$

If $R_1 = R_2 = \cdots = R_n$, then

$$v_o = -\frac{R'}{R_1}(v_1 + v_2 + \cdots + v_n) \tag{13-16b}$$

and the output is proportional to the sum of the inputs.

Many other methods may, of course, be used to combine signals. The present method has the advantage that it may be extended to a very large number of inputs requiring only one additional resistor for each additional input. The result depends, in the limiting case of large amplifier gain, only on the resistors involved, and because of the virtual ground, there is a minimum of interaction between input sources.

Voltage-to-Current Converter   Often it is desirable to convert a voltage signal to a proportional output current. This is required, for example, when we drive a deflection coil in a television tube. If the load impedance has neither side grounded (if it is floating), the simple circuit of Fig. 13-12 with R' replaced by the load impedance Z_L is an excellent *voltage-to-current converter*. For a single input $v_1 = v_s(t)$, the current in Z_L is

$$i_L = \frac{v_s(t)}{R} \tag{13-17}$$

Note that i is independent of the load Z_L, because of the virtual ground of the operational amplifier input. Since the same current flows through the signal source and the load in Fig. 13-12, it is important that the signal source be capable of providing this load current. On the other hand, the amplifier of Fig. 13-13a requires very little current from the signal source due to the very large input resistance seen by the noninverting terminal.

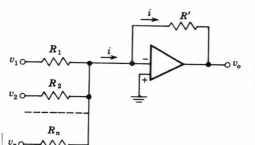

Fig. 13-12   Operational adder, or summing amplifier.

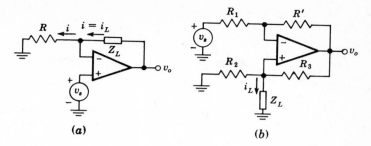

Fig. 13-13 Voltage-to-current converter for (a) a floating load and (b) a grounded load Z_L.

If one side of the load Z_L is grounded, the circuit of Fig. 13-13b can be used. In Prob. 13-15 we show that if $R_3/R_2 = R'/R_1$, then

$$i_L(t) = -\frac{v_s(t)}{R_2} \tag{13-18}$$

The *voltage-to-current converter* is used in the digital-to-analog (D/A) converter discussed in Sec. 14-7.

Current-to-Voltage Converter   Photocells and photomultiplier tubes give an output current which is independent of the load. The circuit in Fig. 13-14 shows an operational amplifier used as a current-to-voltage converter. Due to the virtual ground at the amplifier input, the current in R_s is zero and i_s flows through the feedback resistor R'. Thus the output voltage v_o is $v_o = -i_s R'$. It must be pointed out that the lower limit on current measurement with this circuit is set by the bias current of the inverting input. It is common to parallel R' with a capacitance C' to reduce high-frequency noise.

DC Voltage Follower   The simple configuration of Fig. 13-15 approaches the ideal *voltage follower*. Because the two inputs are tied together (virtually), then $V_o = V_s$ and *the output follows the input*. The LM 102 (National Semiconductor Corporation) is specifically designed for voltage-follower usage

Fig. 13-14 Current-to-voltage converter.

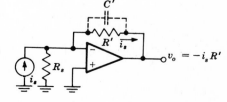

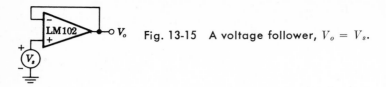

Fig. 13-15   A voltage follower, $V_o = V_s$.

and has very high input resistance (10,000 M), very low input current ($\sim$3 nA), and very low output resistance ($\sim$0 Ω).

13-8   DIFFERENTIAL DC AMPLIFIER

The differential-input single-ended-output instrumentation amplifier is often used to amplify inputs from transducers which convert a physical parameter and its variations into an electric signal. Such transducers are strain-gauge bridges, thermocouples, etc.   The circuit shown in Fig. 13-16 is very simple and uses only one OP AMP.   In Prob. 13-16 we show that if $R_2/R_1 = R_4/R_3$, then

$$V_o = \frac{R_2}{R_1}(V_1 - V_2) \tag{13-19}$$

If the signals V_1 and V_2 have source resistances R_{s1} and R_{s2}, then these resistances add to R_3 and R_1, respectively.   Note that the signal source V_1 sees a resistance $R_3 + R_4 = 101$ K.   If $V_1 = 0$, the inverting input is at ground potential and hence V_2 is loaded by $R_1 = 1$ K.   If this is too heavy a load for the transducer, a voltage follower may be used as a buffer.   If the configuration of Fig. 13-15 precedes each input in Fig. 13-16, the resulting topology of three IC OP AMPS represents a very high-input-resistance, high-performance, low-cost, dc amplifier system.

Bridge Amplifier   A differential amplifier is often used to amplify the output from a transducer bridge, as shown in Fig. 13-17.   Nominally, the four arms of the bridge have equal resistances R.   However, one of the branches has a resistance which changes to $R + \Delta R$ with temperature or some other physical parameter.   The goal of the measurement is to obtain the fractional change δ of the resistance value of the active arm, or $\delta = \Delta R/R$.

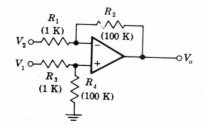

Fig. 13-16   Differential amplifier using one OP AMP.   (The offset-voltage balancing arrangement is not indicated.)

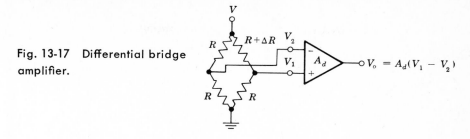

Fig. 13-17   Differential bridge amplifier.

In Prob. 13-19 we find that for the circuit of Fig. 13-17, the output V_o is given by

$$V_o = -\frac{A_d V}{4} \frac{\delta}{1 + \delta/2} \tag{13-20}$$

For small changes in R ($\delta \ll 1$) Eq. (13-20) reduces to

$$V_o = -\frac{A_d V}{4} \delta \tag{13-21}$$

13-9   STABLE AC–COUPLED AMPLIFIER

In some applications the need arises for the amplification of an ac signal, while any dc signal present must be blocked. A very simple and stable ac amplifier is shown in Fig. 13-18a, where capacitor C blocks the dc component of the input signal and together with the resistor R sets the low-frequency 3-dB response for the overall amplifier.

The output voltage V_o as a function of the complex variable s is found from the equivalent circuit of Fig. 13-18b (where the double-ended heavy arrow represents the virtual ground) to be

$$V_o = -IR' = -\frac{V_s}{R + 1/sC} R'$$

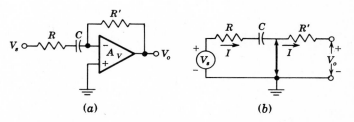

(a)   (b)

Fig. 13-18   (a) AC stable feedback amplifier.   (b) Equivalent circuit when $|A_V| = \infty$.

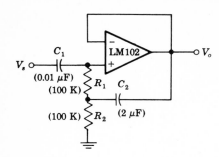

Fig. 13-19 AC voltage follower. (Courtesy of National Semiconductor Corporation.)

and

$$A_{Vf} = \frac{V_o}{V_s} = -\frac{R'}{R}\frac{s}{s + 1/RC} \tag{13-22}$$

From Eq. (13-22) we see that the low 3-dB frequency is

$$f_L = \frac{1}{2\pi RC} \tag{13-23}$$

The high-frequency response is determined by the frequency characteristics of the operational amplifier A_V and the amount of voltage-shunt feedback present.   The midband gain is, from Eq. (13-22), $A_{Vf} = -R'/R$.

AC Voltage Follower   The ac voltage follower is used to provide impedance buffering, that is, to connect a signal source with high internal source resistance to a load of low impedance, which may even be capacitive.   In Fig. 13-19 is shown a practical high-input impedance ac voltage follower using the LM 102 operational amplifier.   We assume that C_1 and C_2 represent short circuits at all frequencies of operation of this circuit.   Resistors R_1 and R_2 are used to provide RC coupling and allow a path for the dc input current into the noninverting terminal.   In the absence of the capacitor C_2, the ac signal source would see an input resistance of only $R_1 + R_2 = 200$ K.   Since the LM 102 is connected as a voltage follower, the voltage gain A_V between the output terminal and the noninverting terminal is very close to unity.   Thus, from our discussion of Miller's theorem in Appendix C-4, the input resistance the source sees becomes, approximately, $R_1/(1 - A_V)$, which is measured to be 12 M at 100 Hz and increases to 100 M at 1 kHz.

13-10   ANALOG INTEGRATION AND DIFFERENTIATION

The analog integrator is very useful in many applications which require the generation or processing of analog signals.   If, in Fig. 13-11, $Z = R$ and a capacitor C is used for Z', as in Fig. 13-20, we can show that the circuit per-

Fig. 13-20 (*a*) Operational integrator. (*b*) Equivalent circuit.

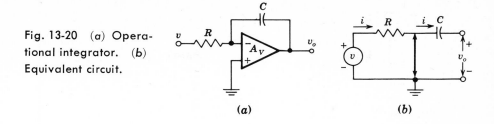

(*a*)   (*b*)

forms the mathematical operation of integration. The input need not be sinusoidal, and hence is represented by the lowercase symbol $v = v(t)$. (The subscript s is now omitted, for simplicity.)   In Fig. 13-20*b*, the double-headed arrow represents a virtual ground.   Hence $i = v/R$, and

$$v_o = -\frac{1}{C} \int i\, dt = -\frac{1}{RC} \int v\, dt \qquad (13\text{-}24)$$

The amplifier therefore provides an output voltage proportional to the integral of the input voltage.

If the input voltage is a constant, $v = V$, then the output will be a ramp, $v_o = -Vt/RC$.   Such an integrator makes an excellent sweep circuit for a cathode-ray-tube oscilloscope, and is called a *Miller integrator,* or *Miller sweep.*

DC Offset and Bias Current   The input stage of the operational amplifier used in Fig. 13-20 is usually a DIFF AMP.   The dc input offset voltage V_{io} appears across the amplifier input, and this voltage will be integrated and will appear at the output as a linearly increasing voltage.   Part of the input bias current will also flow through the feedback capacitor, charging it and producing an additional linearly increasing voltage at the output.   These two ramp voltages continue to increase until the amplifier reaches its saturation point.   We see then that a limit is set on the feasible integration time by the above error components.   The effect of the bias current can be minimized by increasing the feedback capacitor C while simultaneously decreasing the value of R for a given value of the time constant RC.

Practical Circuit   A practical integrator must be provided with an external circuit to introduce initial conditions, as shown in Fig. 13-21.   When switch S is in position 1, the input is zero and capacitor C is charged to the voltage V, setting an initial condition of $v_o = V$.   When swtch S is in position 2, the amplifier is connected as an integrator and its output will be V plus a constant times the time integral of the input voltage v.   In using this circuit, care must be exercised to stabilize the amplifier and $I_{B2}R_2$ must be equal to $I_{B1}R_1$ to minimize the error due to bias current.

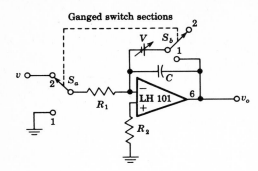

Fig. 13-21 Practical integrator circuit. For minimum offset error due to input bias current it is required that $I_{B1}R_1 = I_{B2}R_2$. (Courtesy of National Semiconductor Corporation.)

Differentiator   If in Fig. 13-11 Z is a capacitor C and if $Z' = R$, we see from the equivalent circuit of Fig. 13-22 that $i = C\,dv/dt$ and

$$v_o = -Ri = -RC\frac{dv}{dt} \tag{13-25}$$

Hence the output is proportional to the time derivative of the input. If the input signal is $v = \sin \omega t$, then the output will be $v_o = -RC\omega \cos \omega t$. Thus the magnitude of the output increases linearly with increasing frequency, and the differentiator circuit has high gain at high frequencies. This results in amplification of the high-frequency components of amplifier noise, and the noise output may completely obscure the differentiated signal.

13-11   ELECTRONIC ANALOG COMPUTATION

The OP AMP is the fundamental building block in an electronic analog computer. As an illustration, let us consider how to program the differential equation

$$\frac{d^2v}{dt^2} + K_1\frac{dv}{dt} + K_2v - v_1 = 0 \tag{13-26}$$

where v_1 is a given function of time, and K_1 and K_2 are real positive constants.

We begin by assuming that d^2v/dt^2 is available in the form of a voltage. Then, by means of an integrator, a voltage proportional to dv/dt is obtained. A second integrator gives a voltage proportional to v. Then an adder (and scale changer) gives $-K_1(dv/dt) - K_2v + v_1$. From the differential equa-

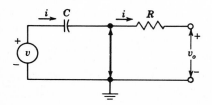

Fig. 13-22 Equivalent circuit of the operational differentiator.

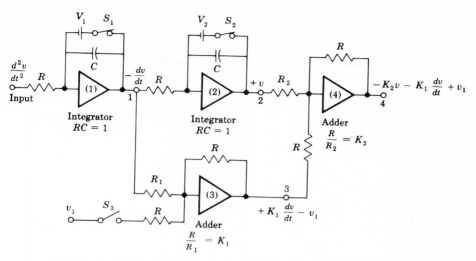

Fig. 13-23 A block diagram of an electronic analog computer. At $t = 0$, S_1 and S_2 are opened and S_3 is closed. Each OP AMP input is as in Fig. 13-21.

tion (13-26), this equals d^2v/dt^2, and hence the output of this summing amplifier is fed to the input terminal, where we had assumed that d^2v/dt^2 was available in the first place.

The procedure outlined above is carried out in Fig. 13-23. The voltage d^2v/dt^2 is assumed to be available at an input terminal. The integrator (1) has a time constant $RC = 1$ s, and hence its output at terminal 1 is $-dv/dt$. This voltage is fed to a similar integrator (2), and the voltage at terminal 2 is $+v$. The voltage at terminal 1 is fed to the inverter and scale changer (3), and its output at terminal 3 is $+K_1(dv/dt)$. This same operational amplifier (3) is used as an adder. Hence, if the given voltage $v_1(t)$ is also fed into it as shown, the output at terminal 3 also contains the term $-v_1$, or the net output is $+K_1(dv/dt) - v_1$. Scale changer–adder (4) is fed from terminals 2 and 3, and hence delivers a resultant voltage $-K_2v - K_1(dv/dt) + v_1$ at terminal 4. By Eq. (13-26) this must equal d^2v/dt^2, which is the voltage that was assumed to exist at the input terminal. Hence the computer is completed by connecting terminal 4 to the input terminal. (This last step is omitted from Fig. 13-23 for the sake of clarity of explanation.)

The specified initial conditions (the value of dv/dt and v at $t = 0$) must now be inserted into the computer. We note that the voltages at terminals 1 and 2 in Fig. 13-23 are proportional to dv/dt and v, respectively. Hence initial conditions are taken care of (as in Fig. 13-21) by applying the correct voltages V_1 and V_2 across the capacitors in integrators 1 and 2, respectively.

The solution is obtained by opening switches S_1 and S_2 and simultaneously closing S_3 (by means of relays) at $t = 0$ and observing the waveform at terminal 2. If the derivative dv/dt is also desired, its waveform is

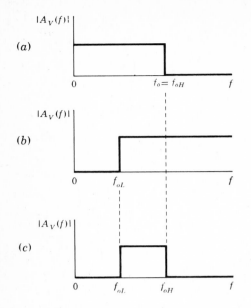

Fig. 13-24   Ideal filter characteristics. (a) Low-pass, (b) high-pass, and (c) bandpass.

available at terminal 1. The indicator may be a cathode-ray tube (with a triggered sweep) or a recorder or, for qualitative analysis with slowly varying quantities, a high-impedance voltmeter.

The solutition of Eq. (13-26) can also be obtained with a computer which contains differentiators instead of integrators. However, integrators are almost invariably preferred over differentiators in analog-computer applications, for the following reasons: Since the gain of an integrator decreases with frequency whereas the gain of a differentiator increases nominally linearly with frequency, it is easier to stabilize the former than the latter with respect to spurious oscillations. As a result of its limited bandwidth, an integrator is less sensitive to noise voltages than a differentiator. Further, if the input waveform changes rapidly, the amplifier of a differentiator may overload. Finally, as a matter of practice, it is convenient to introduce initial conditions in an integrator.

13-12   ACTIVE FILTERS

Consider the ideal low-pass-filter response shown in Fig. 13-24a. In this plot all signals within the band $0 \le f \le f_o$ are transmitted without loss, whereas inputs with frequencies $f > f_o$ give zero output. It is known that such an ideal characteristic is unrealizable with physical elements, and thus it is necessary to approximate it. An approximation for an ideal low-pass filter is of the form

$$A_V(s) = \frac{1}{P_n(s)}$$

(13-27)

where $P_n(s)$ is a polynomial in the variable s with zeros in the left-hand plane. Active filters permit the realization of arbitrary left-hand poles for $A_V(s)$, using the operational amplifier as the active element and only resistors and capacitors for the passive elements.

Since commercially available OP AMPS have unity gain-bandwidth products as high as 100 MHz, it is possible to design active filters up to frequencies of several MHz. The limiting factor for full-power response at those high frequencies is the slewing rate (Sec. 13-6) of the operational amplifier. (Commercial integrated OP AMPS are available with slewing rates as high as 100 V/μs.)

Butterworth Filter   A common approximation of Eq. (13-27) uses the Butterworth polynomials $B_n(s)$, where

$$A_V(s) = \frac{A_{Vo}}{B_n(s)} \tag{13-28}$$

and with $s = j\omega$,

$$|A_V(s)|^2 = A_V(s)A_V(-s) = \frac{A_{Vo}^2}{1 + (\omega/\omega_o)^{2n}} \tag{13-29}$$

From Eqs. (13-28) and (13-29) we note that the magnitude of $B_n(\omega)$ is given by

$$|B_n(\omega)| = \sqrt{1 + \left(\frac{\omega}{\omega_o}\right)^{2n}} \tag{13-30}$$

The Butterworth response [Eq. (13-29)] for various values of n is plotted in Fig. 13-25. Note that the magnitude of A_V is down 3 dB at $\omega = \omega_o$ for

Fig. 13-25 Butterworth low-pass-filter frequency response.

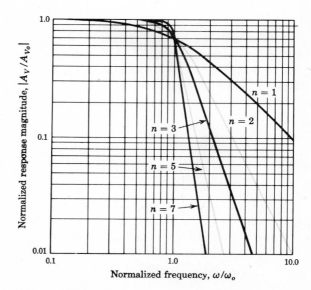

TABLE 13-2   Normalized Butterworth polynomials

n	*Factors of polynomial $B_n(s)$*
1	$(s + 1)$
2	$(s^2 + 1.414s + 1)$
3	$(s + 1)(s^2 + s + 1)$
4	$(s^2 + 0.765s + 1)(s^2 + 1.848s + 1)$
5	$(s + 1)(s^2 + 0.618s + 1)(s^2 + 1.618s + 1)$
6	$(s^2 + 0.518s + 1)(s^2 + 1.414s + 1)(s^2 + 1.932s + 1)$
7	$(s + 1)(s^2 + 0.445s + 1)(s^2 + 1.247s + 1)(s^2 + 1.802s + 1)$
8	$(s^2 + 0.390s + 1)(s^2 + 1.111s + 1)(s^2 + 1.663s + 1)(s^2 + 1.932s + 1)$

all n. The larger the value of n, the more closely the curve approximates the ideal low-pass response of Fig. 13-24a.

If we normalize the frequency by assuming $\omega_o = 1$ rad/s, then Table 13-2 gives the Butterworth polynomials for n up to 8. Note that for n even, the polynomials are the products of quadratic forms, and for n odd, there is present the additional factor $s + 1$.

From the table and Eq. (13-28) we see that the typical second-order Butterworth filter transfer function is of the form

$$\frac{A_V(s)}{A_{Vo}} = \frac{1}{(s/\omega_o)^2 + 2k(s/\omega_o) + 1} \tag{13-31}$$

where $\omega_o = 2\pi f_o$ is the high-frequency 3-dB point and the constant k is called the *damping factor*. Similarly, the first-order filter is

$$\frac{A_V(s)}{A_{Vo}} = \frac{1}{s/\omega_o + 1} \tag{13-32}$$

Practical Realization   Consider the circuit shown in Fig. 13-26a, where the active element is an operational amplifier whose stable midband gain $V_o/V_i = A_{Vo} = (R_1 + R_1')/R_1$ [Eq. (13-2)] is to be determined. We assume that the amplifier input current is zero, and we show in Prob. 13-31 that the transfer function of this network takes the form of Eq. (13-31), where

$$\omega_o = \frac{1}{RC} \tag{13-33}$$

and

$$2k = 3 - A_{Vo} \quad \text{or} \quad A_{Vo} = 3 - 2k \tag{13-34}$$

We are now in a position to synthesize even-order Butterworth filters by cascading prototypes of the form shown in Fig. 13-26a, using identical R's and C's to give the value of ω_o required by Eq. (13-33). The gain A_{Vo} of each operational amplifier must satisfy Eq. (13-34) with the factors $2k$ taken from Table 13-2.

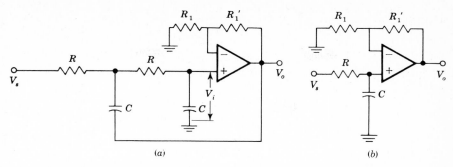

(a)   (b)

Fig. 13-26 (a) Active-filter prototype, second-order low-pass section. (b) First-order low-pass section.

To realize odd-order filters, it is necessary to cascade the first-order filter of Eq. (13-32) with second-order sections such as indicated in Fig. 13-26a. The first-order prototype of Fig. 13-26b has the transfer function of Eq. (13-32) for arbitrary A_{Vo} provided that ω_o is given by Eq. (13-33). For example, a third-order Butterworth active filter consists of the circuit in Fig. 13-26a in cascade with the circuit of Fig. 13-26b, with R and C chosen so that $RC = 1/\omega_o$, with A_{Vo} in Fig. 13-26a selected to give $k = 0.5$ (Table 13-2, $n = 3$), and A_{Vo} in Fig. 13-26b chosen arbitrarily.

EXAMPLE Design a fourth-order Butterworth low-pass filter with a cutoff frequency of 1 kHz.

Solution We cascade two second-order prototypes as shown in Fig. 13-27. For $n = 4$ we have from Table 13-2 and Eq. (13-34)

$$A_{V1} = 3 - 2k_1 = 3 - 0.765 = 2.235$$

and

$$A_{V2} = 3 - 2k_2 = 3 - 1.848 = 1.152$$

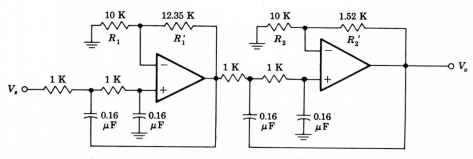

Fig. 13-27 Fourth-order Butterworth low-pass filter with $f_o = 1$ kHz.

From Eq. (13-2), $A_{V1} = (R_1 + R_1')/R_1$. If we arbitrarily choose $R_1 = 10$ K, then for $A_{V1} = 2.235$, we find $R_1' = 12.35$ K, whereas for $A_{V2} = 1.152$, we find $R_2' = 1.520$ K and $R_2 = 10$ K. To satisfy the cutoff-frequency requirement, we have, from Eq. (13-33), $f_o = 1/2\pi RC$. We take $R = 1$ K and find $C = 0.16$ μF. Figure 13-27 shows the complete fourth-order low-pass Butterworth filter.

High-Pass Prototype An idealized high-pass-filter characteristic is indicated in Fig. 13-24b. The high-pass second-order filter is obtained from the low-pass second-order prototype of Eq. (13-31) by applying the transformation

$$\frac{s}{\omega_o}\bigg|_{\text{low-pass}} \rightarrow \frac{\omega_o}{s}\bigg|_{\text{high-pass}} \qquad (13\text{-}35)$$

Thus, interchanging R's and C's in Fig. 13-26a results in a second-order high-pass active filter.

Bandpass Filter A second-order bandpass prototype is obtained by cascading a low-pass second-order section whose cutoff frequency is f_{oH} with a high-pass second-order section whose cutoff frequency is f_{oL}, provided $f_{oH} > f_{oL}$, as indicated in Fig. 13-24c.

Band-Reject Filter Figure 13-28 shows that a band-reject filter is obtained by paralleling a high-pass section whose cutoff frequency is f_{oL} with a low-pass section whose cutoff frequency is f_{oH}. Note that for band-reject characteristics it is required that $f_{oH} < f_{oL}$.

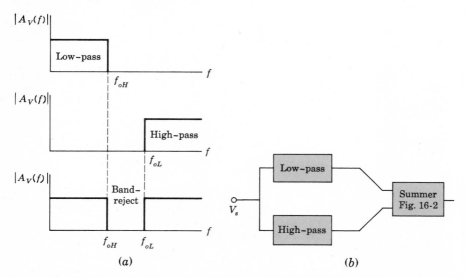

Fig. 13-28 (a) Ideal band-reject-filter frequency response. (a) Parallel combination of low-pass and high-pass filters results in a band-reject filter.

13-13   INTEGRATED CIRCUIT TUNED AMPLIFIER

The differential amplifier stage in monolithic integrated form (Fig. 13-29) is an excellent basic building block for the design of a tuned amplifier (including automatic gain control), an amplitude modulator, or a video amplifier. We now discuss these applications.

Operation of a Tuned Amplifier   This circuit is designed to amplify a signal over a narrow band of frequencies centered at f_o. The *simplified* schematic diagram shown in Fig. 13-30 is used to explain the operation of this circuit. The external leads 1, 2, 3, . . . of the IC in this figure correspond to those in Fig. 13-29. The input signal is applied through the tuned transformer $T1$ to the base of $Q1$. The load R_L is applied across the tuned transformer $T2$ in the collector circuit of $Q3$. The amplification is performed by the transistors $Q1$ and $Q3$, whereas the magnitude of the gain is controlled by $Q2$. The combination of $Q1$-$Q3$ acts as a common-emitter common-base (CE-CB) pair, known as a *cascode* combination. It can be shown that the input resistance and the current gain of a cascode circuit are essentially the same as those of a CE stage, the output resistance is the same as that of a CB stage, and the reverse-open-circuit voltage amplification is given by $h_r \approx 10^{-7}$. The extremely small value of h_r for the cascode transistor pair makes this circuit especially useful in tuned-amplifier design. The reduction in the "reverse internal feedback" of the compound device simplifies tuning, reduces the possibility of oscillation, and results in improved stability of the amplifier.

The voltage V_{AGC} applied to the base of $Q2$ is used to provide automatic gain control. From Fig. 13-7 we see that if V_{AGC} is at least 120 mV greater than V_R, $Q3$ is cut off and all the current of $Q1$ flows through $Q2$. Since $Q3$

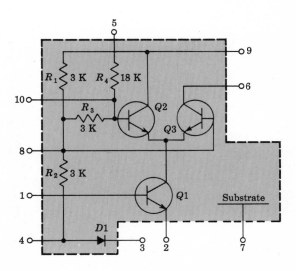

Fig. 13-29   The MC 1550 integrated circuit. (Courtesy of Motorola Semiconductor Inc.)

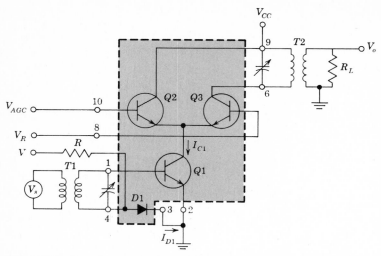

Fig. 13-30   Tuned amplifier consisting of the $Q1$-$Q3$ cascade, with the gain controlled by $Q2$.

is cut off, its transconductance is zero and the gain $A_V = V_o/V_s$ becomes zero. If V_{AGC} is less than V_R by more than 120 mV, $Q2$ is cut off and the collector current of $Q1$ flows through $Q3$, increasing the transconductance of $Q3$ and resulting in maximum voltage gain A_V.

An important advantage of this amplifier is its ability to vary the value of A_V by changing V_{AGC} without detuning the input circuit. This follows from the fact that variations in V_{AGC} cause changes in the division of the current between $Q2$ and $Q3$ without affecting significantly the collector current of $Q1$. Thus the input impedance of $Q1$ remains constant and the input circuit is not detuned.

Biasing of this integrated amplifier is obtained using a technique similar to that discussed in Sec. 10-14. The voltage V and resistor R establish the dc current I_{D1} through the diode $D1$. Since the diode and transistor $Q1$ are on the same silicon chip, very close to each other, and with $V_{D1} = V_{BE1}$, the collector current I_{C1} of $Q1$ is within ± 5 percent of I_{D1}.

A Practical Tuned Amplifier   A hybrid monolithic circuit which embodies the principles discussed above is indicated in Fig. 13-31. (For the moment, assume that the audio generator V_a is not present; $V_a = 0$.) The shaded block is the MC 1550 IC chip of Fig. 13-29. All other components are discrete elements added externally. Resistors R_1 and R_2 bias the diode $D1$ (and hence determine the collector current of $Q1$). These resistors also establish the bias voltage for $Q3$. Resistors R_3 and R_4 serve to "widen" the AGC voltage range from 120 mV to approximately 850 mV, thus rendering the AGC terminal less susceptible to external noise pick-up.

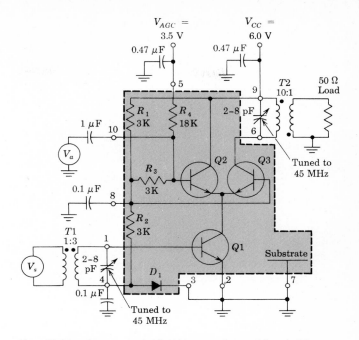

Fig. 13-31 A practical 45-MHz tuned amplifier (with $V_a = 0$), or an RF modulator if $V_a \neq 0$. (Courtesy of Motorola Semiconductor, Inc.)

The source V_s is a 45-MHz RF (radio-frequency) generator whose resistance is 50 Ω. The transformers are wound with No. 32 wire on T12-2 cores; $T1$ with 6:18 turns has a magnetizing inductance $L_M = 1.1$ μH, and $T2$ with 30:3 turns gives $L_M = 2.5$ μH. The variable capacitors in Fig. 13-31 are adjusted so as to resonate with these inductors at 45 MHz.

Amplitude Modulator The RF carrier signal V_s may be varied in amplitude by changing the AGC voltage. Hence, if an audio signal V_a is applied to terminal 10 in Fig. 13-31 so as to modify the AGC voltage, the output will be the amplitude-modulated waveform indicated in Fig. 3-24.

13-14   A CASCODE VIDEO AMPLIFIER

A video amplifier, as opposed to a tuned amplifier, must amplify signals over a wide band of frequencies, say up to 20 MHz. The MC 1550 can be used as a cascode video amplifier (Fig. 13-32a) by avoiding tuned input and output circuits. Between pins 1 and 4, 50 Ω is inserted in order to properly terminate

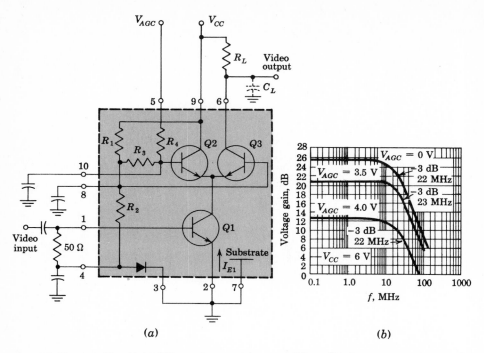

Fig. 13-32 (a) The MC 1550 used as a video amplifier; (b) frequency response for three different values of V_{AGC}.

the coaxial cable carrying the video signal. This small resistance has negligible effect on the biasing of $Q1$. The load R_L is placed directly in the collector lead of $Q3$.

The small-signal analysis of this video amplifier can be made using an approximate equivalent circuit. If both transistors $Q2$ and $Q3$ are operating in their active region, the collector of $Q1$ sees the very small input emitter resistance of two common-base stages in parallel. We can represent $Q1$ with its hybrid-π model. The video output is taken from the collector of $Q3$, which is operating as a common-base stage.

The result of the small-signal analysis of this video circuit is indicated in Fig. 13-32b for three different values of the automatic-gain-control voltage V_{AGC}. The high 3-dB frequency is approximately 22 MHz, and the voltage gain is -25 (for $R_L = 375\ \Omega$ and a load capacitance of 5 pF).

REVIEW QUESTIONS

13-1 (a) Draw the schematic block diagram of the basic OP AMP with inverting and noninverting inputs. (b) Indicate its equivalent circuit.

13-2 List six characteristics of the ideal OP AMP.

13-3 (*a*) Draw the schematic diagram of an ideal inverting OP AMP with voltage-shunt feedback impedances Z and Z'. (*b*) Indicate the virtual-ground model for calculating the gain.

13-4 (*a*) Draw the schematic diagram of an ideal noninverting OP AMP with voltage-shunt feedback. (*b*) Derive the expression for the voltage gain.

13-5 (*a*) Define an ideal DIFF AMP. (*b*) Define difference signal v_d and common-mode signal v_c.

13-6 (*a*) Draw the circuit of an emitter-coupled DIFF AMP. (*b*) Explain why the CMRR $\rightarrow \infty$ for a symmetrical circuit with $R_e \rightarrow \infty$.

13-7 (*a*) Why is R_e in an emitter-coupled DIFF AMP replaced by a constant-current source? (*b*) Draw such a circuit. (*c*) Explain why the network replacing R_e acts as an approximately constant current I_O. (*d*) Explain how I_O is made to be independent of temperature.

13-8 Explain why the CMRR is infinite if a true constant-current source is used in a symmetrical emitter-coupled DIFF AMP.

13-9 (*a*) Sketch the transfer characteristics of a DIFF AMP. (*b*) Over what differential voltage is the DIFF AMP a good limiter? (*c*) Over what differential voltage is the transfer characteristic quite linear? (*d*) How does the transconductance vary (qualitatively) with differential voltage? (*e*) Explain why AGC is possible with the DIFF AMP.

13-10 (*a*) Draw an IC OP AMP in block-diagram form. (*b*) Identify each stage by function.

13-11 Define (*a*) *input bias current*, (*b*) *input offset current*, (*c*) *input offset voltage*, (*d*) *output offset voltage*, (*e*) *power supply rejection ratio*, and (*f*) *slew rate* for an OP AMP.

13-12 Show the balancing arrangement for (*a*) an inverting and (*b*) a noninverting OP AMP.

13-13 Indicate an OP AMP connected as (*a*) an *inverter*, (*b*) a *scale changer*, (*c*) a *phase shifter*, and (*d*) an *adder*.

13-14 Draw the circuit of a *voltage-to-current converter* if the load is (*a*) floating and (*b*) grounded.

13-15 Draw the circuit of a *current-to-voltage converter*. Explain its operation.

13-16 Draw the circuit of a dc *voltage follower* and explain its operation.

13-17 Draw the circuit of a dc differential amplifier having (*a*) low input resistance and (*b*) high input resistance.

13-18 Draw the circuit of an ac *voltage follower* having very high input resistance. Explain its operation.

13-19 Draw the circuit of an OP AMP integrator and indicate how to apply the initial condition. Explain its operation.

13-20 Sketch the idealized characteristics for the following filter types: (*a*) low-pass, (*b*) high-pass, (*c*) bandpass, and (*d*) band-rejection.

13-21 Draw the prototype for a low-pass active-filter section of (*a*) first order, (*b*) second order, and (*c*) third order.

13-22 (*a*) Sketch the basic building block for an IC tuned amplifier. (*b*) Explain how automatic gain control (AGC) is obtained. (*c*) Why does AGC not cause detuning?

13-23 Draw the circuit of an *amplitude modulator* and explain its operation.

13-24 Draw the circuit of an IC video amplifier with AGC.

14 / OPERATIONAL AMPLIFIER NONLINEAR ANALOG SYSTEMS

Among the *nonlinear* analog system configurations discussed in this chapter are the following: comparators, sample-and-hold circuits, ac/dc converters, and square-wave and triangle-waveform generators. The DIFF AMP configuration is used in emitter-coupled logic. Digital-to-analog and analog-to-digital converters are described.

14-1   COMPARATORS

With the exception of amplitude modulation and automatic gain control, all the systems discussed in Chap. 13 operate linearly. This chapter is concerned with nonlinear OP AMP functions.

The comparator is a circuit which compares an input signal $v_i(t)$ with a reference voltage V_R. When the input v_i exceeds V_R, the comparator output v_o takes on a value which is very different from the magnitude of v_o when v_i is smaller than V_R. The DIFF AMP input-output curve of Fig. 13-7 approximates this comparator characteristic. Note that the total input swing between the two extreme output voltages is $\sim 8V_T = 200$ mV. This range may be reduced considerably by cascading two DIFF AMPS. The MC 1530 OP AMP serves as a comparator if connected open-loop, as shown in Fig. 14-1a. The transfer characteristic is given in Fig. 14-1b, and it is now observed that the change in output state takes place with a variation in input of only 2 mV. Note that the input offset voltage contributes an error in the point of comparison between v_i and V_R of the order of 1 mV. The reference V_R may be any voltage, provided that it does not exceed the maximum common-mode range.

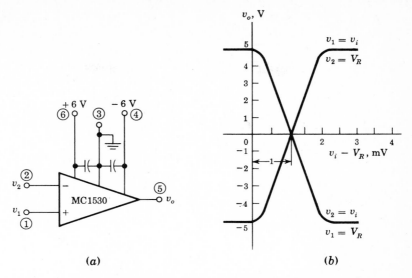

(a)   (b)

Fig. 14-1   (a) The MC 1530 operational amplifier as a comparator.
(b) The transfer characteristic.

Zero-Crossing Detector   If V_R is set equal to zero, the output will respond almost discontinuously every time the input passes through zero. Such an arrangement, called a *zero-crossing detector*, is indicated in Fig. 14-2.

Some of the most important systems using comparators will now be listed. Other applications are discussed in Secs. 14-4 and 14-8.

Square Waves from a Sine Wave   If the input to an OP AMP comparator is a sine wave, the output is a square wave.   If a zero-crossing detector is used, a symmetrical square wave results, as shown in Fig. 14-2c.   This idealized waveform has vertical sides which, in reality, should extend over a range of a few millivolts of input voltage v_i.

Timing-Markers Generator from a Sine Wave   The square-wave output v_o of the preceding application is applied to the input of an RC series circuit. If the time constant RC is very small compared with the period T of the sine-wave input the voltage v' across R is a series of positive and negative pulses, as indicated in Fig. 14-2d.   If v' is applied to a clipper with an ideal diode (Fig. 14-2a), the load voltage v_L contains only positive pulses (Fig. 14-2e). Thus the sinusoid has been converted into a train of positive pulses whose spacing is T.   These may be used for timing markers (on the sweep voltage of a cathode-ray tube, for example).

Phasemeter   The phase angle between two voltages can be measured by a method based on the circuit of Fig. 14-2.   Both voltages are converted into pulses, and the time interval between the pulse of one wave and that obtained from the second sine wave is measured.   This time interval is proportional to the phase difference.   Such a phasemeter can measure angles from 0 to 360°.

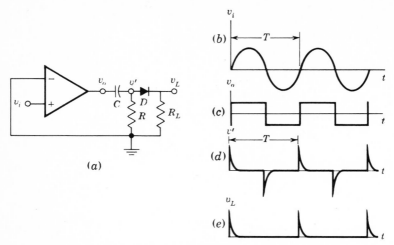

Fig. 14-2 A zero-crossing detector converts a sinusoid v_i into a square wave v_o. The pulse waveforms v' and v_L result if v_o is fed into a short-time-constant RC circuit in cascade with a diode clipper.

Amplitude-Distribution Analyzer A comparator is a basic building block in a system used to analyze the amplitude distribution of the noise generated in an active device or the voltage spectrum of the pulses developed by a nuclear-radiation detector, etc. To be more specific, suppose that the output of the comparator is 10 V if $v_i > V_R$ and 0 V if $v_i < V_R$. Let the input to the comparator be noise. A dc meter is used to measure the average value of the output square wave. For example, if V_R is set at zero, the meter will read 10 V, which is interpreted to mean that the probability that the amplitude is greater than zero is 100 percent. If V_R is set at some value V'_R and the meter reads 7 V, this is interpreted to mean that the probability that the amplitude of the noise is greater than V'_R is 70 percent, etc. In this way the cumulative amplitude probability distribution of the noise is obtained by recording meter readings as a function of V_R.

Pulse-Time Modulation If a periodic sweep waveform is applied to a comparator whose reference voltage V_R is not constant but rather is modulated by an audio signal, it is possible to obtain a succession of pulses whose relative spacing reflects the input information. The result is a *time-modulation system* of communication.

14-2   SAMPLE–AND–HOLD CIRCUITS

A typical data-acquisition system receives signals from a number of different sources and transmits these signals in suitable form to a computer or a commu-

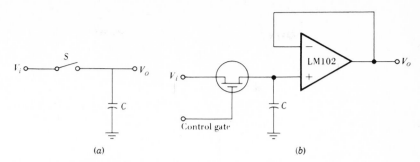

(a)   (b)

Fig. 14-3   Sample-and-hold circuit.   (a) Schematic, (b) practical.

nication channel.   A multiplexer (Sec. 7-4) selects each signal in sequence, and then the analog information is converted into a constant voltage over the gating-time interval by means of a *sample-and-hold circuit*.   The constant output of the sample-and-hold may then be converted to a digital signal by means of an analog-to-digital (A/D) converter (Sec. 14-8) for digital transmission.

A sample-and-hold circuit in its simplest form is a switch S in series with a capacitor, as in Fig. 14-3a.   The voltage across the capacitor tracks the input signal during the time T_g when a logic control gate closes S, and holds the instantaneous value attained at the end of the interval T_g when the control gate opens S.   The switch may be a relay (for very slow waveforms), a sampling diode-bridge gate, a bipolar transistor switch, or an MOSFET controlled by a gating signal.   The MOSFET makes an excellent chopper because its *offset voltage* when ON ($\sim 5~\mu V$) is much smaller than that of a bipolar junction transistor.

The circuit shown in Fig. 14-3b is one of the simplest practical sample-and-hold circuits.   A negative pulse at the gate of the p-channel MOSFET will turn the switch ON, and the holding capacitor C will charge with a time constant $R_{ON}C$ to the instantaneous value of the input voltage.   In the absence of a negative pulse, the switch is turned OFF and the capacitor is isolated from any load through the LM 102 OP AMP.   Thus it will hold the voltage impressed upon it.   It is recommended that a capacitor with polycarbonate, polyethylene, or Teflon dielectric be used.   Most other capacitors do not retain the stored voltage, due to a polarization phenomenon which causes the stored voltage to decrease with a time constant of several seconds.   Even if the polarization phenomenon does not occur, the OFF current of the switch (~ 1 nA) and the bias current of the OP AMP will flow through C.   Since the maximum input bias current for the LM 102 is 10 nA, it follows that with a 10-μF capacitance the drift rate during the HOLD period will be less than 1 mV/s.

Two additional factors influence the operation of the circuit: the reaction time, called *aperture time* (typically less than 100 ns), which is the delay between the time that the pulse is applied to the switch and the actual time the switch closes, and the *acquisition time*, which is the time it takes for the capacitor to change from one level of holding voltage to the new value of input voltage after the switch has closed.

When the hold capacitor is larger than 0.05 μF, an isolation resistor of approximately 10 K should be included between the capacitor and the + input of the OP AMP. This resistor is required to protect the amplifier in case the output is short-circuited or the power supplies are abruptly shut down while the capacitor is charged.

14-3   PRECISION AC/DC CONVERTERS

If a sinusoid whose peak value is less than the threshold or cutin voltage V_γ ($\sim$0.6 V) is applied to the rectifier circuit of Fig. 3-6, we see that the output is zero for all times. In order to be able to rectify millivolt signals, it is clearly necessary to reduce V_γ. By placing the diode in the feedback loop of an OP AMP, the cutin voltage is divided by the open-loop gain A_V of the amplifier. Hence V_γ is virtually eliminated and the diode approaches the ideal rectifying component. If in Fig. 14-4a the input v_i goes positive by at least V_γ/A_V, then v' exceeds V_γ and D conducts. Because of the virtual connection between the noninverting and inverting inputs (due to the feedback with D ON), $v_o \approx v_i$. Therefore the circuit acts as a voltage follower for positive signals (in excess of approximately 0.1 mV). When v_i swings negatively, D is OFF and no current is delivered to the external load except for the small bias current of the LM 101A and the diode reverse saturation current.

Precision Clamp   By modifying the circuit of Fig. 14-4a, as indicated in Fig. 14-4b, an almost ideal clamp (Sec. 3-5) is obtained. If $v_i < V_R$, then v' is positive and D conducts. As explained above, under these conditions the output equals the voltage at the noninverting terminal, or $v_o = V_R$. If $v_i > V_R$, then v' is negative, D is OFF, and $v_o = v_i$. In summary: The output follows the input for $v_i > V_R$ and v_o is clamped to V_R if v_i is less than V_R by about 0.1 mV. When D is reverse-biased in Fig. 14-4a or b, a large differential voltage may appear between the inputs and the OP AMP must be able to

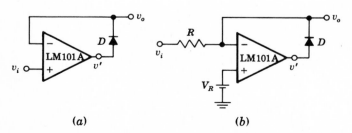

(a)   *(b)*

Fig. 14-4   (a) A precision diode.   (b) A precision clamp.

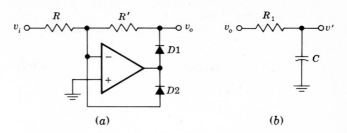

Fig. 14-5   (a) A half-wave rectifier.   (b) A low-pass filter which can be cascaded with the circuit in (a) to obtain an average detector.

withstand this voltage.   Also note that when $v_i > V_R$, the input stage saturates because the feedback through D is missing.

Fast Half-Wave Rectifier   By adding R' and $D2$ to Fig. 14-4b and setting $V_R = 0$, we obtain the circuit of Fig. 14-5a.   If v_i goes negative, $D1$ is ON, $D2$ is OFF, and the circuit behaves as an inverting OP AMP, so that $v_o = -(R'/R)v_i$.   If v_i is positive, $D1$ is OFF and $D2$ is ON.   Because of the feedback through $D2$, a virtual ground exists at the input and $v_o = 0$.   If v_i is a sinusoid, the circuit performs half-wave rectification.   Because the amplifier does not saturate, it can provide rectification at frequencies up to 100 kHz.

An alternative configuration to that in Fig. 14-5a is to ground the left-hand side of R and to impress v_i at the noninverting terminal.   The output now has a value of $(R + R')/R$ times the input for positive voltages and $v_o = v_i$ for negative inputs.   Hence half-wave rectification is obtained if $R' \gg R$.   A full-wave system is indicated in Prob. 14-1.

Active Average Detector   Consider the circuit of Fig. 14-5a to be cascaded with the low-pass filter of Fig. 14-5b.   If v_i is an amplitude-modulated carrier (Fig. 3-24), the R_1C filter removes the carrier and v' is proportional to the average value of the audio signal.   In other words, this configuration represents an *average detector*.

Active Peak Detector   If a capacitor is added at the output of the precision diode of Fig. 14-4a, a peak detector results.   The capacitor in Fig. 14-6a will hold the output at $t = t'$ to the most positive value attained by the input v_i prior to t', as indicated in Fig. 14-6b.   This operation follows from the fact that if $v_i > v_o$, the OP AMP output v' is positive, so that D conducts.   The capacitor is then charged through D (by the output current of the amplifier) to the value of the input because the circuit is a voltage follower.   When v_i falls below the capacitor voltage, the OP AMP output goes negative and the diode becomes reverse-biased.   Thus the capacitor gets charged to the most positive value of the input.

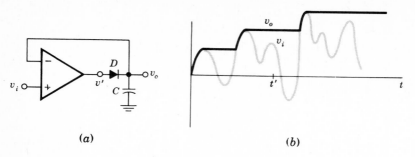

(a)   (b)

Fig. 14-6 (a) A positive peak detector. (b) An arbitrary input waveform v_i and the corresponding output v_o.

The circuit is a special case of a sample-and-hold circuit, and the capacitor
leakage current considerations given in Sec. 14-2 also apply to this configuration. If the output is loaded, a buffer voltage follower should be used to prevent the load resistance from discharging C. To reset the circuit, a low-leakage
switch such as an MOSFET gate must be placed across the capacitor.

14-4   WAVEFORM GENERATORS

The operational amplifier comparator, together with an integrator, can be
used to generate a square wave, a pulse, or a triangle waveform, as we now
demonstrate.

Square-Wave Generator In Fig. 14-7a, the output v_o is shunted to
ground by two Zener diodes connected back to back and is limited to either
$+V_{Z2}$ or $-V_{Z1}$, if $V_\gamma \ll V_Z$ (Fig. 3-12). A fraction $\beta = R_3/(R_2 + R_3)$ of the
output is fed back to the noninverting input terminal. The differential input
voltage v_i is given by

$$v_i = v_c - \beta v_o \qquad\qquad (14\text{-}1)$$

From the transfer characteristic of the comparator given in Fig. 14-1 we see
that if v_i is positive (by at least 1 mV), then $v_o = -V_{Z1}$, whereas if v_i is negative (by at least 1 mV), then $v_o = +V_{Z2}$. Consider an instant of time when
$v_i < 0$ or $v_c < \beta v_o = \beta V_{Z2}$. The capacitor C now charges exponentially toward
V_{Z2} through the integrating $R'C$ combination. The output remains constant
at V_{Z2} until v_c equals $+\beta V_{Z2}$, at which time the comparator output reverses to
$-V_{Z1}$. Now v_c charges exponentially toward $-V_{Z1}$. The output voltage v_o
and capacitor voltage v_c waveforms are shown in Fig. 14-7b for the special case
$V_{Z1} = V_{Z2} = V_Z$. If we let $t = 0$ when $v_c = -\beta V_Z$ for the first half cycle,
we have

$$v_c(t) = V_Z[1 - (1 + \beta)\epsilon^{-t/R'C}] \qquad\qquad (14\text{-}2)$$

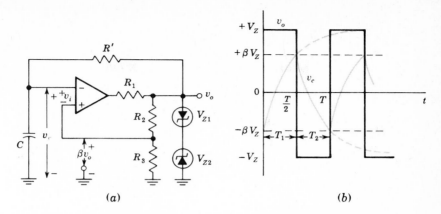

Fig. 14-7 (a) A square-wave generator. (b) Output and capacitor voltage waveforms.

Since at $t = T/2$, $v_c(t) = +\beta V_Z$, we find T, solving Eq. (14-2), to be given by

$$T = 2R'C \ln \frac{1 + \beta}{1 - \beta} \tag{14-3}$$

Note that T is independent of V_Z.

This square-wave generator is particularly useful in the frequency range 10 Hz to 10 kHz. At higher frequencies the delay time of the operational amplifier as it moves out of saturation, through its linear range, and back to saturation in the opposite direction, becomes significant. Also, the slew rate of the operational amplifier limits the slope of the output square wave. The frequency stability depends mainly upon the Zener-diode stability and the capacitor, whereas waveform symmetry depends on the matching of the two Zener diodes. If an unsymmetrical square wave is desired, then $V_{Z1} \neq V_{Z2}$.

The circuit will operate in essentially the same manner as described above if $R_1 = 0$ and the avalanche diodes are omitted. However, now the amplitude of the square wave depends upon the power-supply voltage and the output-stage configuration of the OP AMP.

The circuit of Fig. 14-7 is called an *astable multivibrator* because it has two quasi-stable states. The output remains in one of these states for a time T_1 and then abruptly changes to the second state for a time T_2, and the cycle of period $T = T_1 + T_2$ repeats.

Pulse Generator A *monostable multivibrator* has one stable state and one quasi-stable state. The circuit remains in its stable state until a triggering signal causes a transition to the quasi-stable state. Then, after a time T, the

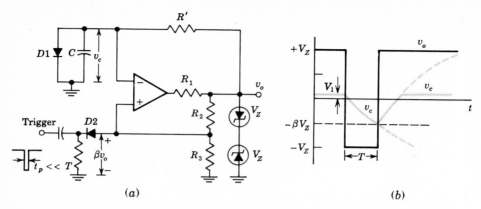

Fig. 14-8   (a) Monostable multivibrator.   (b) Output and capacitor voltage waveforms.

circuit returns to its stable state.   Hence a single pulse has been generated, and the circuit is referred to as a *one-shot*.

The square-wave generator of Fig. 14-7 is modified in Fig. 14-8 to operate as a monostable multivibrator by adding a diode ($D1$) clamp across C and by introducing a narrow negative triggering pulse through $D2$ to the noninverting terminal.   To see how the circuit operates, assume that it is in its stable state with the output at $v_o = +V_Z$ and the capacitor clamped at

$$v_c = V_1 \approx 0.7 \text{ V}$$

(the ON voltage of $D1$ with $\beta V_Z > V_1$).   If the trigger amplitude is greater than $\beta V_Z - V_1$, then it will cause the comparator to switch to an output $v_o = -V_Z$.   The capacitor will now charge through R' toward $-V_Z$ because $D1$ becomes reverse biased.   When v_c becomes more negative than $-\beta V_Z$, the comparator output swings back to $+V_Z$.   The capacitor now starts charging toward $+V_Z$ through R' until v_c reaches V_1 and C becomes clamped again at $v_c = V_1$.   In Prob. 14-5 we find that the pulse width T is given by

$$T = R'C \ln \frac{1 + (V_1/V_Z)}{1 - \beta} \tag{14-4}$$

If $V_Z \gg V_1$ and $R_2 = R_3$, so that $\beta = \frac{1}{2}$, then $T = 0.69R'C$.   For short pulse widths the switching times of the comparator become important and limit the operation of the circuit.   If $R_1 = 0$ and the Zener diodes are omitted, Eq. (14-4) remains valid with V_Z determined by the power supply.

Triangle-Wave Generator   We observe from Fig. 14-7b that v_c has a triangular waveshape but that the sides of the triangles are exponentials rather than straight lines.   To linearize the triangles, it is required that C be charged with a constant current rather than the exponential current supplied through R' in Fig. 14.7a.   In Fig. 14-9 an OP AMP integrator is used to supply constant

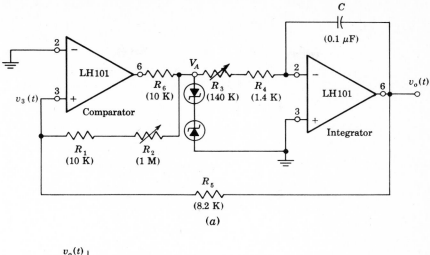

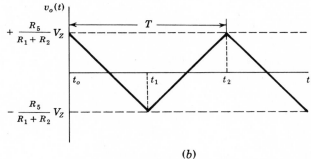

Fig. 14-9 (a) Practical triangle-wave generator. (b) Output waveform. (Courtesy of National Semiconductor Corp.)

current to C so that the output is linear. Because of the inversion through the integrator, this voltage is fed back to the noninverting terminal of the comparator in this circuit rather than to the inverting terminal as in Fig. 14-7.

When the comparator has reached either the positive or negative saturation state, the matched Zener diodes will clamp the voltage V_A at either $+V_Z$ or $-V_Z$. Let us assume that $V_A = +V_Z$ at $t = t_o$. The current flowing into the integrator is

$$I^+ = \frac{V_Z}{R_3 + R_4} \tag{14-5}$$

and the integrator output becomes a negative-going ramp the slope of which is constant and given by

$$\frac{dv_o}{dt} = \frac{-I^+}{C} = \frac{-V_Z}{(R_3 + R_4)C} \tag{14-6}$$

The voltage at pin 3 of the threshold detector is, using superposition,

$$v_3(t) = \frac{R_5 V_Z}{R_1 + R_2 + R_5} + \frac{(R_1 + R_2)v_o(t)}{R_1 + R_2 + R_5} \tag{14-7}$$

When $v_3(t)$ goes through zero and becomes negative, the comparator output changes to the negative-output state and $V_A = -V_Z$. At this time, $t = t_1$, $v_3(t_1) = 0$, or from Eq. (14-7) we find

$$v_o(t_1) = -\frac{R_5}{R_1 + R_2} V_Z \tag{14-8}$$

The current supplied to the integrator for $t_2 > t > t_1$ is

$$I^- = -\frac{V_Z}{R_3 + R_4} = -I^+$$

and the integrator output $v_o(t)$ becomes a positive-going ramp with the same slope as the negative-going ramp. At a time t_2, when

$$v_o(t_2) = +\frac{R_5}{R_1 + R_2} V_Z \tag{14-9}$$

the comparator switches again to its positive output and the cycle repeats.

The frequency of the triangle wave is determined from Eq. (14-6) and Fig. 14-9 to be given by

$$f = \frac{R_1 + R_2}{4(R_3 + R_4)R_5 C} \tag{14-10}$$

The amplitude can be controlled by the ratio $R_5 V_Z/(R_1 + R_2)$. The positive and negative peaks are equal if the Zener diodes are matched. It is possible to offset the triangle with respect to ground if we connect a dc voltage to the inverting terminal of the threshold detector or comparator.

The practical circuit shown in Fig. 14-9 makes use of the LH101 OP AMP, which is internally compensated for unity-gain feedback. This monolithic integrated OP AMP has maximum input offset voltage of 5 mV and maximum input bias current of 500 nA. For symmetry of operation the current into the integrator should be large with respect to I_{bias}, and the peak of the output triangle voltage should be large with respect to the input offset voltage.

14-5   REGENERATIVE COMPARATOR (SCHMITT TRIGGER)

As indicated in Fig. 14-1, the transfer characteristic of the MC1530 DIFF AMP makes the change in output from -5 V to $+5$ V for a swing of 2 mV in input voltage. Hence the average slope of this curve or the large-signal voltage gain A_V is $A_V = 10/(2 \times 10^{-3}) = 5{,}000$. By employing positive (regenerative) voltage-series feedback, as is done in Figs. 14-7 and 14-8 for the astable and monostable multivibrators, the gain may be increased greatly. Consequently the

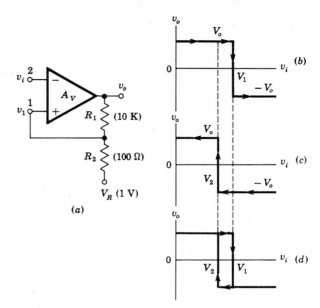

Fig. 14-10   (a) A Schmitt trigger. The transfer characteristic for (b) increasing v_i and (c) decreasing v_i (d) The compositive input-output curve.

total output excursion takes place in a time interval during which the input is changing by much less than 2 mV. Theoretically, if the loop gain $-\beta A_V$ is adjusted to be unity, then the gain with feedback A_{Vf} becomes infinite [Eq. (12-4)]. Such an idealized situation results in an abrupt (zero rise time) transition between the extreme values of output voltage. If a loop gain in excess of unity is chosen, the output waveform continues to be virtually discontinuous at the comparison voltage. However, the circuit now exhibits a phenomenon called *hysteresis*, or *backlash*, which is explained below.

The regenerative comparator of Fig. 14-10a is commonly referred to as a *Schmitt trigger* (after the inventor of a vacuum-tube version of this circuit). The input voltage is applied to the inverting terminal 2 and the feedback voltage to the noninverting terminal 1. The feedback factor is $\beta = R_2/(R_1 + R_2)$. For $R_2 = 100 \ \Omega$, $R_1 = 10$ K, and $A_V = -5{,}000$, the loop gain is

$$-\beta A_V = 0.1 \times \frac{5{,}000}{10.1} = 49.5 \gg 1$$

Assume that $v_i < v_1$, so that $v_o = +V_o$ (+5 V). Then, using superposition, we find from Fig. 14-10a that

$$v_1 = \frac{R_1 V_R}{R_1 + R_2} + \frac{R_2 V_o}{R_1 + R_2} \equiv V_1 \qquad (14\text{-}11)$$

If v_i is now increased, then v_o remains constant at V_o, and $v_1 = V_1 = $ constant until $v_i = V_1$. At this *threshold, critical,* or *triggering voltage,* the output regeneratively switches to $v_o = -V_o$ and remains at this value as long as $v_i > V_1$. This transfer characteristic is indicated in Fig. 14-10b.

The voltage at the noninverting terminal for $v_i > V_1$ is

$$v_1 = \frac{R_1 V_R}{R_1 + R_2} - \frac{R_2 V_o}{R_1 + R_2} \equiv V_2 \tag{14-12}$$

For the parameter values given in Fig. 14-10 and with $V_o = 5$ V,

$$V_1 = 0.99 + 0.05 = 1.04 \text{ V}$$

$$V_2 = 0.99 - 0.05 = 0.94 \text{ V}$$

Note that $V_2 < V_1$, and the difference between these two values is called the *hysteresis* V_H.

$$V_H = V_1 - V_2 = \frac{2R_2 V_o}{R_1 + R_2} = 0.10 \text{ V} \tag{14-13}$$

If we now decrease v_i, then the output remains at $-V_o$ until v_i equals the voltage at terminal 1 or until $v_i = V_2$. At this voltage a regenerative transition takes place and, as indicated in Fig. 14-10c, the output returns to $+V_o$ almost instantaneously. The complete transfer function is indicated in Fig. 14-10d, where the portions without arrows may be traversed in either direction, but the other segments can only be obtained if v_i varies as indicated by the arrows. Note that because of the hysteresis, the circuit triggers at a higher voltage for increasing than for decreasing signals.

We note above that transfer gain increases from 5,000 toward infinity as the loop gain increases from zero to unity, and that there is no hysteresis as long as $-\beta A_V \leq 1$. However, adjusting the gain precisely to unity is not feasile. The DIFF AMP parameters, and hence the gain A_V, are variable over the signal excursion. Hence an adjustment which ensures that the maximum loop gain is unity would result in voltage ranges where this amplification is less than unity, with a consequent loss in speed of response of the circuit. Furthermore, the circuit may not be stable enough to maintain a loop gain of precisely unity for a long period of time without frequent readjustment. In practice, therefore, a loop gain in excess of unity is chosen and a small amount of hysteresis is tolerated. In most cases a small value of V_H is not a matter of concern. In other applications a large backlash range will not allow the circuit to function properly. Thus if the peak-to-peak signal were smaller than V_H, then the Schmitt circuit, having responded at a threshold voltage by a transition in one direction, would never reset itself. In other words, once the output has jumped to, say, V_o, it would remain at this level and never return to $-V_o$.

The most important use made of the Schmitt trigger is to convert a very slowly varying input voltage into an output having an abrupt (almost discontinuous) waveform, occurring at a precise value of input voltage. This regenerative comparator may be used in all the applications listed in Sec. 14-1. For example, the use of the Schmitt trigger as a squaring circuit is illustrated in

Fig. 14-11   Response of the Schmitt trigger
to an arbitrary input signal.

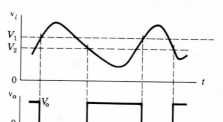

Fig. 14-11.   The input signal is arbitrary except that it has a large enough excursion to carry the input beyond the limits of the hysteresis range V_H.   The output is a square wave as shown, the amplitude of which is independent of the peak-to-peak value of the input waveform.   The output has much faster leading and trailing edges than does the input.

14-6   EMITTER–COUPLED LOGIC (ECL)

The transfer characteristic of the difference amplifier is discussed in Sec. 13-4.   We find that the emitter current remains essentially constant and that this current is switched from one transistor to the other as the signal at the input transistor varies from about 0.1 V below to 0.1 V above the reference voltage V_{B2} at the base of the second transistor (Fig. 13-7).   Except for a very narrow range of input voltage the output voltage takes on only one of two possible values and, hence, behaves as a binary circuit.   Hence the DIFF AMP, which is considered in detail in the chapter on analog systems, is also important as a digital device.   A logic family based upon this basic building block is called *emitter-coupled logic* (ECL) or *current-mode logic* (CML).   Since in the DIFF AMP clipper or comparator neither transistor is allowed to go into saturation, the ECL is the fastest of all logic families; a propagation delay time as low as 1 ns per gate is possible.   The high speed (and high fan-out) attainable with ECL is offset by the increased power dissipation per gate relative to that of the saturating logic families.

A 2-input OR (and also NOR) gate is drawn in Fig. 14-12a.   This circuit is obtained from Fig. 13-5 by using two transistors in parallel at the input. Consider positive logic.   If both A and B are low, then neither $Q1$ nor $Q2$ conducts whereas $Q3$ is in its active region.   Under these circumstances Y is low and Y' is high.   If either A or B is high, then the emitter current switches to the input transistor the base of which is high, and the collector current of $Q3$ drops approximately to zero.   Hence Y goes high and Y' drops in voltage.   Note that OR logic is performed at the output Y and NOR logic at Y', so that $Y' = \bar{Y}$.   The logic symbol for such an OR gate with both true and false outputs is indicated in Fig. 14-12b.   The availability of complementary

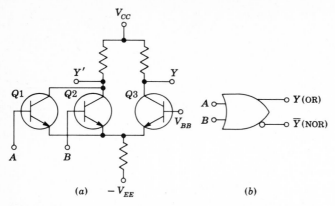

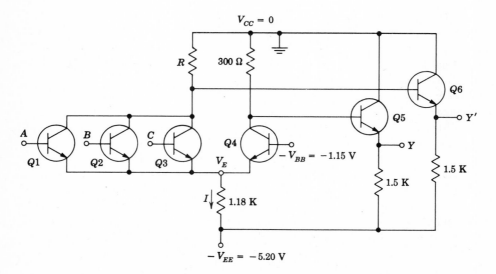

Fig. 14-12 (a) DIFF AMP converted into a 2-input emitter-coupled logic circuit. (b) The symbol for a 2-input OR/NOR gate.

outputs is clearly an advantage to the logic design engineer since it avoids the necessity of adding gates simply as inverters.

One of the difficulties with the ECL topology of Fig. 14-12a is that the $V(0)$ and $V(1)$ levels at the outputs differ from those at the inputs. Hence emitter followers $Q5$ and $Q6$ are used at the outputs to provide the proper dc-level shifts. The basic Motorola ECL 3-input gate is shown in Fig. 14-13. The reference voltage $- V_{BB}$ is obtained from a temperature-compensated network

Fig. 14-13 A 3-input ECL OR/NOR gate, with no dc-level shift between input and output voltages.

(not indicated).   The quantitative operation of the gate is given in the following illustrative problem.

EXAMPLE   (a) What are the logic levels at output Y of the ECL gate of Fig. 14-13? Assume a drop of 0.7 V between base and emitter of a conducting transistor. (b) Calculate the noise margins.   (c) Verify that a conducting transistor is in its active region (*not* in saturation).   (d) Calculate R so that the logic levels at Y' are the complements of those at Y.   (e) Find the average power dissipated by the gate.

Solution   (a) If all inputs are low, then assume transistors $Q1$, $Q2$, and $Q3$ are cut off and $Q4$ is conducting.   The voltage at the common emitter is

$$V_E = -1.15 - 0.7 = -1.85 \text{ V}$$

The current I in the 1.18-K resistance is

$$I = \frac{-1.85 + 5.20}{1.18} = 2.84 \text{ mA}$$

Neglecting the base current compared with the emitter current, I is the current in the 300-Ω resistance and the output voltage at Y is

$$v_Y = -0.3I - V_{BE5} = -(0.3)(2.84) - 0.7 = -1.55 \text{ V} = V(0)$$

If all inputs are at $V(0) = -1.55$ V and $V_E = -1.85$ V, then the base-to-emitter voltage of an input transistor is

$$V_{BE} = -1.55 + 1.85 = 0.30 \text{ V}$$

Since the cutin voltage is $V_{BE,\text{cutin}} = 0.5$ V (Table 4-1), then the input transistors are nonconducting, as was assumed above.

If at least one input is high, then assume that the current in the 1.18-K resistance is switched to R, and $Q4$ is cut off.   The drop in the 300-Ω resistance is then zero.   Since the base and collector of $Q5$ are effectively tied together, $Q5$ now behaves as a diode.   Assuming 0.7 V across $Q5$ as a first approximation, the diode current is $(5.20 - 0.7)/1.5 = 3.0$ mA.   We can assume that the diode voltage for 3.0 mA is 0.75 V.   Hence

$$v_Y = -0.75 \text{ V} = V(1)$$

If one input is at -0.75 V, then $V_E = -0.75 - 0.7 = -1.45$ V, and

$$V_{BE4} = -1.15 + 1.45 = 0.30 \text{ V}$$

which verifies the assumption that $Q4$ is cut off; since $V_{BE,\text{cutin}} = 0.5$ V.

Note that the total output swing between the two logic levels is only $1.55 - 0.75 = 0.80$ V (800 mV).   This voltage is much smaller than the value (in excess of 4 V) obtained with a DTL or TTL gate.

(b) If all inputs are at $V(0)$, then the calculation in part (a) shows that an input transistor is within $0.50 - 0.30 = 0.20$ V of cutin.   Hence a positive noise spike of 0.20 V will cause the gate to malfunction.

If one input is at $V(1)$, then we find in part (a) that $V_{BE4} = 0.30$ V.   Hence a negative noise spike at the input of 0.20 V drops V_E by the same amount and brings V_{BE4} to 0.5 V, or to the edge of conduction. Note that the noise margins are quite small (± 200 mV).

(c) From part (a) we have that, when $Q4$ is conducting, its collector voltage with respect to ground is the drop in the 300-Ω resistance, or $V_{C4} = -(0.3)(2.84) = -0.85$ V. Hence the collector junction voltage is

$$V_{CB4} = V_{C4} - V_{B4} = -0.85 + 1.15 = +0.30 \text{ V}$$

For an n-p-n transistor this represents a reverse bias, and $Q4$ must be in its active region.

If any input, say A, is at $V(1) = -0.75$ V $= V_{B1}$, then $Q1$ is conducting and the output $Y' = \bar{Y} = V(0) = -1.55$ V. The collector of $Q1$ is more positive than $V(0)$ by V_{BE6}, or

$$V_{C1} = -1.55 + 0.7 = -0.85 \text{ V}$$

and

$$V_{CB1} = V_{C1} - V_{B1} = -0.85 + 0.75 = -0.10 \text{ V}$$

For an n-p-n transistor this represents a forward bias, but one whose magnitude is less than the cutin voltage of 0.5 V. Therefore $Q1$ is *not* in saturation; it is in its active region.

(d) If input A is at $V(1)$, then $Q1$ conducts and $Q4$ is OFF. Then

$$V_E = V(1) - V_{BE1} = -0.75 - 0.7 = -1.45 \text{ V}$$

$$I = \frac{V_E + V_{EE}}{1.18} = \frac{-1.45 + 5.20}{1.18} = 3.17 \text{ mA}$$

In part (c) we find that, if $Y' = \bar{Y}$, then $V_{C1} = -0.85$ V. This value represents the drop across R if we neglect the base current of $Q1$. Hence

$$R = \frac{0.85}{3.17} = 0.27 \text{ K} = 270 \ \Omega$$

This value of R ensures that, if an input is $V(1)$, then $Y' = V(0)$. If all inputs are at $V(0) = -1.55$ V, then the current through R is zero and the output is -0.75 V $= V(1)$, independent of R.

Note that, if I had remained constant as the input changed state (true current-mode switching), then R would be identical to the collector resistance (300 Ω) of $Q4$. The above calculation shows that R is slightly smaller than this value.

(e) If the input is low, $I = 2.84$ mA (part a), whereas if the input is high, $I = 3.17$ mA (part d). The average I is $\frac{1}{2}(2.84 + 3.17) = 3.00$ mA. Since $V(0) = -0.75$ V and $V(1) = -1.55$ V, the currents in the two emitter followers are

$$\frac{5.20 - 0.75}{1.50} = 2.96 \text{ mA} \quad \text{and} \quad \frac{5.20 - 1.55}{1.50} = 2.40 \text{ mA}$$

The total power supply current drain is $3.00 + 2.96 + 2.40 = 8.36$ mA and the power dissipation is $(5.20)(8.36) = 43.5$ mW.

Note that the current drain from the power supply varies very little as the input switches from one state to the other. Hence power line spikes (of the type discussed in Sec. 6-10 for TTL gates) are virtually nonexistent.

Fig. 14-14   An implied-OR connection at the output of two ECL gates.

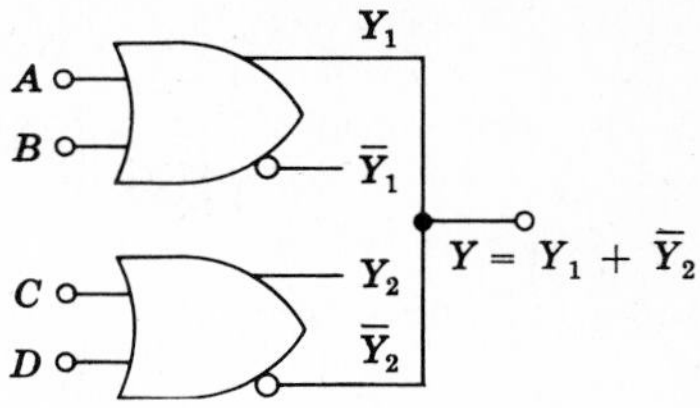

The input resistance can be considered infinite if all inputs are low so that all input transistors are cut off. If an input is high, then $Q4$ is OFF, and the input resistance corresponds to a transistor with an emitter resistor $R_e = 1.18$ K, and from Eq. (10-24) a reasonable estimate is $R_i \approx 100$ K. The output resistance is that of an emitter follower (or a diode) and a reasonable value is $R_o \approx 15\ \Omega$.

If the outputs of two or more ECL gates are tied together as in Fig. 14-14, then wired-OR logic (Sec. 6-9) is obtained (Prob. 14-12). Open-emitter gates are available for use in this application.

Summary   The principal characteristics of the ECL gate are summarized below:

Advantages

1. Since the transistors do not saturate, then the highest speed of any logic family is available.
2. Since the input resistance is very high and the output resistance is very low, a large fan-out is possible.
3. Complementary outputs are available.
4. Current switching spikes are not present in the power-supply leads.
5. Outputs can be tied together to give the implied-OR function.
6. There is little degradation of parameters with variations in temperature.
7. The number of functions available is high.
8. Easy data transmission over long distances by means of balanced twisted-pair 50-Ω lines is possible.

Disadvantages

1. A small voltage difference (800 mV) exists between the two logic levels and the noise margin is only ± 200 mV.
2. The power dissipation is high relative to the other logic families.
3. Level shifters are required for interfacing with other families.
4. The gate is slowed down by heavy capacitive loading.

14-7   DIGITAL–TO–ANALOG CONVERTERS

Many systems accept a digital word as an input signal and translate or convert it to an analog voltage or current. These systems are called *digital-to-analog*, or D/A, *converters*. The digital word is presented in a variety of codes, the most common being pure binary or binary-coded-decimal (BCD).

A D/A converter is indicated schematically in Fig. 14-15. The blocks S_0, S_1, S_2, . . . , S_{N-1} in Fig. 14-15 are electronic switches which are digitally controlled. For example, when a 1 is present on the MSB line, switch S_{N-1} connects the 10-K resistor to the reference voltage $-V_R(-10 \text{ V})$; conversely, when a 0 is present on the MSB line, the switch connects the resistor to the ground line. Thus the switch is a single-pole double-throw (SPDT) electronic switch. The operational amplifier acts as a current-to-voltage converter. The output V_o of an N-bit D/A converter is given by the following equation:

$$V_o = \left(\frac{a_o}{2^{N-1}} + \frac{a_1}{2^{N-2}} + \frac{a_2}{2^{N-3}} + \cdots + \frac{a_{N-2}}{2^1} + \frac{a_{N-1}}{2^0} \right) V \qquad (14\text{-}14)$$

where the coefficients a_n represent the binary word and $a_n = 1(0)$ if the nth bit is 1(0). The most significant bit (MSB) is that corresponding to a_{N-1}, and its weight is V, while the least significant bit (LSB) corresponds to a_0, and its weight is $V/2^{N-1}$, where $V \equiv V_R(R'/R)$ and V_R is a stable reference voltage shown in Fig. 14-15.

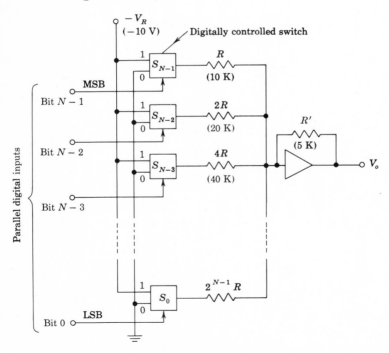

Fig. 14-15   D/A converter with binary weighted resistors.

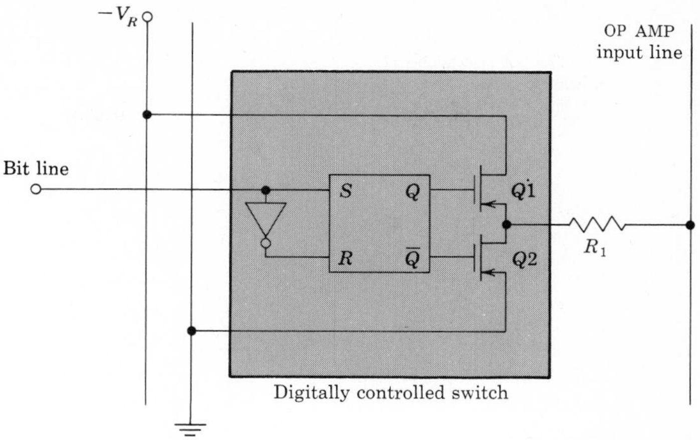

Fig. 14-16 An MOS FLIP-FLOP and a pair of totem-pole MOSFETs implement the single-pole double-throw switch of Fig. 14-15. The resistance R_1 depends upon the bit under consideration. For example for the $N - 3$ bit, $R_1 = 4R$ (Fig. 14-15).

Consider, for example, a 5-bit word ($N = 5$) and for simplicity, choose $R = 2R'$ and $V_R = 32$ V, so that $V = 16$ and Eq. (14-14) becomes

$$V_o = a_o + 2a_1 + 4a_2 + 8a_3 + 16a_4 \qquad (14\text{-}15)$$

Then, if $a_0 = 1$ and all other a's are zero, we have $V_o = 1$. If $a_1 = 1$ and all other a's are zero, we obtain $V_o = 2$. If $a_0 = a_1 = 1$ and all other a's are zero, $V_o = 2 + 1 = 3$ V, etc. Clearly, V_o is an analog voltage proportional to the digital input.

The implementation of the switching devices using p-channel MOS transistors is shown in Fig. 14-16. The S-R FLIP-FLOP is also implemented with MOSFETs and holds the bit on the corresponding bit line. Let us assume that logic 1 corresponds to -10 V and logic 0 corresponds to 0 V (negative logic). A 1 on the bit line sets the FLIP-FLOP at $Q = 1$ and $\bar{Q} = 0$, and thus transistor $Q1$ is ON, connecting the resistor R_1 to the reference voltage $-V_R$, while transistor $Q2$ is kept OFF. Similarly, a 0 at the input bit line will connect the resistor to the ground terminal. The accuracy and stability of this D/A converter depend primarily on the absolute accuracy of the resistors and the tracking of each other with temperature. Since all resistors are different and the largest is $2^{N-1}R$, where R is the smallest resistor, their values become excessively large, and it is very difficult and expensive to obtain stable, precise resistors of such values.

A Ladder-Type D/A Converter A circuit utilizing twice the number of resistors in Fig. 14-15 for the same number of bits (N) but of values R and

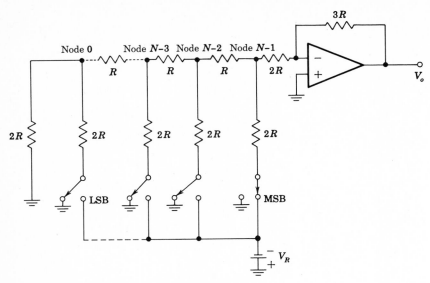

Fig. 14-17   D/A converter using R, $2R$ ladder.

$2R$ only is shown in Fig. 14-17.   The ladder used in this circuit is a current-splitting device, and thus the ratio of the resistors is more critical than their absolute value.   We observe from the figure that at any of the ladder nodes the resistance is $2R$ looking to the left or the right or toward the switch. Hence the current will split equally toward the left and right, and this happens at every node.   Considering node $N - 1$ and assuming the MSB turned on, the voltage at that node will be $-V_R/3$.   Since the gain of the operational amplifier to node $N - 1$ is $-3R/2R$, the weight of the MSB becomes

$$V_o = - \frac{V_R}{3}\left(-\frac{3R}{2R}\right) = \frac{V_R}{2}$$

Similarly, we show in Prob. 14-14 that when the second MSB bit is on and all others are off, the output will be $V_o = +V_R/4$, the third MSB bit gives $+V_R/8$, and the LSB gives $+V_R/2^N$.

The circuits discussed so far use a negative reference voltage and give a positive analog output voltage.   If negative binary numbers are to be converted, the sign-magnitude approach is used; an extra bit is added to the binary word to represent the sign, and this bit can be used to select the polarity of the reference voltage.

A typical 8-bit D/A converter by Zeltex Inc. is packaged in a module measuring 1.9 by 1.7 by 0.4 in. and includes the op amp, reference voltage, ladder network, and the switches.   The 3750 D/A Converter (Fairchild Semiconductor) is a MOS/LSI 10-bit circuit using p-channel enhancement-mode transistors.   The digital word can be entered serially or in parallel, and the output is available through 10 SPDT MOS switches.   The user must provide the resistive ladder network which is connected to the poles of the 10 switches. The 3750 contains an input shift register in which the data are stored and a holding register which retains the state of the previous 10-bit input word and drives the output switches.   The device is available in a 36-pin dual-in-line package.

Multiplying D/A Converter   A D/A converter which uses a varying analog signal instead of a fixed reference voltage is called a *multiplying* D/A *converter*.   From Eq. (14-14) we see that the output is the product of the digital word and the analog voltage V_R and its value depends on the binary word. This arrangement is often referred to as a *programmable attenuator* because the output V_o is a fraction of the input V_R and the attenuator setting can be controlled by a computer.

14-8   ANALOG–TO–DIGITAL CONVERTERS

It is often required that data taken in a physical system be converted into digital form.   Such data would normally appear in electrical analog form. For example, a temperature difference would be represented by the output of a thermocouple, the strain of a mechanical member would be represented by the electrical unbalance of a strain-gauge bridge, etc.   The need therefore arises for a device that converts analog information into digital form.   A very large number of such devices have been invented.   We shall consider below one such A/D converter.

In this system a continuous sequence of equally spaced pulses is passed through a gate.   The gate is normally closed, and is opened at the instant of the beginning of a linear ramp.   The gate remains open until the linear sweep voltage attains the reference voltage of a comparator, the level of which is set equal to the analog voltage to be converted.   The number of pulses in the train that pass through the gate is therefore proportional to the analog voltage.   If the analog voltage varies with time, it will of course not be possible to convert the analog data continuously, but it will be required that the analog data be sampled at intervals.   The maximum value of the analog voltage will be represented by a number of pulses n.   It is clear that n should be made as large as possible consistent with the requirement that the time interval between two successive pulses shall be larger than the timing error of the time modulator.   The recurrence frequency of the pulses is equal, at a minimum, to the product of n and the sampling rate.   Actually, the recurrence rate will be larger in order to allow time for the circuit to recover between samplings.

One form of digital voltmeter uses the above-described analog-to-digital converter.   The number of pulses which pass through the gate is proportional

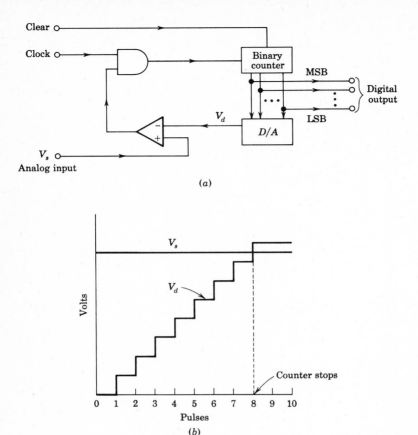

Fig. 14-18 (*a*) A/D converter using a counter; (*b*) counter ramp waveform.

to the voltage being measured. These pulses go to a counter whose reading is indicated visually by means of some form of luminous diplay (Sec. 7-7).

The principles discussed previously are used in the A/D converter shown in Fig. 14-18*a*. The *clear* pulse resets the counter to the zero count. The counter then records in binary form the number of pulses from the clock line. The clock is a source of pulses equally spaced in time. Since the number of pulses counted increases linearly with time, the binary word representing this count is used as the input of a D/A converter whose output is shown in Fig. 14-18*b*. As long as the analog input V_s is greater than V_d, the comparator output is high and the AND gate is open for the transmission of the clock pulses to the counter. When V_d exceeds V_s, the comparator output changes to the low value, and the AND gate is disabled. This stops the counting at the time when $V_s \approx V_d$ and the counter can be read out as the digital word representing the analog input voltage.

Successive-Approximation A/D Converter   The successive-approxima-tion technique is another method to implement an A/D converter.   Instead of a binary counter as shown in Fig. 14-18a, a programmer is used.   The pro-grammer sets the most significant bit (MSB) to 1, with all other bits to 0, and the comparator compares the D/A output with the analog signal.   If the D/A output is larger, the 1 is removed from the MSB, and it is tried in the next most significant bit.   If the analog input is larger, the 1 remains in that bit.   Thus a 1 is tried in each bit of the D/A decoder until, at the end of the process, the binary equivalent of the analog signal is obtained.   The 3751 MOS/LSI circuit (Fairchild Semiconductor) is a 12-bit A/D converter mono-lithic circuit which makes use of the successive-approximation technique.   The ladder network must be supplied externally, and by choosing the appropriate coding of the resistor values in the ladder, the output can be in either binary or BCD digital form.   The device is available in a 36-pin dual-in-line package.

REVIEW QUESTIONS

14-1   (a) What does an IC comparator consist of?   (b) Sketch the transfer characteristic and indicate typical voltage values.

14-2   Sketch the circuit for converting a sinusoid (a) into a square wave and (b) into a series of positive pulses, one per cycle.

14-3   Explain how to measure the phase difference between two sinusoids.

14-4   Sketch a *sample-and-hold* circuit and explain its operation.

14-5   Sketch the circuit of a precision (a) diode and (b) clamp and explain their operation.

14-6   (a) Sketch the circuit of a fast half-wave rectifier and explain its operation. (b) How is this circuit converted into an *average detector?*

14-7   Sketch the circuit of a *peak detector* and explain its operation.

14-8   (a) Draw the circuit of a square-wave generator using an OP AMP.   (b) Explain its operation by drawing the capacitor voltage waveform.   (c) Derive the expression for the period of a symmetrical waveform.

14-9   (a) Draw the circuit of a pulse generator (a monostable multivibrator) using an OP AMP.   (b) Explain its operation by referring to the capacitor waveform.

14-10   (a) Draw the circuit of a triangle generator using a comparator and an integrator.   (b) Explain its operation by referring to the output waveform.   (c) What is the peak amplitude?

14-11   (a) Sketch a regenerative comparator system and explain its operation. (b) What parameters determine the loop gain?   (c) What parameters determine the hysteresis?   (d) Sketch the transfer characteristic and indicate the hysteresis.

14-12   (a) Sketch a 2-input OR (and also NOR) ECL gate.   (b) What parameters determine the noise margin?   (c) Why are the two collector resistors unequal?   (d) Explain why power line spikes are virtually nonexistent.

14-13   List and discuss at least four advantages and four disadvantages of the ECL gate.

14-14 (*a*) Draw a schematic diagram of a D/A converter. Use resistance values whose ratios are multiples of 2. (*b*) Explain the operation of the converter.

14-15 Repeat Rev. 14-14 for a ladder network whose resistances have one of two values, R or $2R$.

14-16 Indicate the circuit of the MOS switch in a D/A converter.

14-17 (*a*) Draw the block diagram for an A/D converter. (*b*) Explain the operation of this system.

PROBABLE VALUES OF GENERAL PHYSICAL CONSTANTS†

Constant	Symbol	Value
Electronic charge	q	1.602×10^{-19} C
Electronic mass	m	9.109×10^{-31} kg
Ratio of charge to mass of an electron	q/m	1.759×10^{11} C/kg
Mass of atom of unit atomic weight (hypothetical)	. . . . . .	1.660×10^{-27} kg
Mass of proton	m_p	1.673×10^{-27} kg
Ratio of proton to electron mass	m_p/m	1.837×10^3
Planck's constant	h	6.626×10^{-34} J-s
Boltzmann constant	$\bar{k}$	1.381×10^{-23} J/°K
	k	8.620×10^{-5} eV/°K
Stefan-Boltzmann constant	σ	5.670×10^{-8} W/(m²)(°K⁴)
Avogadro's number	N_A	6.023×10^{23} molecules/mole
Gas constant	R	8.314 J/(deg)(mole)
Velocity of light	c	2.998×10^8 m/s
Faraday's constant	F	9.649×10^3 C/mole
Volume per mole	V_o	2.241×10^{-2} m³
Acceleration of gravity	g	9.807 m/s²
Permeability of free space	μ_o	1.257×10^{-6} H/m
Permittivity of free space	ϵ_o	8.849×10^{-12} F/m

† E. A. Mechtly, "The International System of Units: Physical Constants and Conversion Factors," National Aeronautics and Space Administration, NASA SP-7012, Washington, D.C., 1964.

CONVERSION FACTORS
AND PREFIXES

1 ampere (A)	$= 1$ C/s
1 angstrom unit (Å)	$= 10^{-10}$ m
	$= 10^{-4}$ μm
1 atmosphere pressure	$= 760$ mm Hg
1 coulomb (C)	$= 1$ A-s
1 electron volt (eV)	$= 1.60 \times 10^{-19}$ J
1 farad (F)	$= 1$ C/V
1 foot (ft)	$= 0.305$ m
1 gram-calorie	$= 4.185$ J
giga (G)	$= \times 10^9$
1 henry (H)	$= 1$ V-s/A
1 hertz (Hz)	$= 1$ cycle/s
1 inch (in.)	$= 2.54$ cm
1 joule (J)	$= 10^7$ ergs
	$= 1$ W-s
	$= 6.25 \times 10^{18}$ eV
	$= 1$ N-m
	$= 1$ C-V
kilo (k)	$= \times 10^3$
1 kilogram (kg)	$= 2.205$ lb
1 kilometer (km)	$= 0.622$ mile
1 lumen	$= 0.0016$ W
	(at 0.55 μm)

1 lumen per square foot	$= 1$ ft-candle (fc)
mega (M)	$= \times 10^6$
1 meter (m)	$= 39.37$ in.
micro (μ)	$= \times 10^{-6}$
1 micron	$= 10^{-6}$ m
	$= 1$ μm
1 mil	$= 10^{-3}$ in.
	$= 25$ μm
1 mile	$= 5,280$ ft
	$= 1.609$ km
milli (m)	$= \times 10^{-3}$
nano (n)	$= \times 10^{-9}$
1 newton (N)	$= 1$ kg-m/s^2
pico (p)	$= \times 10^{-12}$
1 pound (lb)	$= 453.6$ g
1 tesla (T)	$= 1$ Wb/m^2
1 ton	$= 2,000$ lb
1 volt (V)	$= 1$ W/A
1 watt (W)	$= 1$ J/s
1 weber (Wb)	$= 1$ V-s
1 weber per square meter (Wb/m^2)	$= 10^4$ gauss

APPENDIX C / SUMMARY OF NETWORK THEORY

In this book we use linear passive elements such as resistors, capacitors, and inductors in combination with voltage and/or current sources and solid-state devices to form networks. The theorems discussed in this appendix are used frequently in the analysis of these electronic circuits.

C-1   RESISTIVE NETWORKS

Voltage and Current Sources In this section we review some basic concepts and theorems in connection with resistive networks containing voltage and current sources. The circuit symbols and reference directions of independent voltage and current sources are shown in Fig. C-1. An ideal voltage source is defined as a voltage generator whose output voltage $v = v_s$ is independent of the current delivered by the generator. The output voltage is usually specified as a function of time such as, for example, $v_s = V_m \cos \omega t$ or a constant dc voltage. Similarly, an ideal current source delivers an arbitrary current $i = i_s$ independent of the voltage between the two terminals of the current source. The reference polarity for the voltage source v_s means that 1 C of positive charge moving from the negative to the positive terminal through the voltage source acquires v_s joules of energy. In the same way the arrow reference for the current source i_s indicates that i_s coulombs of positive charge per second pass through the source in the direction of the arrow. In a practical voltage or current source there is always some energy converted to heat in an irreversible energy-conversion process. This energy loss can be represented by the loss in a series or parallel source resistance R_s, as shown in Fig. C-1c and d.

A *controlled* or *dependent* source is one whose voltage or current is a function of the voltage or current elsewhere in the circuit. For example, Fig. C-2a represents a small-signal circuit model of a transistor at low frequencies. At the input there is a dependent voltage generator $h_r v_o$ whose *voltage* is proportional to the output *voltage* v_o and the proportionality factor is h_r. At the

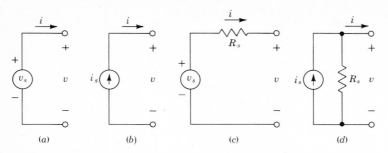

Fig. C-1 (a), (b) ideal and (c), (d) practical voltage and current sources. A circle with a $+$ and $-$ sign is the symbol for the ideal voltage generator. An arrow inside a circle is the symbol for an ideal current generator. The source resistance is designated by R_s.

output there is a dependent current generator $h_f i_1$ whose *current* is proportional to the input *current* i_1 and the proportionality factor is h_f.

Another active device studied in this book is the MOSFET, and its equivalent low-frequency small-signal model is indicated in Fig. C-2b. Note that at the output there is a dependent *current* source $g v_s$ controlled by the input *voltage* v_s and the proportionality factor is g.

Resistance Ohm's law states that the voltage V across a conductor is proportional to the current I in this circuit element. The proportionality factor V/I is called the *resistance* and is expressed in ohms (abbreviated Ω) if V is in volts and I in amperes.

$$V = IR \tag{C-1}$$

In most electronic circuits it is convenient to express the resistance values in kilohms [abbreviated $k\Omega$ (or more simply K)]. Then Eq. (C-1) continues to be valid if I is expressed in milliamperes (mA) and V in volts (V). If a conductor does not obey Eq. (C-1), it is said to be a nonlinear (or nonohmic) resistor.

To find the resistance R seen between two points in a network, an external voltage source V is considered to be applied between these two points and the current I drawn from the source V is determined. The effective resistance is $R = V/I$, provided that in the above procedure each *independent* source in the network is replaced by its internal source resistance R_s; an ideal voltage source by a short circuit, and an ideal current source by an open circuit (Fig. C-1). *All dependent sources, however, must be retained in the network.*

The two basic laws which allow us to analyze electric networks (linear or nonlinear) are known as *Kirchhoff's current law* (KCL) and *Kirchhoff's voltage law* (KVL).

Kirchhoff's Current Law (KCL)   *The sum of all currents toward a node must be zero at any instant of time.*   A *node* is a point where two or more circuit components meet such as points 1 or 2 in Fig. C-2a. When we apply this law, currents directed toward a node are usually taken as positive and those directed away are taken as negative. The opposite convention can also be used as long as we are consistent for all nodes of the network. The positive reference direction of the current through any resistor of the network can be assigned arbitrarily with the understanding that, if the computed current is determined to be negative, the actual current direction is opposite to that assumed. The physical principle on which KCL is based is the law of the conservation of charge, since a violation of KCL would require that some electric charge be "lost" or be "created" at the node.

Kirchhoff's Voltage Law (KVL)   *The sum of all voltage drops around a loop must be zero at all times.*   A closed path in a circuit is called a *loop* or *mesh*. A voltage *drop* V_{12} between two nodes 1 and 2 in a circuit (the potential of point 1 with respect to point 2) is defined as the energy in joules (J) removed from the circuit when a positive charge q of 1 C moves from point 1 to point 2. For example, a voltage drop of $+5$ V across the terminal nodes 1 and 2 of a resistor means that 5 J of energy are removed from the circuit and dissipated as heat in the resistor when a positive charge of 1 C moves *from point 1 to point 2.* If the voltage is -5 V, then point 2 is at a higher voltage than point 1 ($V_{12} = -5$ V represents a voltage *rise*), and a positive charge of 1 C moving from point 1 to point 2 gains 5 J of energy. This, of course, is impossible when a resistor is connected between the two nodes 1 and 2, and is possible only if the negative terminal of a battery is connected to node 1 and the positive terminal to node 2.

It should be clear that KVL is a consequence of the law of conservation of energy. In writing the KVL equations, we go completely around a loop, add all voltage drops, and set the sum equal to zero. Remember these two rules:

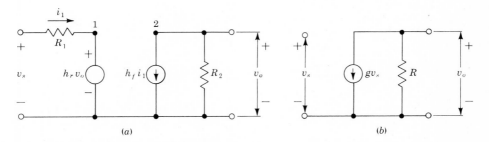

(a)   (b)

Fig. C-2 (a) A transistor circuit model contains a **voltage-controlled voltage generator** $h_r v_o$ and a **current-controlled current generator** $h_f i_1$ (the proportionality factors h_r and h_f are dimensionless). (b) An MOSFET circuit model contains a **voltage-controlled current source** $g v_s$ (the proportional factor g has the dimensions of A/V).

(1) *There is a positive drop in the direction of the current in a resistor.* (2) *There is a positive drop through a battery (or dc source) in the direction from the + to the — terminal, independent of the direction of the current.*

The two fundamental laws (KCL and KVL) will be illustrated in the following examples. Consider first the situation where a resistor R_L is placed directly across the output terminals of a real (nonidealized) voltage source (Fig. C-1c). This added component is called the *load resistor* or, simply, the *load*. The result is that a single mesh is formed as indicated in Fig. C-3. We wish to find the voltage v across R_L.

The current i around the loop flows through R_s and R_L. Traversing this mesh in the assumed direction of current starting at node 2, adding all voltage drops, and setting the sum to zero (as required by KVL) yields

$$-v_s + iR_s + iR_L = 0$$

or

$$i = \frac{v_s}{R_s + R_L} \quad \text{and} \quad v = iR_L = \frac{R_L v_s}{R_s + R_L}. \tag{C-2}$$

Note that under *open-circuit* conditions (defined as $R_L \to \infty$), $v = v_s$. This result is obviously correct since no current can flow in an open circuit so that $i = 0$, $iR_s = 0$, and $v = v_s = $ open-circuit voltage. Also note that under *short-circuit* conditions (defined as $R_L = 0$; an ideal zero-resistance wire), the output voltage drops to zero, $v = 0$. Now the current is a maximum (with respect to variations in the value of R_L) and $i = v_s/R_s = $ *short-circuit current*. The voltage v_s may be a function of time, and then v will also be a function of time.

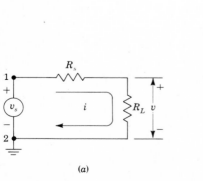
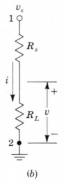

(a)   (b)

Fig. C-3 (a) A load R_L is placed across a voltage source whose internal resistance is R_s. (b) The *same* circuit is redrawn in a different way. The tiny circle at node 1 is used to indicate that a power supply v_s exists between this node and node 2 which has been designated as the reference node. Since one terminal of a generator is usually connected to the metal chassis on which it is built, this terminal is called *ground*. The standard symbol for a ground is shown at node 2.

An equivalent alternative way to draw the circuit of Fig. C-3a is shown in Fig. C-3b.   The caption explains the meaning of the symbols at nodes 1 and 2. This configuration is referred to as a *voltage divider* or an *attenuator*.   Note that v is less than v_s (for any finite R_L), and that the fraction of v_s appearing across R_L is the *attenuation*.

$$\frac{v}{v_s} = \frac{R_L}{R_s + R_L} \tag{C-3}$$

EXAMPLE   Find the current flowing in the loop shown in Fig. C-4.

Solution   We arbitrarily, assign (guess) the positive current direction as shown in Fig. C-4.   As a consequence of the assumed positive current flow, the resistor voltage drops are positive in the directions indicated in the figure.   Let us go around the loop in the direction of the assumed positive current flow (counterclockwise) and sum the voltage drops, starting at ground (node 4).

$$V_{43} + V_{32} + V_{21} + V_{14} = 0 \tag{C-4}$$

Expressing I in milliamperes, the individual voltage drops in volts are: $V_{43} = +14$, $V_{32} = IR_2 = 9I$, $V_{21} = IR_1 = I$, and $V_{14} = +6$.   Substituting the above in Eq. (C-4) we find $14 + 9I + I + 6 = 0$ or $I = -20/10 = -2$ mA.   Thus, the magnitude of the current is 2 mA and the minus sign indicates that the current flows clockwise around the loop, opposite to the assumed direction of positive current.

In the next example we must make use of both KCL and KVL equations because the network has more than one mesh.

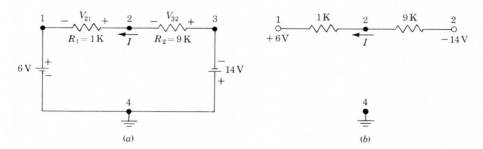

Fig. C-4   (a) A single-loop resistive circuit.   (b) The same circuit drawn differently. The voltage +6 V shown at node 1 indicates that there is a dc generator (or battery) between this node and the reference (ground) node 4, giving a 6-V drop between nodes 1 and 4.   Similarly, the −14 V at node 3 means that there is a 14-V rise (negative drop) between nodes 3 and 4.

EXAMPLE (a) Find the currents I_1, I_2, and I_3 in the circuit shown in Fig. C-5. (b) Find the voltage drop V_{24}.

Solution We assign the arbitrary reference directions of positive currents as shown in the figure. We must sum the voltage drops in each loop by going around each mesh in the arbitrary direction shown by the loop arrows, called the *mesh currents*. Note that the current in R_1 is the mesh current I_1 and that in R_2 is the mesh current I_2. However, the current in R_3 is a combination of I_1 and I_2. Applying KVL we obtain the following equations:

Loop 1 $V_{12} + V_{24} + V_{41} = 0$ (C-5)

Loop 2 $V_{32} + V_{24} + V_{43} = 0$ (C-6)

where the individual voltage drops are given below:

$V_{12} = I_1 R_1 = I_1$ $V_{24} = -I_3 R_3 = -2I_3$ $V_{41} = -6$

$V_{32} = I_2 R_2 = 9I_2$ $V_{43} = 14$

Substituting these values into Eqs. (C-5) and (C-6) gives

$I_1 - 2I_3 - 6 = 0$

$9I_2 - 2I_3 + 14 = 0$

Since we have only two equations for our three unknowns, we must use the KCL equation at node 2 to obtain the additional equation

$I_1 + I_2 + I_3 = 0$ or $I_3 = -(I_1 + I_2)$

Substituting this value of I_3 into the equations for I_1 and I_2 gives

$3I_1 + 2I_2 = 6$

$2I_1 + 11I_2 = -14$

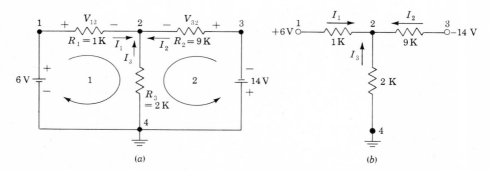

Fig. C-5 (a) Two-loop resistive network. (b) The same network with the voltages (with respect to ground) at nodes 1 and 3 indicated, but with the battery symbols omitted.

Solving these simultaneous algebraic equations, we find

$I_1 = 3.242 \qquad I_2 = -1.862 \qquad$ and $\qquad I_3 = -1.379$ mA

b. The voltage drop V_{24} is

$V_{24} = -I_3R_3 = 1.379 \times 2 = 2.758$ V

The voltage drop between any two nodes in a network is independent of the path chosen between the nodes. For example, V_{24} may be found by going from 2 to 1 to 4 and adding all voltage drops along this path. Thus,

$V_{24} = -I_1R_1 + 6 = -3.242 + 6 = 2.758$ V

which agrees with the value found by going directly from 2 to 4 through R_3.

In solving the above illustrative problem, we chose the two internal meshes 1 and 2. There is a third mesh in this network; the one around the outside loop 4-1-2-3-4. However, this outside mesh is not independent of the other two meshes. *An independent loop is one whose KVL equation includes at least one voltage not included in the other equations.* The number of independent KVL equations is equal to the number of independent loops.

A *junction* is defined as a point where three or more circuit elements meet. Of the four nodes in Fig. C-5, nodes 2 and 4 are junctions. *The number of independent KCL equations is equal to one less than the number of junctions.* Hence, in solving the above problem only one KCL equation is required.

Series and Parallel Combinations of Resistors The circuit of Fig. C-6*a* consists of three resistors in *series*, which means that the same current flows in each resistor. From KVL,

$-V + IR_1 + IR_2 + IR_3 = 0$

The equivalent resistance R between nodes 1 and 2 is given by

$$R \equiv \frac{V}{I} = R_1 + R_2 + R_3 \tag{C-7}$$

To find the total resistance in a series circuit, add the individual values of resistance.

Resistors are in parallel when the same voltage appears across each resistor. Hence, Fig. C-6*b* shows three resistors in parallel:

$$I_1 = \frac{V}{R_1} = G_1V \qquad I_2 = \frac{V}{R_2} = G_2V \qquad I_3 = \frac{V}{R_3} = G_3V$$

where $G \equiv 1/R$ is called the *conductance*. Its dimensions are A/V or reciprocal ohms, called *mhos* (℧). Applying KCL to Fig. C-6*b* yields

$$I = I_1 + I_2 + I_3 = (G_1 + G_2 + G_3)V$$

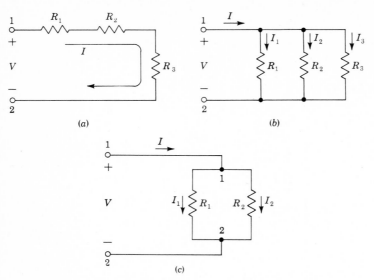

Fig. C-6 (a) Resistors in series. (b) Resistors in parallel. (c) A current divider.

The equivalent conductance between nodes 1 and 2 is, by definition,

$$G \equiv \frac{I}{V} = G_1 + G_2 + G_3 \tag{C-8}$$

To find the total conductance in a parallel circuit, add the individual values of conductance. Equation (C-8) is equivalent to

$$\frac{1}{R} = \frac{1}{R_1} + \frac{1}{R_2} + \frac{1}{R_3} \tag{C-9}$$

Of course, the number of resistors in the series or parallel circuits of Fig. C-6 is not limited to three; it can be any number, two or more. For the special case of two resitors, Eq. (C-9) is equivalent to

$$R = R_1 \| R_2 = \frac{R_1 R_2}{R_1 + R_2} \tag{C-10}$$

where the symbol $\|$ is to be read "in parallel with." It follows from this equation that two resistors in parallel have an effective resistance which is *smaller* than either resistor.

Just as a series circuit gives voltage attenuation [Fig. C-3b and Eq. (C-3)] so a parallel circuit gives current attenuation. In Fig. C-6c the current I_1 in R_1 (or I_2 in R_2) is less than the current I entering node 1. Thus, using Eq. (C-10), we have

$$V = IR = \frac{IR_1R_2}{R_1 + R_2} = I_1R_1$$

or

$$I_1 = \frac{R_2I}{R_1 + R_2} \tag{C-11}$$

Note that if $R_1 = 0$, $I_1 = I$. This result is intuitively correct since all the current should flow in the short circuit. On the other hand if $R_1 \to \infty$, then $I_1 = 0$, which is certainly true because no current can flow in an open circuit.

C-2 NETWORK THEOREMS

The currents and voltages in any network regardless of its complexity may be obtained by a systematic application of KCL and KVL. However, the analysis may often be simplified by using one or more of the additional network theorems discussed in this section.

Superposition Theorem *The response of a linear network containing several independent sources is found by considering each generator separately and then adding the individual responses.* When evaluating the response due to one source, each of the other independent generators is replaced by its internal resistance, that is, set $v_s = 0$ for a voltage source and $i_s = 0$ for a current generator.

EXAMPLE Find the currents I_1, I_2, and I_3 in the circuit of Fig. C-5, using the superposition theorem.

Solution First consider the currents I_1', I_2', and I_3' due to the 6-V supply. Then node 3 must be short-circuited to node 4, so as to eliminate the response due to the -14-V source. This connection puts R_2 and R_3 in parallel, as indicated in Fig. C-7a. From Eq. (C-10) this parallel combination has a resistance of

$$\frac{R_2R_3}{R_2 + R_3} = \frac{9 \times 2}{9 + 2} = 1.636 \text{ K}$$

The resistance seen by the 6-V supply is R_1 plus the above value, hence

$$I_1' = \frac{6}{1 + 1.636} = 2.276 \text{ mA}$$

From the current attenuation formula, Eq. (C-11),

$$I_2' = \frac{-I_1'R_3}{R_2 + R_3} = \frac{-2.276 \times 2}{9 + 2} = -0.414 \text{ mA}$$

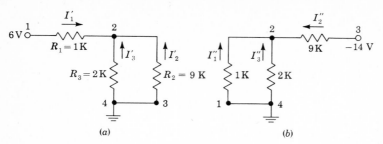

(a)   (b)

Fig. C-7 Superposition is applied to the network of Fig. C-5.
The circuit from which to calculate the response due to (a) the
6-V supply and (b) the −14-V supply.

Similarly,

$$I_3' = \frac{-I_1'R_2}{R_2 + R_3} = \frac{-2.276 \times 9}{2 + 9} = -1.862 \text{ mA}$$

We now find the currents I_1'', I_2'', and I_3'' due to the −14-V supply. To eliminate the effect of the 6-V source, nodes 1 and 4 are connected together as shown
in Fig. C-7b. Proceeding as above, we find

$$I_2'' = \frac{-14}{9 + (1 \times 2)/3} = -1.448 \text{ mA}$$

$$I_1'' = +1.448 \times \tfrac{2}{3} = 0.9655 \text{ mA}$$

$$I_3'' = +1.448 \times \tfrac{1}{3} = 0.4826 \text{ mA}$$

The net current is the algebraic sum of the currents due to each excitation. Thus

$$I_1 = I_1' + I_1'' = 2.276 + 0.966 = 3.242 \text{ mA}$$

$$I_2 = I_2' + I_2'' = -0.414 - 1.448 = -1.862 \text{ mA}$$

$$I_3 = I_3' + I_3'' = -1.862 + 0.483 = -1.379 \text{ mA}$$

These are the same values obtained in the previous section. Note that for this
particular network the analysis in Sec. C-1 using KVL and KCL is simpler than
that given here using superposition.

Thévenin's Theorem *Any linear network may, with respect to a pair of
terminals, be replaced by a voltage generator V_{Th} (equal to the open-circuit voltage)
in series with the resistance R_{Th} seen between these terminals.* To find R_{Th} all
independent voltage sources are short-circuited and all *independent* current
sources are open-circuited. This theorem is often used to reduce the number
of meshes in a network. For example, the two-mesh circuit of Fig. C-5 may
be reduced to a single loop by replacing the components to the left of terminals 2 and 4 (including R_3) by the Thévenin equivalent. For convenience,
the circuit of Fig. C-5 is redrawn in Fig. C-8a. The components in the shaded

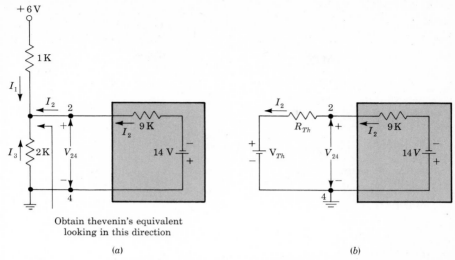

Obtain thevenin's equivalent
looking in this direction

(a)   (b)

Fig. C-8 (a) **The circuit of Fig. C-5 redrawn.** (b) **Thévenin's theorem applied to the circuit in (a) looking to the left of nodes 2 and 4.**

box are those to the right of nodes 2 and 4, and they are redrawn unaltered in Fig. C-8b. The other circuit elements do not appear in Fig. C-8b but are replaced by V_{Th} and R_{Th}. Thévenin's theorem states that I_2 and V_{24} calculated from this reduced circuit are identical to the corresponding values in Fig. C-5.

The open-circuit voltage V_{Th} is obtained by disconnecting the components in the box from Fig. C-8a. From the voltage-attenuator formula Eq. (C-2),

$$V_{Th} = \frac{6 \times 2}{1 + 2} = 4 \text{ V}$$

To find the resistance seen to the left of nodes 2 and 4, the 6-V supply is imagined reduced to zero, which is equivalent to connecting the top of the 1-K resistor to ground. Hence, this resistor is now placed in parallel with the 2-K resistor, and

$$R_{Th} = \frac{1 \times 2}{1 + 2} = 0.667 \text{ K}$$

From the equivalent circuit of Fig. C-8b we obtain

$$I_2 = \frac{-(14 + V_{Th})}{9 + R_{Th}} = \frac{-18}{9.667} = -1.862 \text{ mA}$$

and

$$V_{24} = -9I_2 - 14 = 9 \times 1.862 - 14 = 2.758 \text{ V}$$

These two values agree with the numerical values found in Sec. C-1. The currents I_3 and I_1 do not appear in Fig. C-8b and must be found from Fig. C-8a. Thus,

$$I_3 = \frac{-V_{24}}{2} = \frac{-2.758}{2} = -1.379 \text{ mA}$$

and

$$I_1 = \frac{6 - V_{24}}{1} = 6 - 2.758 = 3.242 \text{ mA}$$

which are the same currents found previously.

Norton's Theorem *Any linear network may, with respect to a pair of terminals, be replaced by a current generator (equal to the short-circuit current) in parallel with the resistance seen between the two terminals.*

From Thévenin's and Norton's theorems it follows that a voltage source V in series with a resistance R is equivalent to a current source I in parallel with R, provided that $I = V/R$. These equivalent circuits are indicated in Figs. C-1c d with $v_s = V$, $R_s = R$, and $i_s = I = V/R_s$.

Open-Circuit-Voltage—Short-Circuit-Current Theorem As corollaries to Thévenin's and Norton's theorems we have the following relationships. If V represents the *open-circuit voltage*, I the *short-circuit current*, and $R(G)$ the resistance (conductance) between two terminals in a network, then

$$V = IR = \frac{I}{G} \qquad I = \frac{V}{R} = GV \qquad R = \frac{V}{I} \tag{C-12}$$

In spite of their disarming simplicity, these equations (reminiscent of Ohm's law) should not be overlooked because they are most useful in analysis. For example, the first equation, which states "open-circuit voltage equals short-circuit current divided by conductance," is often the simplest way to find the voltage between two points in a network, as the following problem illustrates.

EXAMPLE Find the voltage between nodes 2 and 4 in the circuit of Fig. C-5, using the *open-circuit-voltage—short-circuit-current theorem.*

Solution If a zero-resistance wire connects node 2 to 4, then the current in this short circuit due to the 6-V battery is $\frac{6}{1} = 6$ mA, and due to the other battery, it is $-\frac{14}{9} = -1.556$ mA. Hence, the total short-circuit current is $I = 6 - 1.556 = 4.444$ mA.

The conductance between 2 and 4 with nodes 1 and 3 grounded, corresponds to R_1, R_2, and R_3 all in parallel. Hence,

$$G = \tfrac{1}{1} + \tfrac{1}{2} + \tfrac{1}{9} = 1.611 \text{ K}$$

Therefore,

$$V_{24} = \frac{I}{G} = \frac{4.444}{1.611} = 2.759 \text{ V} \tag{C-13}$$

which agrees with the value found previously. The currents I_1 and I_2 can now be found by applying KVL to each loop in Fig. C-5. For example,

$$I_2 = \frac{-(14 + V_{24})}{9} = \frac{-(14 + 2.759)}{9} = -1.862 \text{ mA}$$

which is the same value obtained previously. Note that this method, or the use of Thévenin's theorem, results in a simpler solution than the straightforward analysis, using KVL and KCL.

Nodal Method of Analysis   When the number of junction voltages (with respect to the reference, or ground, node) is less than the number of independent meshes, then the choice of nodal voltages as the unknowns leads to a simpler solution than considering the mesh currents as the unknowns. For example, the circuit of Fig. C-5 has two independent meshes but only one independent junction voltage. In terms of the one unknown independent voltage V_{24}, the currents are

$$I_1 = \frac{6 - V_{24}}{1} \qquad I_2 = \frac{-14 - V_{24}}{9} \qquad I_3 = \frac{-V_{24}}{2}$$

By KCL the sum of these three currents (which enter node 2) must equal zero. Hence,

$$\frac{6}{1} - \frac{V_{24}}{1} - \frac{14}{9} - \frac{V_{24}}{9} - \frac{V_{24}}{2} = 0$$

or

$$V_{24}\left(\frac{1}{1} + \frac{1}{9} + \frac{1}{2}\right) = \frac{6}{1} - \frac{14}{9} = 4.444 \text{ mA}$$

which is equivalent to Eq. (C-13) namely, $V_{24}G = I$, where I is the short-circuit current. Hence, the nodal method in this simple case is equivalent to the analysis using the open-circuit-voltage—short-circuit-current theorem.

C-3   THE SINUSOIDAL STEADY STATE

If a sinusoidal excitation (a voltage or current) is applied to a linear network, then the response (the voltage between any two nodes or the current in any branch of the network) will also be sinusoidal. (It is assumed that all transients have died down so that a steady state is reached.) Let us verify this general statement for the simple parallel combination of resistor R and capacitor C in Fig. C-9 to which has been applied the sinusoidal source voltage

$$v = V_m \cos \omega t = V_m \cos 2\pi f t \qquad (C\text{-}14)$$

where f is the *frequency* of the source in hertz (cycles per second), $\omega = 2\pi f$ is called the *angular frequency*, and V_m is the *maximum* or *peak* value of voltage. We shall now prove that the generator current i is also a sinusoidal waveform.

A capacitor C is a component (say, two metals separated by a dielectric) which stores charge q (coulombs) proportional to the applied voltage v (volts) so that

$$q = Cv \qquad (C\text{-}15)$$

where the proportionality factor C is called the *capacitance*. The dimensions of C are coulombs per volt, which is abbreviated as *farads* (F). The capacitor current i_C is therefore

$$i_C = \frac{dq}{dt} = C\frac{dv}{dt} \qquad (C\text{-}16)$$

or, using Eq. (C-14),

$$i_C = -\omega C V_m \sin \omega t \qquad (C\text{-}17)$$

From Ohm's law, the resistor current i_R is given by

$$i_R = \frac{v}{R} = \frac{V_m}{R} \cos \omega t \qquad (C\text{-}18)$$

From Kirchhoff's current law (KCL), $i = i_R + i_C$, or

$$i = \frac{V_m}{R} \cos \omega t - \omega C V_m \sin \omega t \qquad (C\text{-}19)$$

which has the form

$$i = I_m \cos \theta \cos \omega t - I_m \sin \theta \sin \omega t \qquad (C\text{-}20)$$

where

$$I_m \cos \theta \equiv \frac{V_m}{R} \qquad \text{and} \qquad I_m \sin \theta \equiv \omega C V_m \qquad (C\text{-}21)$$

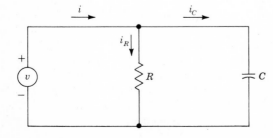

Fig. C-9 A parallel RC combination excited by a sinusoidal voltage.

From the trigonometric identity,

$$\cos (\theta + \alpha) = \cos \theta \cos \alpha - \sin \theta \sin \alpha \qquad \text{(C-22)}$$

then Eq. (C-20), with $\alpha \equiv \omega t$, is equivalent to

$$i = I_m \cos (\omega t + \theta) \qquad \text{(C-23)}$$

We have thus verified that the generator current is indeed a sinusoid. The peak current is I_m, and i is shifted in phase by the angle θ with respect to the source voltage $V_m \cos \omega t$. We say that "the generator current *leads* its voltage by the phase angle θ."

The peak current I_m and the phase θ are obtained from Eqs. (C-21). If the two equations are each squared and then added, we obtain

$$I_m{}^2 \cos^2 \theta + I_m{}^2 \sin^2 \theta = \frac{V_m{}^2}{R^2} + \omega^2 C^2 V_m{}^2 \qquad \text{(C-24)}$$

Since $\cos^2 \theta + \sin^2 \theta = 1$, then

$$I_m = V_m \sqrt{\frac{1}{R^2} + \omega^2 C^2} \qquad \text{(C-25)}$$

Dividing the second equation in Eq. (C-21) by the first yields

$$\frac{I_m \sin \theta}{I_m \cos \theta} = \frac{\omega C V_m}{V_m/R}$$

or

$$\tan \theta = \omega C R \qquad \text{(C-26)}$$

For a more complicated network than the one in Fig. C-9, the analysis would involve a prohibitive amount of trigonometric manipulation. Hence, we now present a simpler alternative general method for solving sinusoidal networks in the steady state. Some important concepts (such as phasors, complex plane, and impedance) are first introduced.

Phasors   Each current (or voltage) in a network is a sinusoid, which has a peak value and a phase angle. Hence it can be represented by a vector, which is a directed line segment, having a length and direction. For a sinusoid this vector is called a *phasor*. Its magnitude represents the effective or rms value and is given by the peak value divided by $\sqrt{2}$ [Eq. (3-17)]. The direction of the phasor is the phase θ in the sinusoidal waveform $I_m \cos (\omega t + \theta)$, and the angle θ is measured counterclockwise with respect to the horizontal axis. In this section we use boldface **I** (**V**) to denote a phasor current (voltage). In phasor notation, the current in Eq. (C-23) is written

$$\mathbf{I} = I \angle \theta \qquad \text{(C-27)}$$

where $I = I_m/\sqrt{2}$. This phasor is indicated in Fig. C-10a.

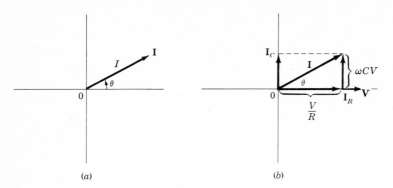

Fig. C-10   (a) The current represented as a phasor of magnitude
I and phase *θ*.   (b) Phasor addition, representing I = I$_R$ + I$_C$.

The applied voltage phasor is, from Eq. (C-14), V = V∠0, where
$V = V_m/\sqrt{2}$, and the current in the resistor is, from Eq. (C-18) I$_R$ = V/R∠0.
These phasors are indicated in Fig. C-10b.   Note that *the current in a resistor
is in phase with the voltage across the resistor.*

Since Eq. (C-17) may be written $i_C = \omega CV_m \cos{(\omega t + 90°)}$, the phasor
representing the capacitor current is

$$I_C = \omega CV \angle 90° \tag{C-28}$$

where $V = V_m/\sqrt{2}$ is the rms voltage.   Note that *the current in a capacitor
leads the voltage across the capacitor by 90°.*   The phasor I$_C$ is plotted in Fig.
C-10b.   The generator current is the sum of the resistor current and the
capacitor current, or in phasor notation

$$I = I_R + I_C = \frac{V}{R} \angle 0 + \omega CV \angle 90° \tag{C-29}$$

This vector sum is indicated in Fig. C-10b, where it is found that

$$|I|^2 = \frac{V^2}{R^2} + \omega^2 C^2 V^2 \qquad \text{and} \qquad \tan{\theta} = \omega CR$$

in agreement with Eqs. (C-25) and (C-26).   Note how simple the phasor
method of analysis is compared with the above solution, with the use of instantaneous values of current and voltage and manipulating the equations with
the aid of trigonometric identities.   By introducing the concept of the complex
plane, the analysis may be further simplified.   An essentially algebraic, rather
than a trigonometric, solution is now obtained.

The *j* Operator   A useful convention is to take the symbol *j* to represent a *phase lead* of 90°.   In place of Eq. (C-28) we now write I$_C$ = *jωCV*,

and for the total current in Eq. (C-29) we have

$$| = \frac{V}{R} + j\omega CV \qquad\qquad (C\text{-}30)$$

This equation is interpreted to mean that $|$ is a phasor formed by combining the phasor V/R horizontally (at zero phase) with ωCV plotted vertically (at a phase of 90°). Hence, the vertical axis is also called the j axis. The current $|$, shown in Fig. C-10b, is identical with that found above.

From the definition of j it follows that $j|$ is a phasor whose magnitude is that of $|$ but whose phase is 90° greater than the phase of $|$. In other words, j "multiplying" a phasor $|$ is an operator which rotates $|$ in the counterclockwise direction by 90°. Consider $| = 1$, a phasor of magnitude 1 and phase 0. Then $j| = j1$ has a magnitude 1 and phase 90°, as indicated in Fig. C-11. Then $j(j1)$ represents a rotation of $j1$ by 90°, which results in a phasor of unit magnitude pointing along the negative horizontal axis, as indicated in Fig. C-11. In a purely formal manner we may write

$$j(j1) = j^2 1 = -1 \qquad \text{or} \qquad j = \sqrt{-1} \qquad\qquad (C\text{-}31)$$

Because of this formalism the vertical axis is called the j or *imaginary axis* and the horizontal axis is designated as the *real axis*. The plane of Fig. C-11 is now called the *complex plane*.

Note that higher powers of j are easily found. Thus,

$$j^3 = j(j^2) = j(-1) = -j \qquad\qquad (C\text{-}32)$$

which represents a phasor of magnitude 1 and phase $-90°$. The reciprocal of j is $-j$, as is easily verified. Thus

$$\frac{1}{j} = \frac{1}{j}\frac{j}{j} = \frac{j}{j^2} = -j \qquad\qquad (C\text{-}33)$$

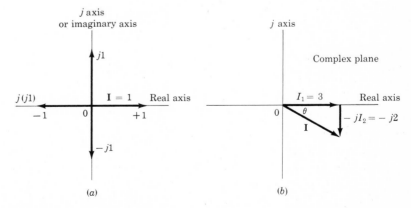

(a)   (b)

Fig. C-11   (a) Concerning the operator j.   (b) A phasor current plotted in the complex plane.

because $j^2 = -1$ from Eq. (C-31).   A point in the complex plane is called a
complex number, and it is evident that a phasor is a complex number.   Hence,
the analysis of sinusoidal circuits is carried out most simply by treating currents and voltages as complex numbers representing phasors.

Assume that a complicated circuit is analyzed (by the general method
outlined in Sec. C-4) and that the following complex current is obtained

$$ I = I_1 - jI_2 = 3 - j2 \qquad \text{mA} \tag{C-34} $$

This phasor is indicated in the complex plane in Fig. C-11*b*.   From this
diagram it follows that the rms current $|I|$ and the phase angle θ are given by

$$ |I| = \sqrt{I_1{}^2 + I_2{}^2} = \sqrt{13} = 3.61 \text{ mA} $$

and

$$ \theta = -\arctan \frac{I_2}{I_1} = -\arctan \frac{2}{3} = -33.7° = -0.588 \text{ rad} $$

If the frequency f is 1 kHz, then the instantaneous current is, from Eq. (C-23),
$i = 3.61 \sqrt{2} \cos (6{,}280t - 0.588)$ mA.

C-4   SIMPLIFIED SINUSOIDAL NETWORK ANALYSIS

Consider a linear network containing resistors, capacitors, inductors, and sinusoidal sources.   The steady-state response is desired.   A straightforward
method of solution is possible which is analogous to that used with networks
containing only resistive components and constant (dc) supply voltages (or
currents).   The analysis consists of writing the KVL and KCL equations
for the network and then solving for the complex (phasor) currents and voltages.   In order to carry out such an analysis, it is first necessary to introduce
the concept of *complex resistance* or *reactance*.   After defining reactance, a
number of specific circuits are solved by using this simple method of analysis.

Reactance   The ratio of the voltage† V across a passive circuit component to the current I through the element for each of the three basic components is as follows:

$$ \text{Resistance} \qquad \frac{V}{I} = R $$

$$ \text{Capacitance} \qquad \frac{V}{I} = \frac{1}{j\omega C} = \frac{-j}{\omega C} \tag{C-35} $$

$$ \text{Inductance} \qquad \frac{V}{I} = j\omega L $$

† In this section, and throughout the text, bold face **V** (or **I**) used to designate a
complex (phasor) voltage (or current) is replaced by an italic symbol V (or I).

The first equation of Eqs. (C-35) is Ohm's law.   The second equation follows from Eq. (C-28).   An inductor is a component (say, a coil of wire) whose terminal voltage v is proportional to the rate of change of current.   The proportionality factor L (henrys, H) is called the *inductance*.   From $v = L \, di/dt$, the third equation of Eq. (C-35) can be obtained in a manner analogous to that used in the preceding section to obtain Eq. (C-28).

From Eqs. (C-35) it follows that a capacitor behaves as a "complex resistance" $-j/\omega C$ and an inductor acts like a "complex resistance" $j\omega L$.   A more commonly used phrase for complex resistance is *reactance*, denoted by the real positive symbol X.

$$\text{Capacitive reactance} = -jX_C \text{ where } X_C \equiv 1/\omega C$$

and

$$\text{Inductive reactance} = +jX_L \text{ where } X_L \equiv \omega L$$

In applying KVL to a circuit containing reactive elements, it must be remembered that the drop across a capacitor is $-jX_C I = -jI/\omega C$ and the drop across an inductor is $jX_L I = j\omega L I$.   It follows from the above considerations that KVL applied to the series circuit of Fig. C-12 yields

$$V = RI + j\omega L I - \frac{j}{\omega C} I \tag{C-36}$$

or

$$I = \frac{V}{R + j(\omega L - 1/\omega C)} = \frac{V}{R + jX} \tag{C-37}$$

where the total series reactance is $X \equiv \omega L - 1/\omega C$.   The current may be expressed in standard complex number form $I = I_1 + jI_2$ by multiplying both the numerator and the denominator by the complex conjugate (change j to $-j$) of the denominator.   Thus,

$$I = \frac{V}{R + jX} \frac{R - jX}{R - jX} = \frac{V}{R^2 + X^2} (R - jX) \tag{C-38}$$

From this equation we find that the magnitude and phase of I are given by

$$|I| = \frac{V}{\sqrt{R^2 + X^2}} \quad \text{and} \quad \tan \theta = -\frac{X}{R} \tag{C-39}$$

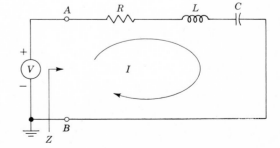

Fig. C-12   An *RLC* series circuit.

Impedance   The ratio of the voltage between any two points A and B of a network to the current in this portion of the circuit is called the *impedance* Z between A and B.   For the circuit of Fig. C-12

$$Z \equiv \frac{V}{I} = R + j\left(\omega L - \frac{1}{\omega C}\right) \tag{C-40}$$

from Eq. (C-36). Since the generator V is placed directly between A and B, then Z is the impedance "seen" by the source V.   Note that for a series circuit the impedance equals the sum of the resistances plus reactances in the loop.   This statement is analogous to the law for a dc series circuit, which states that the total resistance is the sum of the resistances in series.   It should be emphasized that, whereas Z is a complex quantity, it is not a phasor, since it does *not* represent a current or a voltage varying sinusoidally with time.

Two impedances Z_1 and Z_2 in parallel represent an equivalent impedance Z given by

$$Z = \frac{Z_1 Z_2}{Z_1 + Z_2} \tag{C-41}$$

corresponding to Eq. (C-10) for two resistors in parallel.   For the parallel combination of a resistor R and a capacitor C as in Fig. C-9, $Z_1 = R$, $Z_2 = -j/\omega C = 1/j\omega C$, and from Eq. (C-41),

$$Z = \frac{R(1/j\omega C)}{R + 1/j\omega C} = \frac{R}{1 + j\omega CR} \tag{C-42}$$

This same result is obtained by applying KCL to Fig. C-9.   Using phasor notation, we have

$$I = I_R + I_C = \frac{V}{R} + \frac{V}{1/j\omega C} = \frac{V}{R} + j\omega CV \tag{C-43}$$

and $Z = V/I$ gives the result in Eq. (C-42).

As another example of a one-mesh circuit, consider the high-pass RC configuration of Fig. 11-1, which is repeated in Fig. C-13 for easy reference.   The output voltage is

$$V_o = IR = \frac{V_i}{Z}R = \frac{R}{Z}V_i = \frac{RV_i}{R + 1/j\omega C} \tag{C-44}$$

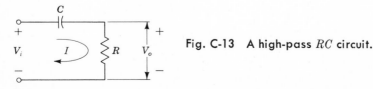

Fig. C-13   A high-pass RC circuit.

If $j\omega$ is replaced by s, then

$$V_o = \frac{V_i}{1 + 1/sRC} = \frac{sV_i}{s + 1/RC} \tag{C-45}$$

which is Eq. (11-1) with $R = R_1$ and $C = C_1$.

Loop or Mesh Equations   This general method of analysis described in Sec. C-1 in connection with Fig. C-5 is to apply KVL around each independent loop.   We illustrate this method with reference to the two-mesh circuit of Fig. C-14.   For the assumed directions of the loop currents it is clear that the current in R_1 is $I_1 - I_2$ so that KCL is satisfied at node A.   The sum of the voltage *drops* around loop 1 *in the direction of* I_1 is

$$-V_1 + I_1\left(\frac{-j}{\omega C_1} + R_1\right) - I_2 R_1 = 0 \tag{C-46}$$

and KVL around mesh 2 *in the direction of* I_2 is

$$+V_2 - I_1 R_1 + I_2\left(R_1 + R_2 - \frac{j}{\omega C_2}\right) = 0 \tag{C-47}$$

Equations (C-46) and (C-47) are solved simultaneously for the two currents I_1 and I_2.   It is assumed that the two voltage sources have the same frequency. If this is not true, then the principle of superposition (Sec. C-2) must be used. In Eqs. (C-46) and (C-47) set $V_2 = 0$, $\omega = \omega_1$, and solve the resulting simultaneous equations for the currents.   The result will be a sinusoid of frequency f_1 for I_1 (and also for I_2).   Then set $V_1 = 0$ and $\omega = \omega_2$ and again solve Eqs. (C-46) and (C-47) for the currents, which will now be sinusoids of frequency f_2.   The total current i_1 will be obtained by *adding the two instantaneous values of time*, and will be of the form

$$i_1 = A_1 \cos(\omega_1 t + \theta_1) + A_2 \cos(\omega_2 t + \theta_2)$$

Note that *it is meaningless to add phasors of different frequencies*.

Admittance   The reciprocal of the impedance is called the admittance and is designated by Y, so that

$$Y \equiv \frac{1}{Z} = G + jB \tag{C-48}$$

Fig. C-14   A two-mesh sinusoidal network.

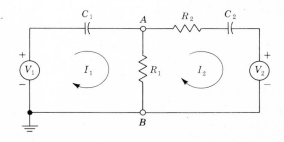

The real part of Y is the *conductance* G and the imaginary part is the *susceptance* B. If a resistor is under consideration then $Z = R$, $G = 1/R$, and $B = 0$. On the other hand if the circuit element is a capacitor $Z = 1/j\omega C$ and $Y = j\omega C$ so that $B = \omega C$ and $G = 0$.

Since $I = V/Z$, then $I = YV$. For a resistor $I_R = GV_R$ and for a capacitor $I_C = j\omega CV_C$. For the circuit of Fig. C-9, with R and C in parallel, $V_R = V_C = V$, and the total current is

$$I = I_R + I_C = (G + j\omega C)V$$

The admittance of this combination is $Y = I/V = G + j\omega C$, which agrees with Eq. (C-43) if $G = 1/R$.

Nodal Equations   This general method of solution is to apply KCL at each independent node in the network. We illustrate this type of analysis with the single-stage high-frequency transistor of Fig. (11-11) which is repeated in Fig. C-15 for easy reference. There are two independent junction voltages which, *with respect to ground*, are designated V_1 and V_2 respectively. We introduce the conductances $G = 1/R$ and $g = 1/r$ for convenience.

The currents away from node 1 are gV_1 in r, $j\omega C_e V_1$ in C_e, $G(V_1 - V_s)$ in R, and $j\omega C_c(V_1 - V_2)$ in C_c. Hence, KCL at this node yields

$$0 = (g + j\omega C_e)V_1 + G(V_1 - V_s) + j\omega C_c(V_1 - V_2) \qquad \text{(C-49)}$$

This equation agrees with Eq. (11-45), with the following change of notation: $V_1 = V_{b'e}$, $V_2 = V_o$, $G = G'_s$, $g = g_{b'e}$, and $s = j\omega$.

The currents away from node 2 are the dependent current $g_m V_1$, $j\omega C_c(V_2 - V_1)$ in C_c and V_2/R_L in R_L. At this node KCL is, with $s = j\omega$,

$$0 = (g_m - sC_c)V_1 + \left(sC_c + \frac{1}{R_L}\right)V_2 \qquad \text{(C-50)}$$

which is equivalent to Eq. (11-46).

In passing, note that the nodal equations implicitly satisfy Kirchhoff's voltage laws. For example, KVL around the left-hand loop in Fig. C-15 is

$$-V_s - I_R R + V_1 = 0$$

or

$$I_R = \frac{V_1 - V_s}{R} = G(V_1 - V_s)$$

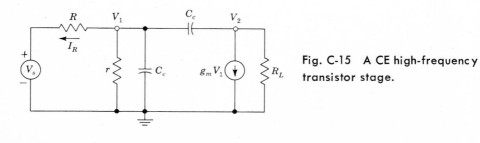

Fig. C-15   A CE high-frequency transistor stage.

which is precisely the expression used, in applying KCL to node V_1, for the current in R.

Network Theorems　The network theorems discussed in Sec. C-2 in connection with resistive networks apply equally well to networks containing capacitors and inductors as well as resistors in the sinusoidal steady state. These include *superposition, Thévenin's* theorem, *Norton's* theorem, and the *open-circuit-voltage—short-circuit-current theorem*. However, in applying these theorems, the impedance Z must be used in place of the resistance R. For example, Eq. (C-12) is now generalized to

$$V = IZ = \frac{I}{Y} \qquad I = \frac{V}{Z} = VY \qquad Z = \frac{V}{I} \qquad (C-51)$$

Miller's Theorem　This theorem is particularly useful in connection with transistor high-frequency amplifiers. Consider an arbitrary circuit configurations with N distinct nodes 1, 2, 3, . . . , N, as indicated in Fig. C-16. Let the node voltages be V_1, V_2, V_3, . . ., V_N, where $V_N = 0$, since N is the reference, or ground, node. Nodes 1 and 2 (referred to as N_1 and N_2) are interconnected with an impedance Z'. We postulate that we know the ratio V_2/V_1. Designate this ratio V_2/V_1 by K, which in the sinusoidal steady state will be a complex number. We shall now show that the current I_1 drawn from N_1 through Z' can be obtained by disconnecting Z' from terminal 1 and by bridging an impedance $Z'/(1 - K)$ from N_1 to ground, as indicated in Fig. C-16b.

The current I_1 is given by

$$I_1 = \frac{V_1 - V_2}{Z'} = \frac{V_1(1 - K)}{Z'} = \frac{V_1}{Z'/(1 - K)} = \frac{V_1}{Z_1} \qquad (C-52)$$

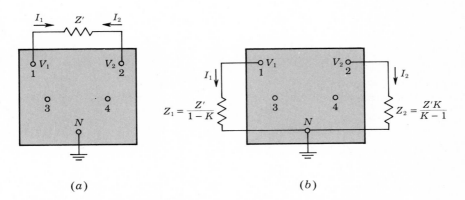

(a)　　　　　　　　　　　　　　　　(b)

Fig. C-16　Pertaining to Miller's theorem. By definition, $K \equiv V_2/V_1$. The networks in (a) and (b) have identical node voltages. Note that $I_1 = -I_2$.

Therefore, if $Z_1 \equiv Z'/(1 - K)$ were shunted across terminals N_1-N, the current I_1 drawn from N_1 would be the same as that from the original circuit. Hence the same expression is obtained for I_1 in terms of the node voltages for the two configurations (Fig. C-16a and b).

In a similar way, it may be established that the correct current I_2 drawn from N_2 may be calculated by removing Z' and by connecting between N_2 and ground an impedance Z_2, given by

$$Z_2 \equiv \frac{Z'}{1 - 1/K} = \frac{Z'K}{K - 1} \tag{C-53}$$

Since identical nodal equations (KCL) are obtained from the configurations of Fig. C-16a and b, these two networks are equivalent. It must be emphasized that this theorem will be useful in making calculations only if it is possible to find the value of K by some independent means.

The first application of Miller's theorem in this text is made in Sec. 11-12. Miller's theorem is also valid for resistive networks with Z' replaced by a resistor R'.

C-5   STEP RESPONSE OF AN RC CIRCUIT

The most common transient problem encountered in electronic circuits is that resulting from a step change in dc excitation applied to a series combination of a resistor and capacitor. Consider the high-pass RC circuit of Fig. C-17 to which is applied a step of voltage v_i. The output voltage v_o is taken across the resistor.

Step-Voltage Input   A *step voltage* is one which maintains the value zero for all times $t < 0$ and maintains the value V for all times $t > 0$. The transition between the two voltage levels takes place at $t = 0$ and is accomplished in an arbitrarily short time interval. Thus in Fig. C-18, $v_i = 0$ immediately before $t = 0$ (to be referred to as time $t = 0-$), and $v_i = V$ immediately after $t = 0$ (to be referred to as time $t = 0+$).

From elementary considerations, the response of the network is exponential, with a time constant $RC \equiv \tau$, and the output voltage is of the form

$$v_o = B_1 + B_2\epsilon^{-t/\tau} \tag{C-54}$$

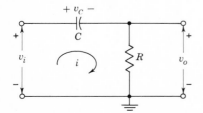

Fig. C-17   The high-pass RC circuit.

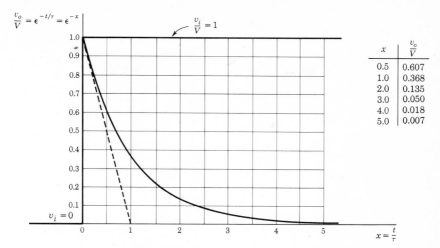

Fig. C-18 Step-voltage response of the high-pass RC circuit. The dashed line is tangent to the exponential at $t = 0+$.

The constant B_1 is equal to the steady-state value of the output voltage because as $t \rightarrow \infty$, $v_o \rightarrow B_1$. If this final value of output voltage is called V_f, then $B_1 = V_f$. The constant B_2 is determined by the initial output voltage, say V_i, because at $t = 0$, $v_o = V_i = B_1 + B_2$ or $B_2 = V_i - V_f$. Hence the general solution for a single-time-constant circuit having initial and final values V_i and V_f, respectively, is

$$v_o = V_f + (V_i - V_f)\epsilon^{-t/\tau} \tag{C-55}$$

This basic equation is used many times throughout this text.

The constants V_f and V_i must now be determined for the circuit of Fig. C-17. The input is a constant ($v_i = V$) for $t > 0$. Since $i = C(dv_C/dt)$, then in the steady state $i = 0$, $v_C = V$, and the final output voltage is zero, or $V_f = 0$.

The above result may also be obtained by the following argument: We have already emphasized that a capacitor C behaves as an open circuit at zero frequency (because the reactance of C varies inversely with f). Hence, any constant (dc) input voltage is "blocked" and cannot reach the output. Hence, $V_f = 0$.

The value of V_i is determined from the following basic considerations. If the instantaneous current through a capacitor is i, then the change in voltage across the capacitor in time t_1 is $(1/C) \int_0^{t_1} i\,dt$. Since the current is always of finite magnitude, the above integral approaches zero as $t_1 \rightarrow 0$. Hence, it follows that *the voltage across a capacitor cannot change instantaneously.*

Applying the above principle to the network of Fig. C-17, we must conclude that since, at $t = 0$, the input voltage changes discontinuously by an

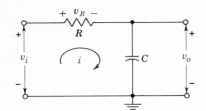

Fig. C-19   The low-pass RC circuit.

amount V, the output must also change abruptly by this same amount.   If we assume that the capacitor is initially uncharged, then the output at $t = 0+$ must jump to V.   Hence, $V_i = V$ and since $V_f = 0$, Eq. (C-55) becomes

$$v_o = V\epsilon^{-t/\tau} \tag{C-56}$$

Input and output are shown in Fig. C-18.   Note that the output is 0.61 of its initial value at 0.5τ, 0.37 at 1τ, and 0.14 at 2τ.   The output has completed more than 95 percent of its total change after 3τ and more than 99 percent of its swing if $t > 5\tau$.   Hence, although the steady state is approached asymptotically, we may assume for most applications that the final value has been reached after 5τ.

Discharge of a Capacitor Through a Resistor   Consider a capacitor C charged to a voltage V.   At $t = 0$, a resistor R is placed across C.   We wish to obtain the capacitor voltage v_o as a function of time.   Since the action of shunting C by R cannot instantaneously change the voltage, $v_o = V$ at $t = 0+$.   Hence, $V_i = V$.   Clearly at $t = \infty$, the capacitor will be completely discharged by the resistor and, therefore, $V_f = 0$.   Substituting these values of V_i and V_f into Eq. (C-55), we obtain Eq. (C-56), and the capacitor discharge is indicated in Fig. C-18.

The Low-Pass RC Circuit   The response of the circuit of Fig. C-19 to a step input is exponential with a time constant RC.   Since the capacitor voltage cannot change instantaneously, the output starts from zero and rises toward the steady-state value V, as shown in Fig. C-20.   The output is given

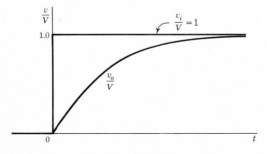

Fig. C-20   Step-voltage response of the low-pass RC circuit.

by Eq. (C-55), or

$$v_o = V(1 - \epsilon^{-t/RC}) \tag{C-57}$$

Note that the circuits of Figs. C-17 and C-19 are identical except that the output $v_o = v_R$ is taken across R in Fig. C-17, whereas the output in Fig. C-19 is $v_o = v_C$.   From Fig. C-19,

$$v_C = v_i - v_R = V - V\epsilon^{-t/RC}$$

where v_R is given by Eq. (C-56).   This result for v_C agrees with Eq. (C-57).

APPENDIX

D / PROBLEMS

CHAPTER 1

1-1   (*a*) An electron is emitted from an electrode with a negligible initial velocity and is accelerated by a potential of 1,000 V. Calculate the final velocity of the particle.

  (*b*) Repeat the problem for the case of a deuterium ion (heavy hydrogen ion—atomic weight 2.01) that has been introduced into the electric field with an initial velocity of 10^5 m/s.

1-2   An electron having an initial kinetic energy of 10^{-16} J at the surface of one of two parallel-plane electrodes and moving normal to the surface is slowed down by the retarding field caused by a 400-V potential applied between the electrodes.

  (*a*) Will the electron reach the second electrode?

  (*b*) What retarding potential would be required for the electron to reach the second electrode with zero velocity?

1-3   The essential features of the displaying tube of an oscilloscope are shown in the accompanying figure. The voltage difference between K and A is V_a and between P_1 and P_2 is V_p. Neither electric field affects the other one. The electrons are emitted from the electrode K with initial zero velocity, and they pass through a hole in the middle of electrode A. Because of the field between P_1 and P_2 they change direction while they pass through these plates and, after that, move with constant velocity toward the screen S. The distance between plates is d. The distance from the center of the plates to the screen is l_s.

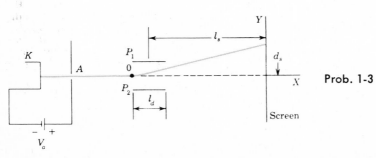

Prob. 1-3

(a) Find the velocity v_x of the electrons as a function of V_a as they cross A.

(b) Find the Y-component of velocity v_y of the electrons as they come out of the field of plates P_1 and P_2 as a function of V_p, l_d, d, and v_x.

(c) Find the distance from the middle of the screen (d_s), when the electrons reach the screen, as a function of tube distances and applied voltages.

(d) For $V_a = 1.0$ kV, and $V_p = 10$ V, $l_d = 1.27$ cm, $d = 0.475$ cm, and $l_s = 19.4$ cm, find the numerical values of v_x, v_y, and d_s.

(e) If we want to have a deflection of $d_s = 10$ cm of the electron beam, what must be the value of V_a?

1-4 A diode consists of a plane emitter and a plane-parallel anode separated by a distance of 0.5 cm. The anode is maintained at a potential of 10 V negative with respect to the cathode.

(a) If an electron leaves the emitter with a speed of 10^6 m/s, and is directed toward the anode, at what distance from the cathode will it intersect the potential-energy barrier?

(b) With what speed must the electron leave the emitter in order to be able to reach the anode?

1-5 A particle when displaced from its equilibrium position is subject to a linear restoring force $f = -kx$, where x is the displacement measured from the equilibrium position. Show by the energy method that the particle will execute periodic vibrations with a maximum displacement which is proportional to the square root of the total energy of the particle.

1-6 A particle of mass m is projected vertically upward in the earth's gravitational field with a speed v_o.

(a) Show by the energy method that this particle will reverse its direction at the height of $v_o^2/2g$, where g is the acceleration of gravity.

(b) Show that the point of reversal corresponds to a "collision" with the potential-energy barrier.

1-7 Prove that the concentration n of free electrons per cubic meter of a metal is given by

$$n = \frac{dv}{AM} = \frac{A_o dv \times 10^3}{A}$$

where d = density, kg/m³
$\quad v$ = valence, free electrons per atom
$\quad A$ = atomic weight
$\quad M$ = weight of atom of unit atomic weight, kg (Appendix A)
$\quad A_o$ = Avogadro's number, molecules/mole

1-8 The specific density of tungsten is 18.8 g/cm³ and its atomic weight is 184.0. Assume that there are two free electrons per atom. Calculate the concentration of free electrons.

1-9 (a) Compute the conductivity of copper for which $\mu = 34.8$ cm²/V-s and $d = 8.9$ g/cm³. The atomic weight is 63.54 and there is one free electron per atom.

(b) If an electric field is applied across such a copper bar with an intensity of 10 V/cm, find the average velocity of the free electrons.

1-10 Compute the mobility of the free electrons in aluminum for which the density is 2.70 g/cm³ and the resistivity is 3.44×10^{-6} Ω-cm. Assume that aluminum has three valence electrons per atom. The atomic weight is 26.98.

1-11 The resistance of No. 18 copper wire (diameter = 1.03 mm) is 6.51 Ω per 1,000 ft. The concentration of free electrons in copper is 8.4×10^{28} electrons/m³. If the current is 2 A, find the (a) drift velocity, (b) mobility, (c) conductivity.

1-12 (a) Calculate the electric field required to give an electron in silicon an average energy of 1 eV.
(b) Is it practical to generate electron-hole pairs by applying a voltage across a bar of silicon? Explain.

1-13 (a) Determine the concentration of free electrons and holes in a sample of germanium at 300°K which has a concentration of donor atoms equal to 2×10^{14} atoms/cm³ and a concentration of acceptor atoms equal to 3×10^{14} atoms/cm³. Is this p- or n-type germanium? In other words, is the conductivity due primarily to holes or to electrons?
(b) Repeat part a for equal donor and acceptor concentrations of 10^{15} atoms/cm³. Is this p- or n-type germanium?
(c) Repeat part a for donor concentration of 10^{16} atoms/cm³ and acceptor concentration of 10^{14} atoms/cm³.

1-14 (a) Find the concentration of holes and of electrons in p-type germanium at 300°K if the conductivity is 100 (Ω-cm)$^{-1}$.
(b) Repeat part a for n-type silicon if the conductivity is 0.1 (Ω-cm)$^{-1}$.

1-15 (a) Show that the resistivity of intrinsic germanium at 300°K is 45 Ω-cm.
(b) If a donor-type impurity is added to the extent of 1 atom per 10^8 germanium atoms, prove that the resistivity drops to 3.7 Ω-cm.

1-16 (a) Find the resistivity of intrinsic silicon at 300°K.
(b) If a donor-type impurity is added to the extent of 1 atom per 10^8 silicon atoms, find the resistivity.

1-17 Consider intrinsic germanium at room temperature (300°K). By what percent does the conductivity increase per degree rise in temperature?

1-18 Repeat Prob. 1-17 for intrinsic silicon.

1-19 Repeat Prob. 1-13a for a temperature of 400°K, and show that the sample is essentially intrinsic.

1-20 A sample of germanium is doped to the extent of 10^{14} donor atoms/cm³ and 7×10^{13} acceptor atoms/cm³. At the temperature of the sample the resistivity of pure (intrinsic) germanium is 60 Ω-cm. If the applied electric field is 2 V/cm, find the total conduction current density.

1-21 The hole concentration in a semiconductor specimen is shown.
(a) Find an expression for and sketch the hole current density $J_p(x)$ for the case in which there is no externally applied electric field.
(b) Find an expression for and sketch the built-in electric field that must exist if there is to be no net hole current associated with the distribution shown.

(c) Find the value of the potential between the points $x = 0$ and $x = W$ if $p(0)/p_o = 10^3$.

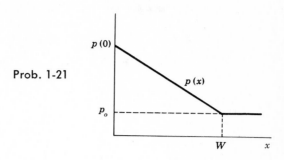

Prob. 1-21

CHAPTER 2

2-1 (a) Sketch logarithmic and linear plots of carrier concentration vs. distance for an abrupt silicon junction if $N_D = 10^{15}$ atoms/cm³ and $N_A = 10^{16}$ atoms/cm³. Give numerical values for ordinates. Label the n, p, and depletion regions. (b) Sketch the space-charge electric field and potential as a function of distance for this case (Fig. 2-1).

2-2 Repeat Prob. 2-1 for an abrupt germanium junction.

2-3 (a) For what voltage will the reverse current in a p-n junction germanium diode reach 90 percent of its saturation value at room temperature? (b) What is the ratio of the current for a forward bias of 0.05 V to the current for the same magnitude of reverse bias? (c) If the reverse saturation current is 10 μA, calculate the forward currents for voltages of 0.1, 0.2, and 0.3 V, respectively.

2-4 (a) Calculate the anticipated factor by which the reverse saturation current of a germanium diode is multiplied when the temperature is increased from 25 to 80°C. (b) Repeat part a for a silicon diode over the range 25 to 150°C.

2-5 It is predicted that, for germanium, the reverse saturation current should increase by 0.11°C⁻¹. It is found experimentally in a particular diode that at a reverse voltage of 10 V, the reverse current is 5 μA and the temperature dependence is only 0.07°C⁻¹. What is the leakage resistance shunting the diode?

2-6 A diode is mounted on a chassis in such a manner that, for each degree of temperature rise above ambient, 0.1 mW is thermally transferred from the diode to its surroundings. (The "thermal resistance" of the mechanical contact between the diode and its surroundings is 0.1 mW/°C.) The ambient temperature is 25°C. The diode temperature is not to be allowed to increase by more than 10°C above ambient. If the reverse saturation current is 5.0 μA at 25°C and increases at the rate of 0.07°C⁻¹, what is the maximum reverse-bias voltage which may be maintained across the diode?

2-7 A silicon diode operates at a forward voltage of 0.4 V. Calculate the factor by which the current will be multiplied when the temperature is increased from 25 to 150°C.

2-8 An ideal germanium p-n junction diode has at a temperature of 125°C a reverse saturation current of 30 μA. At a temperature of 125°C find the dynamic resistance for a 0.2 V bias in (a) the forward direction, (b) the reverse direction.

2-9 (a) Derive equations for the electric field intensity and potential as a function of x in Fig. 2-7.
   HINT: Use Eqs. (1-2) and (2-1).
   (b) From the potential calculated above, verify Eq. (2-11).

2-10 Prove that for an alloy p-n junction (with $N_A \ll N_D$), the width W of the depletion layer is given by

$$W = \left(\frac{2\epsilon \mu_p V_j}{\sigma_p} \right)^{\frac{1}{2}}$$

where V_j is the junction potential under the condition of an applied diode voltage V_d.

2-11 (a) Prove that for an alloy silicon p-n junction (with $N_A \ll N_D$), the depletion-layer capacitance in picofarads per square centimeter is given by

$$C = 2.9 \times 10^{-4} \left(\frac{N_A}{V_j} \right)^{\frac{1}{2}}$$

(b) If the resistivity of the p material is 3.5 Ω-cm, the barrier height V_o is 0.35 V, the applied reverse voltage is 5 V, and the cross-sectional area is circular of 40 mils diameter, find C_T.

2-12 Reverse-biased diodes are frequently employed as electrically controllable variable capacitors. The transition capacitance of an abrupt junction diode is 20 pF at 5 V. Compute the decrease in capacitance for a 1.0-V increase in bias.

2-13 Calculate the barrier capacitance of a germanium p-n junction whose area is 1 mm by 1 mm and whose space-charge thickness is 2×10^{-4} cm. The dielectric constant of germanium (relative to free space) is 16.

2-14 The zero-voltage barrier height at an alloy germanium p-n junction is 0.2 V. The concentration N_A of acceptor atoms in the p side is much smaller than the concentration of donor atoms in the n material, and $N_A = 3 \times 10^{20}$ atoms/m³. Calculate the width of the depletion layer for an applied reverse voltage of (a) 10 V and (b) 0.1 V and (c) for a forward bias of 0.1 V. (d) If the cross-sectional area of the diode is 1 mm², evaluate the space-charge capacitance corresponding to the values of applied voltage in (a) and (b).

2-15 (a) Consider a grown junction for which the uncovered charge density ρ varies linearly with distance. If $\rho = ax$, prove that the barrier voltage V_j is given by

$$V_j = \frac{aW^3}{12\epsilon}$$

(b) Verify that the barrier capacitance C_T is given by Eq. (2-13).
   HINT: Draw the charge variation with x, and use Eqs. (1-2) and (2-1).

2-16 (a) Carry out the integral indicated in Sec. 2-7 and verify that

$$Q = AqL_pp'(0)$$

(b) Show that

$$I_p(x) = \frac{AqD_pp'(0)}{L_p} \epsilon^{-x/L_p}$$

(c) From the equations in parts a and b, verify Eq. (2-17).

2-17 Given a forward-biased silicon diode with $I = 1$ mA. If the diffusion capacitance is $C_D = 1$ μF, what is the diffusion length L_p? Assume that the doping of the p side is much greater than that of the n side.

2-18 (a) Prove that the maximum electric field $\mathcal{E}_m$ at a step-graded junction with $N_A \gg N_D$ is given by

$$\mathcal{E}_m = \frac{2V_j}{W}$$

HINT: Use Eq. (2-1) with $\rho = qN_D$ (Sec. 2-6) to find $\mathcal{E}(x)$.
(b) It is found that Zener breakdown occurs when $\mathcal{E}_m = 2 \times 10^7$ V/m $\equiv \mathcal{E}_Z$. Prove that Zener voltage V_Z is given by

$$V_Z = \frac{\epsilon \mathcal{E}_Z^2}{2qN_D}$$

Note that the Zener breakdown voltage can be controlled by controlling the concentration of donor ions.

2-19 (a) Zener breakdown occurs in germanium at a field intensity of 2×10^7 V/m. Prove that the breakdown voltage is $V_Z = 51/\sigma_p$, where σ_p is the conductivity of the p material in $(\Omega\text{-cm})^{-1}$. Assume that $N_A \ll N_D$. Use the equation for V_Z in Prob. 2-18(b).
(b) If the p material is essentially intrinsic, calculate V_Z.
(c) For a doping of 1 part in 10^8 of p-type material, the resistivity drops to 3.7 Ω-cm. Calculate V_Z.
(d) For what resistivity of the p-type material will $V_Z = 1$ V?

2-20 A series combination of a 15-V avalanche diode and a forward-biased silicon diode is to be used to construct a zero-temperature-coefficient voltage reference. The temperature coefficient of the silicon diode is -1.7 mV/°C. Express in percent per degree celsius the required temperature coefficient of the Zener diode.

2-21 A 10-V battery, a 2-K resistance, and a silicon p-n diode are in series. The Zener diode voltage is 7 V.
(a) Find the approximate current if the diode is put into the circuit so that it is forward-biased. State the approximation you are making.
(b) Repeat part a if the diode is reversed.
(c) The 2-K resistance is replaced by another identical diode. The two diodes are placed in the circuit in series opposing (the two anodes are connected together). Explain what will happen. Will the current be very small (of the order of I_o) or extremely large? Make no quantitative calculations.

(d) If in part c the battery voltage is reduced to 4 V, what is the current and the voltage of each diode?

CHAPTER 3

3-1 Each diode is described by a linearized volt-ampere characteristic, with incremental resistance r and offset voltage V_γ. Diode $D1$ is germanium with $V_\gamma = 0.2$ V and $r = 20\ \Omega$, whereas $D2$ is silicon with $V_\gamma = 0.6$ V and $r = 15\ \Omega$. Find the diode currents if (a) $R = 10$ K, (b) $R = 1$ K.

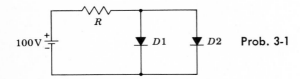

$100V$ R $D1$ $D2$ **Prob. 3-1**

3-2 (a) For the application in Sec. 3-3, plot the voltage across the diode for one cycle of the input voltage v_i. Let $V_m = 2.4$ V, $V_\gamma = 0.6$ V, $R_f = 10\ \Omega$, and $R_L = 100\ \Omega$.

(b) By direct integration find the average value of the diode voltage and the load voltage. Note that these two answers are numerically equal and explain why.

3-3 For the diode clipping circuit of Fig. 3-10a assume that $V_R = 10$ V, $v_i = 20 \sin \omega t$, and that the diode forward resistance is $R_f = 100\ \Omega$ while $R_r = \infty$ and $V_\gamma = 0$. Neglect all capacitances. Draw to scale the input and output waveforms and label the maximum and minimum values if (a) $R = 100\ \Omega$, (b) $R = 1$ K, and (c) $R = 10$ K.

3-4 Repeat Prob. 3-3 for the case where the reverse resistance is $R_r = 10$ K.

3-5 In the diode clipping circuit of Fig. 3-10a and d, $v_i = 20 \sin \omega t$, $R = 1$ K, and $V_R = 10$ V. The reference voltage is obtained from a tap on a 10-K divider connected to a 100-V source. Neglect all capacitances. The diode forward resistance is 50 Ω, $R_r = \infty$, and $V_\gamma = 0$. In both cases draw the input and output waveforms to scale. Which circuit is the better clipper? HINT: Apply Thévenin's theorem (Sec. C-2) to the reference-voltage divider network.

3-6 A symmetrical 5-kHz square wave whose output varies between $+10$ and -10 V is impressed upon the clipping circuit shown. Assume $R_f = 0$, $R_r = 2$ M, and $V_\gamma = 0$. Sketch the steady-state output waveform, indicating numerical values of the maximum, minimum, and constant portions.

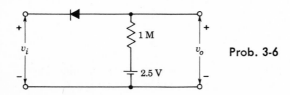

v_i 1 M v_o **Prob. 3-6**

2.5 V

3-7 For the clipping circuits shown in Fig. 3-10b and d derive the transfer characteristic v_o versus v_i, taking into account R_f and V_γ and considering $R_r = \infty$.

3-8 The clipping circuit shown employs temperature compensation. The dc voltage source V_γ represents the diode offset voltage; otherwise the diodes are assumed to be ideal with $R_f = 0$ and $R_r = \infty$.
(*a*) Sketch the transfer curve v_o versus v_i.
(*b*) Show that the maximum value of the input voltage v_i so that the current in $D2$ is always in the forward direction is

$$v_{i,\max} = V_R + \frac{R}{R'}(V_R - V_\gamma)$$

(*c*) What is the temperature dependence of the point on the input waveform at which clipping occurs?

Prob. 3-8

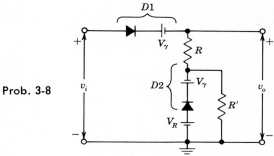

3-9 (*a*) In the clipping circuit shown, $D2$ compensates for temperature variations. Assume that the diodes have infinite back resistance, a forward resistance of 50 Ω, and a break point at the origin ($V_\gamma = 0$). Calculate and plot the transfer characteristic v_o against v_i. Show that the circuit has an extended break point, that is, two break points close together.
(*b*) Find the transfer characteristic that would result if $D2$ were removed and the resistor R were moved to replace $D2$.
(*c*) Show that the double break of part *a* would vanish and only the single break of part *b* would appear if the diode forward resistances were made vanishingly small in comparison with R.

Prob. 3-9

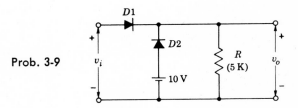

3-10 (*a*) In the peak clipping circuit shown, add another diode $D2$ and a resistor R' in a manner that will compensate for drift of the peak point with temperature.

(b) Show that the break point of the transmission curve occurs at V_R. Assume $R_r \gg R \gg R_f$.

(c) Show that if $D2$ is always to remain in conduction it is necessary that

$$v_i < v_{i,\max} = V_R + \frac{R}{R'}(V_R - V_\gamma)$$

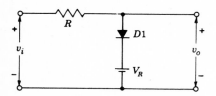

Prob. 3-10

3-11 (a) The input voltage v_i to the two-level clipper shown in part a of the figure varies linearly from 0 to 150 V. Sketch the output voltage v_o to the same time scale as the input voltage. Assume ideal diodes.

(b) Repeat (a) for the circuit shown in part b of the figure.

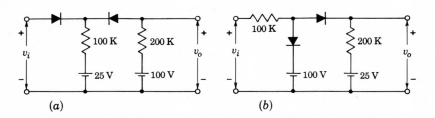

(a)   (b)

Prob. 3-11

3-12 The circuit of Fig. 3-11a is used to "square" a 10-kHz input sine wave whose peak value is 50 V. It is desired that the output voltage waveform be flat for 90 percent of the time. Diodes are used having a forward resistance of 100 Ω and a backward resistance of 100 K.

(a) Find the values of V_{R1} and V_{R2}.

(b) What is a reasonable value to use for R?

3-13 (a) The diodes are ideal. Write the transfer characteristic equations (v_o as a function of v_i).

(b) Plot v_o against v_i, indicating all intercepts, slopes, and voltage levels.

(c) Sketch v_o if $v_i = 40 \sin \omega t$. Indicate all voltage levels.

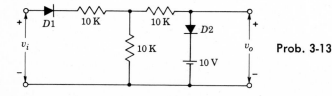

Prob. 3-13

3-14 (a) Repeat Prob. 3-13 for the circuit shown.
(b) Repeat for the case where the diodes have an offset voltage $V_\gamma = 1$ V.

Prob. 3-14

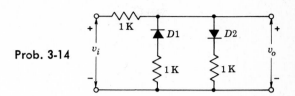

3-15 Assume that the diodes are ideal. Make a plot of v_o against v_i for the range of v_i from 0 to 50 V. Indicate all slopes and voltage levels. Indicate, for each region, which diodes are conducting.

Prob. 3-15

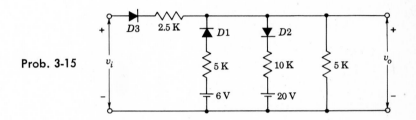

3-16 The triangular waveform shown is to be converted into a sine wave by using clipping diodes. Consider the dashed waveform sketched as a first approximation to the sinusoid. The dashed waveform is coincident with the sinusoid at 0°, 30°, 60°, etc. Devise a circuit whose output is this broken-line waveform when the input is the triangular waveform. Assume ideal diodes and calculate the values of all supply voltages and resistances used. The peak value of the sinusoid is 50 V.

Prob. 3-16

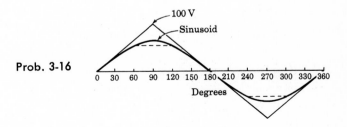

3-17 Construct circuits which exhibit terminal characteristics as shown in parts a and b of the figure.

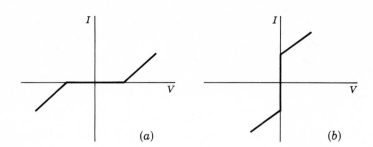

Prob. 3-17   (a)   (b)

3-18 (a) Find the transfer characteristic of the circuit shown.
   (b) Assume that v_i increases linearly with time, sketch the output waveform.
      NOTE: The above circuit is called a comparator. It is used to mark the instant when an arbitrary waveform attains some reference level.

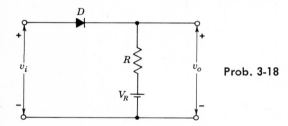

Prob. 3-18

3-19 The diode-resistor comparator of Prob. 3-18 is connected to a device which responds when the comparator output attains a level of 0.1 V. The input is a ramp which rises at the rate of 10 V/μs. The germanium diode has a reverse saturation current of 1 μA. Initially, $R = 1$ K and the 0.1-V output level is attained at a time $t = t_1$. If we now set $R = 100$ K, what will be the corresponding change in t_1? $V_R = 0$.

3-20 (a) Two p-n germanium diodes are connected in series opposing. A 5-V battery is impressed upon this series arrangement. Find the voltage across each junction at room temperature. Assume that the magnitude of the Zener voltage is greater than 5 V.
      Note that the result is independent of the reverse saturation current. Is it also independent of temperature?
      HINT: Assume that reverse saturation current flows in the circuit, and then justify this assumption.
   (b) If the magnitude of the Zener voltage is 4.9 V, what will be the current in the circuit? The reverse saturation current is 5 μA.

3-21 (a) In the circuit of Prob. 3-20, the Zener breakdown voltage is 2.0 V. The reverse saturation current is 5 μA. If the silicon diode resistance could be neglected, what would be the current?
   (b) If the ohmic resistance is 100 Ω, what is the current?
      NOTE: Answer part b by plotting Eq. (2-2) and drawing a load line. Verify your answer analytically by a method of successive approximations.

3-22 A *p-n* germanium junction diode at room temperature has a reverse satura-tion current of 10 μA, negligible ohmic resistance, and a Zener breakdown voltage of 100 V. A 1-K resistor is in series with this diode, and a 30-V battery is impressed across this combination. Find the current (*a*) if the diode is forward-biased, (*b*) if the battery is inserted into the circuit with the reverse polarity. (*c*) Repeat parts *a* and *b* if the Zener breakdown voltage is 10 V.

3-23 The Zener diode can be used to prevent overloading of sensitive meter move-ments without affecting meter linearity. The circuit shown represents a dc voltmeter which reads 20 V full scale. The meter resistance is 560 Ω, and $R_1 + R_2 = 99.5$ K. If the diode is a 16-V Zener, find R_1 and R_2 so that, when $V_i > 20$ V, the Zener diode conducts and the overload current is shunted away from the meter.

Prob. 3-23

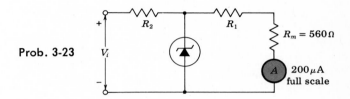

3-24 The saturation currents of the two diodes are 1 and 2 μA. The breakdown voltages of the diodes are the same and are equal to 100 V.
(*a*) Calculate the current and voltage for each diode if $V = 90$ V and $V = 110$ V.
(*b*) Repeat part *a* if each diode is shunted by a 10-M resistor.

Prob. 3-24

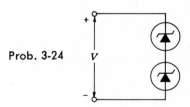

3-25 (*a*) In Fig. 3-14, $V = 300$ V, $V_z = 220$ V, the value of the Zener current is 15 mA and the value of load current is 25 mA. Calculate the value of R which must be used.
(*b*) If the load decreases by 5 mA, what will be the Zener current?
(*c*) The load is as in part *a*. If the supply voltage changes to 340 V, what is I_z?
(*d*) The normal operating range of the avalanche diode is from 3 to 50 mA. If $R = 1.5$ K and $V = 340$ V, over what load current can the output be varied?

3-26 A diode whose internal resistance is 20 Ω is to supply power to a 1,000-Ω load from a 110-V (rms) source of supply. Calculate (*a*) the peak load current, (*b*) the dc load current, (*c*) the ac load current, (*d*) the dc diode voltage, (*e*) the total input power to the circuit, (*f*) the percentage regulation from no load to the given load.

3-27 Show that the maximum dc output power $P_{dc} \equiv V_{dc}I_{dc}$ in a half-wave single-phase circuit occurs when the load resistance equals the diode resistance R_f.

3-28 The efficiency of rectification η_r is defined as the ratio of the dc output power $P_{dc} \equiv V_{dc}I_{dc}$ to the input power $P_i = (1/2\pi)\int_0^{2\pi} v_i i \, d\alpha$.

(a) Show that, for the half-wave-rectifier circuit,

$$\eta_r = \frac{40.6}{1 + R_f/R_L} \%$$

(b) Show that, for the full-wave rectifier, η_r has twice the value given in part a.

3-29 Prove that the regulation of both the half-wave and the full-wave rectifier is given by

$$\% \text{ regulation} = \frac{R_f}{R_L} \times 100\%$$

3-30 (a) Prove Eqs. (3-21) and (3-22) for the dc voltage of a full-wave-rectifier circuit.
(b) Find the dc voltage across a diode by direct integration.

3-31 A full-wave single-phase rectifier circuit uses two diodes as shown in Fig. 3-18. The internal resistance of each diode may be considered to be constant and equal to 500 Ω. These feed into a pure resistance load of 2,000 Ω. The secondary transformer voltage to center tap is 280 V. Calculate (a) the dc load current, (b) the direct current in each diode, (c) the ac voltage across each diode, (d) the dc output power, (e) the percentage regulation.

3-32 (a) v_i is a sinusoidal voltage. Trace the paths of the current when v_i is positive and when v_i is negative. Sketch the output waveform v_o.
(b) In this circuit, can the transformer and the load be exchanged? Explain carefully.
(c) For $v_i = V_m \sin \omega t$, calculate the rms value of v_o.
    NOTE: The above circuit is known as a full-wave bridge rectifier.
(d) What happens when D_3 and D_4 are each replaced by a large capacitor C? Verify that $v_o \approx 2V_m$ as $R_L \to \infty$.
    NOTE: This circuit is known as a full-wave voltage doubler.

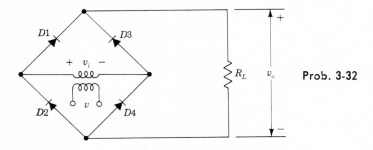

Prob. 3-32

3-33 A 1-mA dc meter whose resistance is 10 Ω is calibrated to read rms volts when used in a *bridge* circuit (see Prob. 3-32) with semiconductor diodes. The meter is used as the load in the bridge circuit and the input signal is applied at v_i without the transformer. The effective resistance of each element may be considered to be zero in the forward direction and infinite in the inverse direction. The sinusoidal input voltage is applied in series with a 5-K resistance. What is the full-scale reading of this meter?

3-34 The circuit shown is a *half-wave voltage doubler*. Analyze the operation of this circuit. Calculate (*a*) the maximum possible voltage across each capacitor, (*b*) the peak inverse voltage of each diode. In this circuit the output voltage is negative with respect to ground. Show that if the connections to the cathode and anode of each diode are interchanged, the output voltage will be positive with respect to ground.

Prob. 3-34

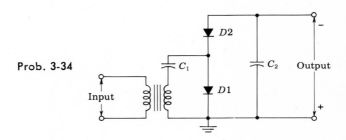

CHAPTER 4

4-1 The transistor of Fig. 4-3*a* has the characteristics given in Figs. 4-6 and 4-7. Let $V_{CC} = 6$ V, $R_L = 200$ Ω, and $I_E = 15$ mA.
(*a*) Find I_C and V_{CB}.
(*b*) Find V_{EB} and V_L.
(*c*) If I_E changes by $\Delta I_E = 10$ mA symmetrically around the point of part *a* and with constant V_{CC}, find the corresponding change in I_C.

4-2 The CB transistor used in the circuit of Fig. 4-3*a* has the characteristics given in Figs. 4-6 and 4-7. Let $I_C = -20$ mA, $V_{CB} = -4$ V, and $R_L = 200$ Ω.
(*a*) Find V_{CC} and I_E.
(*b*) If the supply voltage V_{CC} decreases from its value in part *a* by 2 V while I_E retains its previous value, find the new values of I_C and V_{CB}.

4-3 The CE transistor used in the circuit shown has the characteristics given in Figs. 4-9 and 4-10.
(*a*) Find V_{BB} if $V_{CC} = 10$ V, $V_{CE} = -1$ V, and $R_L = 250$ Ω.
(*b*) If $V_{CC} = 10$ V, find R_L so that $I_C = -20$ mA and $V_{CE} = -4$ V. Find V_{BB}.

Prob. 4-3

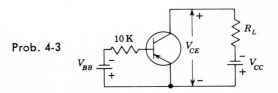

4-4 If $\alpha = 0.98$ and $V_{BE} = 0.7$ V, find R_1 in the circuit shown for an emitter current $I_E = -2$ mA. Neglect the reverse saturation current.

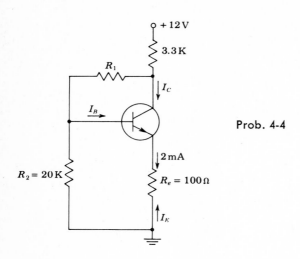

Prob. 4-4

4-5 (a) Find R_c and R_b in the circuit of Fig. 4-11a if $V_{CC} = 10$ V and $V_{BB} = 5$ V, so that $I_C = 10$ mA and $V_{CE} = 5$ V. A silicon transistor with $\beta = 100$, $V_{BE} = 0.7$ V, and negligible reverse saturation current is under consideration.
(b) Repeat part a if a 100-Ω emitter resistor is added to the circuit.

4-6 In the circuit shown, $V_{CC} = 24$ V, $R_c = 10$ K, and $R_e = 270$ Ω. If a silicon transistor is used with $\beta = 45$ and if $V_{CE} = 5$ V, find R. Neglect the reverse saturation current.

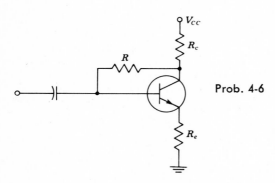

Prob. 4-6

4-7 For the circuit shown, transistors $Q1$ and $Q2$ operate in the active region with $V_{BE1} = V_{BE2} = 0.7$ V, $\beta_1 = 100$, and $\beta_2 = 50$. The reverse saturation currents may be neglected.

(a) Find the currents I_{B2}, I_1, I_2, I_{C2}, I_{B1}, I_{C1}, and I_{E1}.
(b) Find the voltages V_{o1} and V_{o2}.

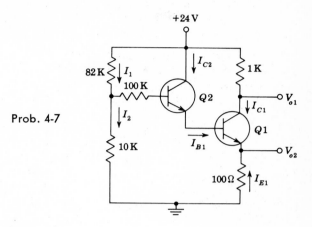

Prob. 4-7

4-8   (a) The reverse saturation current of the germanium transistor in Fig. 4-12 is
2 μA at room temperature (25°C) and increases by a factor of 2 for each temperature increase of 10°C.   The bias is $V_{BB} = 5$ V.   Find the maximum allowable
value for R_B if the transistor is to remain cut off at a temperature of 75°C.
(b) If $V_{BB} = 1.0$ V and $R_B = 50$ K, how high may the temperature increase
before the transistor comes out of cutoff?

4-9   From the characteristic curves for the type 2N404 transistor given in Fig. 4-13,
find the voltages V_{BE}, V_{CE}, and V_{BC} for the circuit shown.

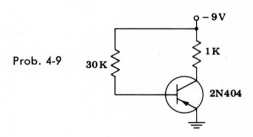

Prob. 4-9

4-10   The silicon transistor shown has the parameters given in Table 4-1 and $\beta = h_{FE} = 50$.
(a) Assume that Q is in the active region and find I_B and I_C.
(b) Verify that the assumption in part a is *not* correct.   Explain briefly.
(c) Assume that Q is in saturation and find I_B and I_C.
(d) Verify that the assumption in part c is justified.   Explain briefly.

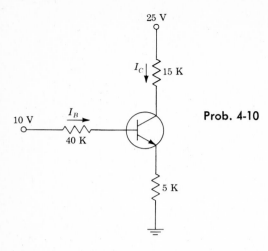

Prob. 4-10

4-11 A silicon transistor with $V_{BE,\text{sat}} = 0.8$ V, $\beta = h_{FE} = 100$, $V_{CE,\text{sat}} = 0.2$ V is used in the circuit shown. Find the minimum value of R_c for which the transistor remains in saturation.

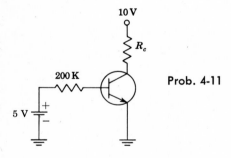

Prob. 4-11

4-12 For the circuit shown, assume $\beta = h_{FE} = 100$.

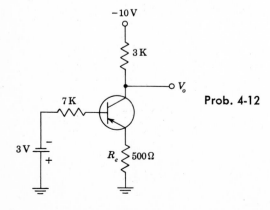

Prob. 4-12

(a) Find if the silicon transistor is in cutoff, saturation, or in the active region

(b) Find V_o.

(c) Find the minimum value for the emitter resistor R_e for which the transistor operates in the active region.

4-13 (a) The circuit shown uses a silicon transistor with $R = 50$ K and $\beta = h_{FE} = 100$. Find the base and collector currents. Use the parameters in Table 4-1.

(b) Is the transistor in the cutoff, active, or saturation region? Make a numerical calculation to justify your answer.

(c) Find the smallest value of R so that the transistor is in its active region.

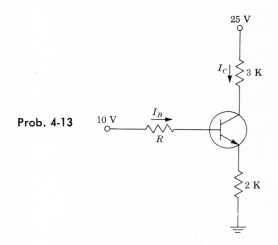

Prob. 4-13

4-14 (a) The silicon transistor shown has $\beta = h_{FE} = 100$. Find the value of R_c so that the transistor is just in saturation. Use the parameters in Table 4-1.

(b) It is desired to take the transistor out of saturation into the active region. Make no numerical calculations, but explain briefly and clearly whether R_c should be increased or decreased.

(c) If $R_c = 5$ K, find the minimum value of h_{FE} so that the transistor is in saturation.

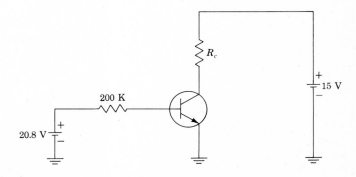

Prob. 4-14

4-15 If the silicon transistor used in the circuit shown has a minimum value of $\beta = h_{FE}$ of 30 and if $I_{CBO} = 10$ nA at 25°C:

(a) Find V_o for $V_i = 12$ V and show that Q is in saturation.

(b) Find the minimum value of R_1 for which the transistor in part a is in the active region.

(c) If $R_1 = 15$ K and $V_i = 1$ V, find V_o and show that Q is at cutoff.

(d) Find the maximum temperature at which the transistor in part c remains at cutoff.

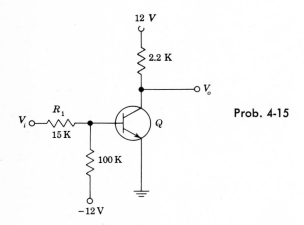

Prob. 4-15

4-16 Silicon transistors with $h_{FE} = 100$ are used in the circuit shown. Neglect the reverse saturation current.

(a) Find V_o when $V_i = 0$ V. Assume $Q1$ is OFF and justify the assumption.

(b) Find V_o when $V_i = 6$ V. Assume $Q2$ is OFF and justify this assumption.

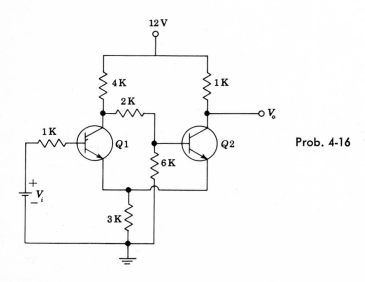

Prob. 4-16

4-17 For the circuit shown, $\alpha_1 = 0.98$, $\alpha_2 = 0.96$, $V_{CC} = 24$ V, $R_c = 120$ Ω, and $I_E = -100$ mA. Neglecting the reverse saturation currents, determine (a) the currents I_{C1}, I_{B1}, I_{E1}, I_{C2}, I_{B2}, and I_C; (b) V_{CE}; (c) I_C/I_B, I_C/I_E.

Prob. 4-17

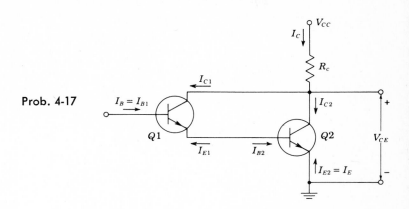

CHAPTER 5

5-1 A silicon wafer is uniformly doped with phosphorus to a concentration of 10^{15} cm^{-3}. Refer to Table 1-1. At room temperature (300°K) find
(a) The conductivity and resistivity.
(b) The concentration of boron, which, if added to the phosphorus-doped wafer, would halve the conductivity.

5-2 List in order the steps required in fabricating a monolithic silicon integrated transistor by the epitaxial-diffused method. Sketch the cross section after each oxide growth. Label materials clearly.

5-3 Sketch *to scale* the cross section of a monolithic transistor fabricated on a 5-mil-thick silicon substrate. HINT: Refer to Sec. 5-1 and Figs. 5-6 and 5-7 for typical dimensions.

5-4 Sketch the five basic diode connections (in circuit form) for the monolithic integrated circuits.

5-5 If the base sheet resistance can be held to within ± 10 percent and resistor line widths can be held to ± 0.1 mil, plot approximate tolerance of a diffused resistor as a function of line width w in mils over the range $0.5 \leq w \leq 5.0$. (Neglect contact-area and contact-placement errors.)

5-6 A 1-mil-thick silicon wafer has been doped uniformly with phosphorus to a concentration of 10^{16} cm^{-3}, plus boron to a concentration of 2×10^{15} cm^{-3}. Find its sheet resistance.

5-7 (a) Calculate the resistance of a diffused resistor 4 mils long, 1 mil wide, and 2 μm thick, given that its sheet resistance is 2.2 Ω/square.
(b) Repeat part a for an aluminum metalizing layer 0.5 μm thick of resistivity 2.8×10^{-6} Ω-cm.

5-8 An integrated junction capacitor has an area of 1,000 mils² and is operated at a reverse barrier potential of 1 V. The acceptor concentration of 10^{15} atoms/cm³ is much smaller than the donor concentration. Calculate the capacitance.

5-9 A thin-film capacitor has a capacitance of 0.4 pF/mil². The relative dielectric constant of silicon dioxide is 3.5. What is the thickness of the SiO_2 layer in angstroms?

5-10 The n-type epitaxial isolation region shown is 8 mils long, 6 mils wide, and 1 mil thick and has a resistivity of 0.1 Ω-cm. The resistivity of the p-type substrate is 10 Ω-cm. Find the parasitic capacitance between the isolation region and the substrate under 5-V reverse bias. Assume that the sidewalls contribute 0.1 pF/mil².

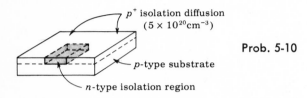

p^+ isolation diffusion
$(5 \times 10^{20} cm^{-3})$

Prob. 5-10

p-type substrate

n-type isolation region

NOTE: *In the problems that follow, indicate your answer by giving the letter of the statement you consider correct.*

5-11 The typical number of diffusions used in making epitaxial-diffused silicon integrated circuits is (*a*) 1, (*b*) 2, (*c*) 3, (*d*) 4, (*e*) 5.

5-12 Epitaxial growth is used in integrated circuits (ICs)
(*a*) To grow selectively single-crystal p-doped silicon of one resistivity on a p-type substrate of a different resistivity.
(*b*) To grow single-crystal n-doped silicon on a single-crystal p-type substrate.
(*c*) Because it yields back-to-back isolating p-n junctions.
(*d*) Because it produces low parasitic capacitance.

5-13 Silicon dioxide (SiO_2) is used in ICs
(*a*) Because it facilitates the penetration of diffusants.
(*b*) Because of its high heat conduction.
(*c*) To control the location of diffusion and to protect and insulate the silicon surface.
(*d*) To control the concentration of diffusants.

5-14 The p-type substrate in a monolithic circuit should be connected to
(*a*) The most positive voltage available in the circuit.
(*b*) The most negative voltage available in the circuit.
(*c*) Any dc ground point.
(*d*) Nowhere, i.e., be left floating.

5-15 Monolithic integrated circuit systems offer greater reliability than discrete-component systems because
(*a*) There are fewer interconnections.
(*b*) High-temperature metalizing is used.
(*c*) Electric voltages are low.
(*d*) Electric elements are closely matched.

5-16 The sheet resistance of a semiconductor is
(a) An undesirable parasitic element.
(b) An important characteristic of a diffused region, especially when used to form diffused resistors.
(c) A characteristic whose value determines the required area for a given value of integrated capacitance.
(d) A parameter whose value is important in a thin-film resistance.

5-17 Isolation in ICs is required.
(a) To make it simpler to test circuits.
(b) To protect the components from mechanical damage.
(c) To protect the transistor from possible thermal effects.
(d) To minimize electrical interaction between circuit components.

5-18 Almost all resistors are made in a monolithic IC
(a) During the emitter diffusion.
(b) While growing the epitaxial layer.
(c) During the base diffusion.
(d) During the collector diffusion.

5-19 Increasing the yield of an integrated circuit
(a) Reduces individual circuit cost.
(b) Increases the cost of each good circuit.
(c) Results in a lower number of good chips per wafer.
(d) Means that more transistors can be fabricated on the same size wafer.

5-20 In a monolithic-type IC
(a) All isolation problems are eliminated.
(b) Resistors and capacitors of any value may be made.
(c) All components are fabricated into a single crystal of silicon.
(d) Each transistor is diffused into a separate isolation region.

5-21 The main purpose of the metalization process is
(a) To interconnect the various circuit elements.
(b) To protect the chip from oxidation.
(c) To act as a heat sink.
(d) To supply a bonding surface for mounting the chip.

CHAPTER 6

6-1 Convert the following decimal numbers to binary form: (a) 671, (b) 325, (c) 152.

6-2 The parameters in the diode OR circuit of Fig. 6-3 are $V(0) = +12$ V, $V(1) = -2$ V, $R_s = 600$ Ω, $R = 10$ K, $R_f = 0$, $R_r = \infty$, and $V_\gamma = 0.6$ V. Calculate the output levels if one input is excited and if (a) $V_R = +12$ V, (b) $V_R = +10$ V, (c) $V_R = +14$ V, and (d) $V_R = 0$ V. For which of these cases is the OR function satisfied (except possibly for a shift in level between input and output)? (e) Repeat part a if three inputs are excited.

6-3 Consider a two-input positive-logic diode OR gate (Fig. 6-3 with the diodes reversed) with $V_R = 0$. The inputs are the square waves v_1 and v_2 indicated.

Sketch the output waveform if the ratio of the amplitude of v_2 to v_1 is (a) 2 and (b) $\frac{1}{2}$. Assume ideal diodes ($R_f = 0$, $R_r = \infty$, and $V_\gamma = 0$) and $R_s = 0$.

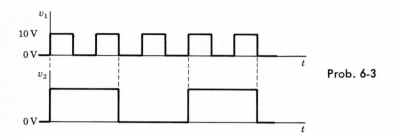

Prob. 6-3

6-4 Consider two signals, a 1-kHz sine wave and a 10-kHz square wave of zero average value, applied to the OR circuit of Fig. 6-3 with $V_R = 0$. Draw the output waveform if the sine-wave amplitude (a) exceeds the square-wave amplitude, (b) is less than the square-wave amplitude.

6-5 Consider a two-input positive-logic diode AND circuit (Fig. 6-5b) with $V_R = 15$ V, $R = 10$ K, $R_s = 1$ K. Assume ideal diodes and neglect all capacitances. A square wave v_i, extending from -5 to $+5$ V with respect to ground, is applied simultaneously to both inputs. (a) Sketch the output v_o and calculate the maximum and minimum voltages with respect to ground. (b) If $v_1 = v_i$ and $v_2 = -v_i$, calculate the voltage levels of v_o and plot.

6-6 Consider a two-input positive-logic diode AND circuit (Fig. 6-5b) with $V_R = 10$ V, $R = 10$ K, and $R_s = 0$. Assume ideal diodes. The input waveforms are v_1 and v_2 sketched in Prob. 6-3. Sketch the output waveform if the ratio of the amplitude of v_2 to v_1 is (a) 2, (b) 1, and (c) $\frac{1}{2}$. Repeat part b if $R_s = 1$ K.

6-7 The binary input levels for the AND circuit shown are $V(0) = 0$ V and $V(1) = 25$ V. Assume ideal diodes. If $v_1 = V(0)$ and $v_2 = V(1)$, then v_o is to be at 5 V. However, if $v_1 = v_2 = V(1)$, then v_o is to rise above 5 V.
(a) What is the maximum value of V_R which may be used?
(b) If $V_R = 20$ V, what is v_o at a coincidence [$v_1 = v_2 = V(1)$]? What are the diode currents?
(c) Repeat part b if $V_R = 40$ V.

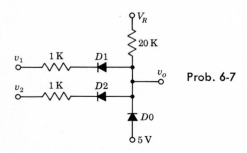

Prob. 6-7

6-8 (a) Verify that the circuit shown is an inverter by calculating the output levels corresponding to input levels of 0 and -6 V. What minimum value of h_{FE} is required? Neglect junction saturation voltages and assume an ideal diode.
(b) If the reverse collector saturation current at 25°C is 5 μA, what is the maximum temperature at which this inverter will operate properly?

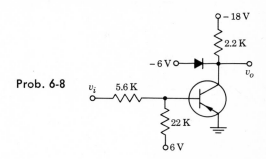

Prob. 6-8

6-9 For the circuit shown in Fig. 6-7, $V_{CC} = 8$ V, $V_{BB} = 8$ V, $V_{EE} = 0$, and $R_c = 2.2$ K. The inverter is to operate properly in the temperature range -25 to 125°C. The silicon transistor used has $(h_{FE})_{min} = 65$ at 25°C, 55 at -25°C, 85 at 125°C, and $I_{CBO} = 5$ nA at 25°C. The desired logic levels are $V(1) = 8 \pm 2$ V, $V(0) = 0.2 \pm 0.2$ V.
(a) Find the maximum value of R_1 if $R_2 = 100$ K.
(b) If the desired logic levels are $V(1) = 4 \pm 1$ V and $V(0) = 0.2 \pm 0.2$ V, what modification should you make to this circuit?

6-10 A half adder is a combination of OR, NOT, and AND gates. It has two inputs and two outputs and the following truth table:

	Input 1	Input 2	Output 1	Output 2
	0	0	0	0
Prob. 6-10	0	1	1	0
	1	0	1	0
	1	1	0	1

Draw the logic block diagram for a half adder.

6-11 The four inputs v_1, v_2, v_3, and v_4 are voltages from zero-impedance sources whose values are either $V(0) = 10$ V or $V(1) = 20$ V. The diodes are ideal. $V_R = 25$ V, $R_1 = 5$ K, and $R_2 = 10$ K.
(a) If $v_1 = v_2 = 10$ V and $v_3 = v_4 = 20$ V, find v_o and the currents in each diode.
(b) If $v_1 = v_3 = 10$ V and $v_2 = v_4 = 20$ V, find v_o and the currents in each diode.
(c) Sketch in block-diagram form the logic performed by this circuit.
(d) Verify that in order for the circuit to operate properly the following inequality must be satisfied:

$$R_2 > \frac{V_R - V(0)}{V(0)} R_1$$

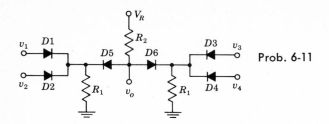

Prob. 6-11

6-12 (a) In block-diagram form indicate the logic performed by the diode system shown. The input levels are $V(0) = -8$ V and $V(1) = +2$ V. Neglect source resistance and assume that the diodes are ideal. Justify your answer by calculating the voltages v_A, v_B, and v_o (and indicating which diodes are conducting) under the following circumstances: (i) all inputs are at $V(0)$; (ii) some but not all inputs in A are at $V(1)$ and all inputs in B are at $V(0)$; (iii) all inputs in A are at $V(1)$ and some inputs in B are at $V(1)$; and (iv) all inputs are at $V(1)$. (b) If the 10-K resistance were increased, at what maximum value would the circuit no longer operate in the manner described above?

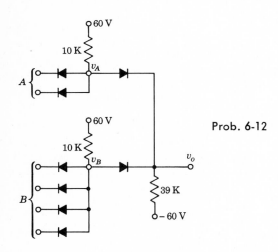

Prob. 6-12

6-13 (a) Verify De Morgan's law [Eq. (6-26)] in a manner analogous to that given in the text in connection with the proof of Eq. (6-24).
(b) Prove Eq. (6-26) by constructing a truth table for each side and verifying that these two tables have the same outputs.

6-14 Verify the auxiliary Boolean identities in Table 6-4.

6-15 Using Boolean algebra, verify
(a) $\overline{\bar{A} + B} + \overline{\bar{A} + \bar{B}} = A$
(b) $AB + AC + B\bar{C} = AC + B\bar{C}$
   HINT: Multiply the first term on the left-hand side by $C + \bar{C} = 1$.
(c) $\overline{AB + BC + CA} = \bar{A}\bar{B} + \bar{B}\bar{C} + \bar{C}\bar{A}$

6-16 Using Boolean algebra, verify
  (a) $(A + B)(B + C)(C + A) = AB + BC + CA$
  (b) $(A + B)(\bar{A} + C) = AC + \bar{A}B$
  (c) $AB + \bar{B}\bar{C} + A\bar{C} = AB + \bar{B}\bar{C}$
    HINT: A term may be multiplied by $B + \bar{B} = 1$.

6-17 Given two N-bit characters which are available in parallel form. Indicate in block-diagram form a system whose output is 1 if and only if *all* corresponding bits are equal, that is, only if the two characters are equal.

6-18 A, B, and C represent the presence of pulses. The logic statement "A or B and C" can have two interpretations. Which are they? In block-diagram form draw the circuit to perform each of the two logic operations.

6-19 A circuit has three input and one output terminals. The output is 1 if any two of the three inputs are 1 and is 0 for any other combination of inputs. Draw a block diagram of this logic circuit.

6-20 In block-diagram form draw a circuit to perform the following logic: If pulses A_1, A_2, and A_3 occur simultaneously or if pulses B_1 and B_2 occur simultaneously, an output pulse is delivered, provided that pulse C does not occur at the same time. No output is to be obtained if A_1, A_2, A_3, B_1, and B_2 occur simultaneously.

6-21 A single-pole double-throw switch is to be simulated with AND, OR, and INHIBITOR circuits. Call the two signal inputs A and B. A third input C receives the switching instructions in the form of a code: 1 (a pulse is present) or 0 (no pulse exists). It is desired that $C = 1$ set the switch to A and $C = 0$ set the switch to B, as indicated schematically. In block-diagram form show the circuit for this switch.

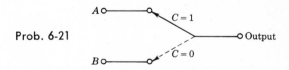

Prob. 6-21

6-22 In block-diagram form draw a circuit which satisfies simultaneously the conditions a, b, and c as follows:
  (a) The output is excited if any pair of inputs A_1, A_2, and A_3 is excited, provided that B is also excited.
  (b) The output is 1 if any one (and only one) of the inputs A_1, A_2, or A_3 is 1, provided that $B = 0$.
  (c) No output is excited if A_1, A_2, and A_3 are simultaneously excited.

6-23 (a) For the illustrative NAND gate of Fig. 6-15 calculate the minimum value of h_{FE} taking junction voltages into account.
  (b) What is the maximum noise voltage (superimposed upon the logic level) which will still permit the circuit to operate properly? Consider the following two cases: (i) a complete coincidence and (ii) all inputs but one in the 1 state.
  (c) What is the maximum value of the source resistance which will still permit proper circuit operation? Assume a 0.7-V drop across a conducting diode.

6-24 Verify that the circuit shown is an EXCLUSIVE OR.

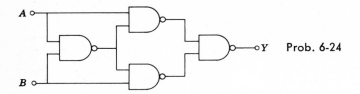

Prob. 6-24

6-25 Verify that the NOR-NOR topology is equivalent to an OR-AND system.

6-26 Verify that the logic operations OR, AND, and NOT may be implemented by using only NOR gates.

6-27 What logic operation is performed by the circuit shown, which consists of interconnected NOR gates?

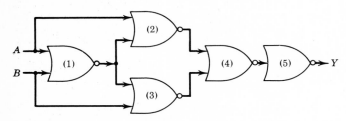

Prob. 6-27

6-28 (*a*) Implement the EXCLUSIVE OR gate using (i) NOR gates, (ii) NAND gates.
(*b*) Repeat part *a* for the half adder of Prob. 6-10.

6-29 The DTL gate shown uses silicon devices with $V_{BE,\text{sat}} = 0.8$ V, $V_{CE,\text{sat}} = 0.2$ V, $V_\gamma = 0.5$ V, and the drop across a conducting diode $= 0.7$ V. The inputs to this switch are obtained from the outputs of similar gates.
(*a*) Verify that the circuit functions as a positive NAND and calculate $h_{FE,\text{min}}$. Assume that the transistor is essentially cut off if the base-to-emitter voltage is at least 0.1 V smaller than the cutin voltage V_γ.

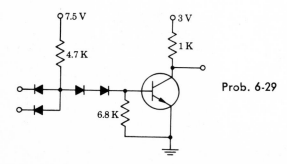

Prob. 6-29

(b) Assume that the diode reverse saturation current is equal to the transistor reverse saturation collector current. Find $I_{CBO,max}$.

(c) If all inputs are high, what is the magnitude of noise voltage at the input which will cause the gate to malfunction?

(d) Repeat part c if at least one input is low.

6-30 (a) Analyze the DTL circuit shown. Use the voltage drops given in Prob. 6-29.

(b) Find $h_{FE,min}$ if two similar gates are to be driven by this circuit.

(c) Find the noise margins.

Prob. 6-30

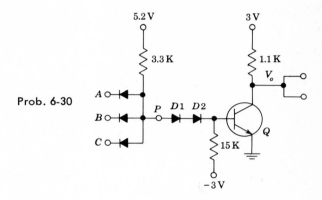

6-31 (a) Analyze the DTL circuit shown. Use the voltage drops given in Prob. 6-29.

(b) If $h_{FE} = 25$, calculate the fan-out N.

(c) For a fan-out of 10 and assuming a diode reverse saturation current of 15 μA, what is $V(1)$?

Prob. 6-31

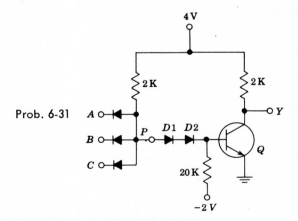

6-32 For the integrated positive DTL NAND gate shown, prove that

(a) The maximum number of diodes in series that can be used is given by $n_{max} = (V_{CC} - V_{BE,sat})/V_D$, where V_D is the voltage drop across a diode.

(b) The maximum fan-out is given by

$$N_{\max} \approx h_{FE} - h_{FE}\left(n + \frac{R_1}{R_2}\right)\frac{V_D}{V_{CC}} - \frac{R_1}{R_c}\left(1 + \frac{V_D}{V_{CC}}\right)$$

Assume that $V_{BE,\text{sat}} \approx V_D$ and $V_{CC} - V_D \gg V_{CE,\text{sat}}$.

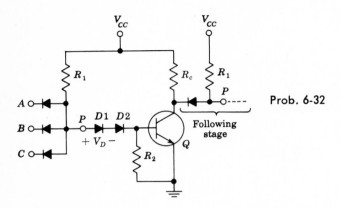

Prob. 6-32

6-33 (a) The discrete-components circuit of a DTL gate shown uses a silicon transistor with worst-case values of $V_{BE,\text{sat}} = 1.0$ V and $V_{CE,\text{sat}} = 0.5$ V. The voltage across any silicon diode (when conducting) is 0.7 V. Assume that $D1$ consists of two diodes in series. The circuit parameters are $V_{CC} = V_{BB} = 12$ V, $R = 15$ K, $R_2 = 100$ K, and $R_c = 2.2$ K. The inputs to this switch are obtained from the outputs of similar gates. Verify that the circuit functions as a positive NAND. In particular, for proper operation, calculate the minimum value of the clamping voltage V' and h_{FE}. HINT: $V(1)$ equals $V' + V_D$. Why?
(b) Will the circuit operate properly if $D1$ is (i) a single diode or (ii) three diodes in series?
(c) Replace $D1$ by a 15-K resistance and repeat part a. Compare the binary levels in parts a and c.
(d) What is the maximum allowed fan-in, assuming that the diodes are ideal? What is a practical limitation on fan-in?

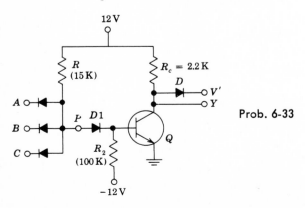

Prob. 6-33

6-34 Verify that this wired circuit performs the EXCLUSIVE OR function.

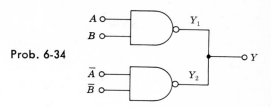

Prob. 6-34

6-35 (a) For the IC positive TTL NAND gate shown in Fig. 6-21, calculate $h_{FE,\min}$ for proper operation of the circuit.
(b) Calculate noise margins.
(c) Calculate the fan-out if $h_{FE,\min} = 30$.

6-36 For the RTL IC positive NOR gate shown in Fig. 6-23 prove that the maximum fan-out can be approximated by the formula

$$N_{\max} = h_{FE,\min} - h_{FE,\min} \frac{0.6}{V_{CC}} - \frac{R_b}{R_c}$$

6-37 The inputs of the RTL IC positive NOR gate shown in Fig. 6-23 are obtained from the outputs of similar gates and the outputs drive similar gates. If the supply voltage of the system is 5 V and the temperature range for proper operation of the gate is -50 to $150°C$, calculate the maximum permissible values of the resistances. Assume $h_{FE} = 30$ at $-50°C$, $I_{CBO} = 10$ nA at $25°C$, and the desired fan-out is 10.

CHAPTER 7

7-1 Indicate how to implement S_n of Eq. (7-1) with AND, OR, and NOT gates.

7-2 Verify that the sum S_n in Eq. (7-1) for a full adder can be put in the form

$$S_n = A_n \oplus B_n \oplus C_{n-1}$$

7-3 (a) For convenience, let $A_n = A$, $B_n = B$, $C_{n-1} = C$, and $C_n = C^1$. Using Eq. (7-4) for C^1, verify Eq. (7-5) with the aid of the Boolean identities in Table 6-4; in other words, prove that

$$\bar{C}^1 = \bar{B}\bar{C} + \bar{C}\bar{A} + \bar{A}\bar{B}$$

(b) Evaluate $D \equiv (A + B + C)\bar{C}^1$ and prove that S_n in Eq. (7-1) is given by

$$S_n = D + ABC$$

7-4 (a) Make a truth table for a binary half subtractor A minus B (corresponding to the half adder of Fig. 7-3). Instead of a carry C, introduce a *borrow* P.
(b) Verify that the digit D is satisfied by an EXCLUSIVE OR gate and that P follows the logic "B but not A."

7-5 The system shown is called a true/complement–zero/one element. Verify the truth table.

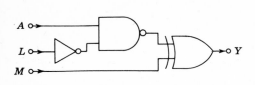

| Control inputs | | Output |
L	M	Y
0	0	$\bar{A}$
0	1	A
1	0	1
1	1	0

Prob. 7-5

7-6 Draw a logic diagram of a 4-to-10-line decoder using OR gates instead of AND gates.

7-7 Draw a logic diagram for a 3-to-8-line decoder.

7-8 Explain how to convert a 4-to-10-line decoder unit into a 3-to-8-line decoder.

7-9 Draw a logic diagram for an 8-to-1-line multiplexer.

7-10 Write the Boolean expression for the output Y of a 4-to-1-line multiplexer with an enable input (Fig. 7-14).

7-11 The block diagram shows two data selectors being used to select 1 out of 32 data inputs. Explain the operation of the system.

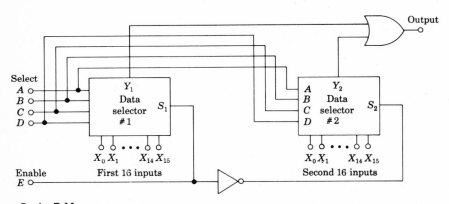

Prob. 7-11

7-12 Design an encoder satisfying the following truth table, using a diode matrix.

Inputs				Outputs			
W_3	W_2	W_1	W_0	Y_3	Y_2	Y_1	Y_0
0	0	0	1	0	1	1	1
0	0	1	0	1	1	0	0
0	1	0	0	1	1	0	1
1	0	0	0	0	0	1	0

Prob. 7-12

7-13 (a) Design an encoder, using multiple-emitter transistors, to satisfy the following truth table.
(b) How many transistors are needed and how many emitters are there in each transistor?

Prob. 7-13

Inputs			Outputs				
W_2	W_1	W_0	Y_4	Y_3	Y_2	Y_1	Y_0
0	0	1	1	1	0	1	0
0	1	0	1	0	0	0	1
1	0	0	0	1	1	1	1

7-14 A block diagram of a three-input (A, B, and C) and eight-output (Y_0 to Y_7) decoder matrix is indicated. The bit Y_6 is to be 1 (5 V) if the input code is 110 corresponding to decimal 6.
(a) Indicate how diodes are to be connected to line Y_6.
(b) Repeat for Y_0, Y_1, and Y_7.

Prob. 7-14

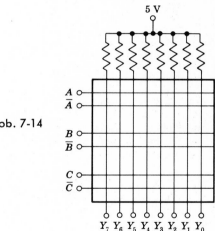

7-15 (a) Write the expressions for Y_1 and Y_3 in the binary-to-Gray-code converter.
(b) Indicate how to implement the relationship for Y_1 with diodes.

7-16 (a) Give the relationships between the output and input bits for the Gray-to-binary-code translator for Y_1 and Y_2.
(b) Indicate how to implement the equation for Y_1 with transistors.

7-17 (a) Write the sum-of-products canonical form for Y_4 of Table 7-5 for the seven-segment indicator code.
(b) Verify that this expression can be minimized to $Y_4 = A + C\bar{B}$.

7-18 How many AND, OR, and NOT gates are required if a three-input adder is implemented with an ROM? Compare these numbers with those used in a full-adder chip.

7-19 Implement the code conversion indicated below using a Read-Only Memory (ROM). Indicate *all* connections between the X input and the Y outputs. Use the standard symbols for inverters, for AND gates, and for multiple-emitter transistors. (Do not use diodes.)

X_1	X_0	Y_3	Y_2	Y_1	Y_0
0	0	1	0	1	1
0	1	0	1	0	1
1	0	0	1	1	1
1	1	1	1	0	0

Prob. 7-19

7-20 Consider an 8-bit comparator. Justify the connections $C' = C_L$, $D' = D_L$, and $E' = E_L$ for the chip handling the more significant bits. HINT: Add 4 to each subscript in Fig. 7-21. Extend Eq. (7-18) for E and Eq. (7-19) for C to take all 8 bits into account.

7-21 (*a*) By means of a truth table verify the Boolean identity

$$Y = (A \oplus B) \oplus C = A \oplus (B \oplus C)$$

(*b*) Verify that $Y = 1(0)$ if an odd (even) number of variables equals 1. This result is *not* limited to three inputs, but is true for any number of inputs. It is used in Sec. 7-9 to construct a parity checker.

7-22 Construct the truth table for the EXCLUSIVE OR tree of Fig. 7-22 for all possible inputs A, B, C, and D. Include $A \oplus B$ and $C \oplus D$ as well as the output Z. Verify that $Z = 1(0)$ for odd (even) parity.

7-23 (*a*) Draw the logic circuit diagram for an 8-bit parity check/generator system.
(*b*) Verify that the output is $0(1)$ for odd (even) parity.

7-24 (*a*) Verify that if $P' = 1$ in Fig. 7-22, this system is an even-parity check. In other words, demonstrate that with $P' = 1$, the output is $P = 0(1)$ for even (odd) parity of the inputs A, B, C, and D.
(*b*) Also verify that P generates the correct even-parity bit.

7-25 (*a*) Indicate an 8-bit parity checker as a block having 8 input bits (collectively designated A_1), an output P_1, and an input control P'_1. Consider a second 8-bit unit with inputs A_2, output P_2, and control P'_2. Show how to cascade the two packages in order to check for odd parity of a 16-bit word. Verify that the system operates properly if $P'_1 = 1$. Consider the four possible parity combinations of A_1 and A_2.
(*b*) Show how to cascade three units to obtain the parity of a 24-bit word. Should $P'_1 = 0$ or 1 for odd parity?
(*c*) Show how to cascade units to obtain the parity of a 10-bit word.

CHAPTER 8

8-1 (*a*) Verify that it is not possible for both outputs in Fig. 8-2*a* to be in the same state.
(*b*) Verify that if $S = 0$ and $R = 1$ in Fig. 8-2*b*, the latch is reset to $Q = 0$.
(*c*) If $S = R = 0$, verify that the state of the latch is undetermined (it could be either $Q = 1$ or $Q = 0$).
(*d*) If $S = R = 1$, verify that both outputs would go to 1. Is this a valid situation?

8-2 Draw the logic diagram for an *S-R* FLIP-FLOP using AOI gates instead of NAND gates.

8-3 The excitation table for a *J-K* FLIP-FLOP is shown. An X in the table is to be interpreted to mean that it does not matter whether this entry is a 1 or a 0. It is referred to as a "don't care" condition. Thus the second row indicates that if the output is to change from 0 to 1, the J input must be 1, whereas K can be either 1 or 0. Verify this excitation table by referring to the truth table of Fig. 8-4*b*.

Prob. 8-3

Q_n	Q_{n+1}	J_n	K_n
0	0	0	X
0	1	1	X
1	0	X	1
1	1	X	0

8-4 Verify that the *J-K* FLIP-FLOP truth table is satisfied by the difference equation

$$Q_{n+1} = J_n \bar{Q}_n + \bar{K}_n Q_n$$

8-5 (*a*) For the *J-K* FLIP-FLOP of Fig. 8-5, verify that for $Cr = 1$, $Pr = 0$, and $Ck = 0$, the 1 state is preset independent of the values of J_n and K_n.
(*b*) Repeat part *a* for $Ck = 1$, provided that $J_n = K_n = 0$.
(*c*) Verify that $Cr = Pr = Ck = 0$ leads to an indeterminate state; i.e., it may be 0 or 1.

8-6 (*a*) Verify that there is no race-around difficulty in the *J-K* circuit of Fig. 8-5 for any data input combination except $J = K = 1$.
(*b*) Explain why the race-around condition does not exist (even for $J = K = 1$) provided that $t_p < \Delta t < T$.

8-7 (*a*) For the master-slave *J-K* FLIP-FLOP of Fig. 8-6 assume $Q = 1$, $\bar{Q} = 0$, $Ck = 1$, $K = 0$, and J arbitrary. What is Q_M?
(*b*) If K changes to 1, what is Q_M?
(*c*) If K returns to 0, what is Q_M? Note that Q_M does *not* return to its initial value. Hence K (and J) must not vary during the pulse.

8-8 The indicated waveforms J, K, and Ck are applied to a *J-K* FLIP-FLOP. Plot the output waveform for Q and $\bar{Q}$ lined up with respect to the clock pulses.

NOTE: Assume that the output $Q = 0$ when the first clock pulse is applied and that $Pr = Cr = 1$.

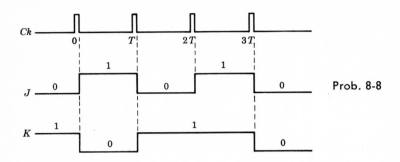

Prob. 8-8

8-9 (a) Verify that an S-R FLIP-FLOP is converted to a T type if S is connected to $\bar{Q}$ and R to Q.
(b) Verify that a D-type FLIP-FLOP becomes a T type if D is tied to $\bar{Q}$.

8-10 Augment the shift register of Fig. 8-9 with a four-input NOR gate whose output is connected to the *serial input* terminal. The NOR-gate inputs are Q_4, Q_3, Q_2, and Q_1.
(a) Verify that regardless of the initial state of each FLIP-FLOP, when power is applied, the register will assume correct operation as a ring counter after P clock pulses, where $P \le 4$.
(b) If initially $Q_4 = 1$, $Q_3 = 1$, $Q_2 = 0$, $Q_1 = 0$, and $Q_0 = 1$, sketch the waveform at Q_0 for the first 16 pulses.
(c) Repeat part b if $Q_4 = 0$, $Q_3 = 1$, $Q_2 = 0$, $Q_1 = 0$, and $Q_0 = 0$.

8-11 (a) Draw a waveform chart for the twisted-ring counter; i.e., indicate the waveforms Q_4, Q_3, Q_2, Q_1, and Q_0 for, say, 12 pulses. Assume that initially $Q_0 = Q_1 = Q_2 = Q_3 = Q_4 = 0$.
(b) Write the truth table after each pulse.
(c) By inspection of the table show that two-input AND gates can be used for decoding. For example, pulse 1 is decoded by $Q_4 \bar{Q}_3$. Why?

8-12 (a) For the modified ring counter shown, assume that initially $Q_0 = 1$, $Q_1 = 0$, and $Q_2 = 0$. Make a table of the readings Q_0, Q_1, Q_2, J_2, and K_2 after each

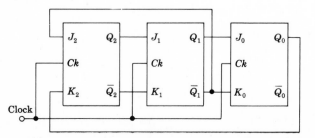

Prob. 8-12

clock pulse. How many pulses are required before the system begins to operate as a divide-by-N counter? What is N?
(b) Repeat part a if initially $Q_0 = 1$, $Q_1 = 0$, and $Q_2 = 1$.

8-13 A 50:1 ripple counter is desired.
(a) How many FLIP-FLOPS are required?
(b) If 4-bit FLIP-FLOPS are available on a chip, how many chips are needed? How are these interconnected?
(c) Indicate the feedback connections to the clear terminals.

8-14 (a) Indicate a divide-by-14 ripple-counter block diagram. Include a latch in the clear input.
(b) What are the inputs to the feedback NAND gate for a 153:1 ripple counter?

8-15 Consider the operation of the latch in Fig. 8-13. Make a table of the quantities Ck, Q_1, Q_3, P_1, $\overline{Ck}$, and $P_2 = Cr$ for the following conditions:
(a) Immediately after the tenth pulse.
(b) After the tenth pulse and assuming Q_1 has reset before Q_3.
(c) During the eleventh pulse.
(d) After the eleventh pulse.
This table should demonstrate that
(a) The tenth pulse sets the latch to clear the counter.
(b) The latch remains set until all FLIP-FLOPS are cleared.
(c) The positive edge of the eleventh pulse resets the latch so that $Cr = 1$.
(d) The negative edge of the eleventh pulse initiates the new counting cycle.

8-16 (a) The circuit shown is a *programmable* ripple counter. It is understood that $J = K = Cr = 1$ and that the latch in Fig. 8-13b exists between P_1 and P_2. If $Pr_0 = Pr_1 = 1$ and $Pr_2 = Pr_3 = 0$, what is the count N? Explain the operation of the system carefully.
(b) Why is the latch required?
(c) Generalize the result of part a as follows. The counter has n stages and is to divide by N, where $2^n > N > 2^{n-1}$. How must the preset inputs be programmed?

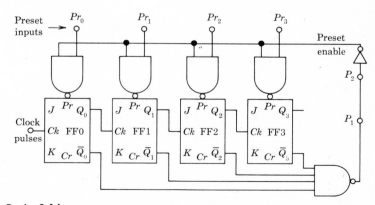

Prob. 8-16

8-17 From the logic diagram of the synchronous counter shown, write the truth table of Q_0, Q_1, and Q_2 after each pulse and verify that this is a 5:1 counter.

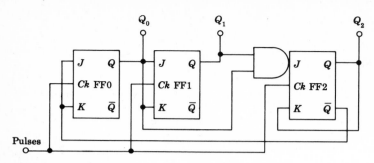

Prob. 8-17

8-18 Consider a two-stage synchronous counter (both stages receive the pulses at the Ck input). In each counter $K = 1$. If $J_0 = \bar{Q}_1$ and $J_1 = Q_0$, draw the circuit. From a truth table of Q_0 and Q_1 after each pulse, demonstrate that this is a $3:1$ counter.

CHAPTER 9

9-1 The drain resistance R_d of an n-channel FET with the source grounded is 2 K. The FET is operating at a quiescent point $V_{DS} = 10$ V, and $I_{DS} = 3$ mA, and its characteristics are given in Fig. 9-3.
(a) To what value must the gate voltage be changed if the drain current is to change to 5 mA?
(b) To what value must the voltage V_{DD} be changed if the drain current is to be brought back to its previous value? The gate voltage is maintained constant at the value found in part a.

9-2 Draw the circuit of an MOSFET negative AND gate and explain its operation.

9-3 Consider the FLIP-FLOP circuit shown. Assume $V_T = V_{ON} = 0$ and $|V_{DD}| \gg V_T$.
(a) Assume $v_{i1} = v_{i2} = 0$. Verify that the circuit has two possible stable states; either $v_{o1} = 0$ and $v_{o2} = -V_{DD}$ or $v_{o1} = -V_{DD}$ and $v_{o2} = 0$.
(b) Show that the state of the FLIP-FLOP may be changed by momentarily allowing one of the inputs to go to $-V_{DD}$; in other words by applying a negative input pulse.

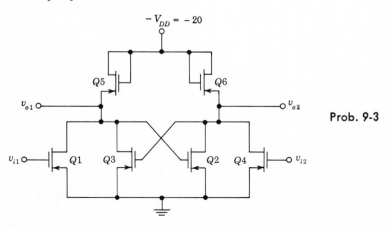

Prob. 9-3

9-4 Draw a CMOS inverter using positive logic.

9-5 (a) The complementary MOS negative NAND gate is indicated. Explain its operation.
(b) Draw the corresponding positive NAND gate.

Prob. 9-5

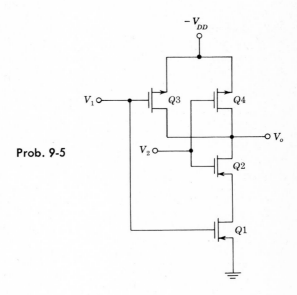

9-6 The circuit of a CMOS positive NOR gate is indicated. Explain its operation.

Prob. 9-6

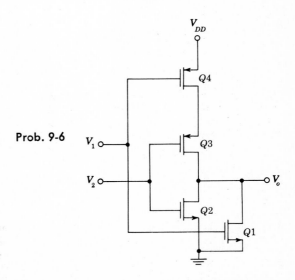

9-7 Draw an MOSFET circuit satisfying the logic equation, $Y = \overline{A + BC}$, where Y is the output corresponding to the three inputs A, B, and C.

9-8 (a) Verify that the circuit of Fig. 9-17 performs the function of a NAND gate. Let the voltage levels of V_1 and V_2 be 0 V or $-V_{DD}$.

(b) Verify that this circuit dissipates less power than the corresponding circuit of Fig. 9-13.

(c) Draw the circuit of a dynamic NOR gate corresponding to Fig. 9-14. Repeat parts a and b for this circuit.

9-9 Show that in the four-phase shift-register stage there is no dc current path to ground even when clocks ϕ_1 and ϕ_2 overlap or when clocks ϕ_3 and ϕ_4 overlap. Assume that the input terminal is maintained at zero volts. Explain the operation of this shift register by referring to the timing waveforms indicated in the figure.

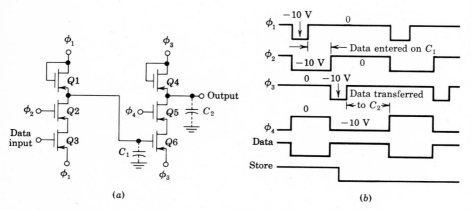

(a)   (b)

Prob. 9-9

9-10 Draw the logic diagrams of a recirculating or refresh memory to store 512 words each 4 bits long, using the TI 3309JC as the basic building block. Input data are available in parallel form, and data output must be presented also in parallel form. (Refer also to timing diagram Prob. 9-9.)

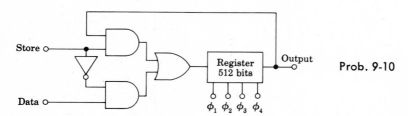

Prob. 9-10

9-11 Draw a 16-word 4-bit RAM matrix using the basic 1-bit RAM of Fig. 9-23 and using linear selection.

9-12 The figure shows a 64-word 1-bit RAM with on-chip decoding. The memory accepts a 6-bit address word. Using the above memory unit as a building block, construct a 64-word by 4-bit memory.

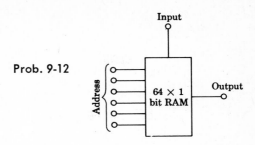

Prob. 9-12

9-13   The figure shows a 16-bit coincident memory matrix.   A specific bit is selected
by applying a logic 1 to the coincident X and Y address lines.
(a) Draw the diagram of a 16-word by N-bit memory (each word N bits long)
using the above RAM as the basic building block.
(b) What determines the maximum value of N in this configuration?

Prob. 9-13

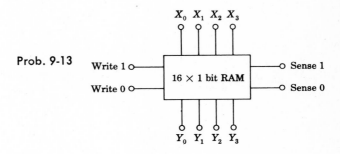

CHAPTER 10

10-1   (a) Describe how to obtain h_{ie} from the CE characteristics.
(b) Repeat part a for h_{re}.   Explain why this procedure, although correct in
principle, is inaccurate in practice.

10-2   The transistor whose characteristics are shown in Fig. 10-2 is biased at $V_{CE} =$
-8 V and $I_B = -300$ μA.   Compute graphically h_{fe} and h_{oe} at the quiescent
point specified above.

10-3   The circuit of the given figure represents the small-signal model of a CE
transistor-amplifier stage.   Show that

(a) $A_I \equiv \dfrac{I_o}{I_b} = -\dfrac{h_{fe}}{1 + h_{oe}R_L}$

(b) $R_i \equiv \dfrac{V_i}{I_b} = h_{ie} + h_{re}A_I R_L$

(c) $A_V \equiv \dfrac{V_o}{V_i} = \dfrac{A_I R_L}{R_i}$

(d) $Y_o = \dfrac{1}{R_o} = h_{oe} - \dfrac{h_{fe}h_{re}}{h_{ie} + R_s}$

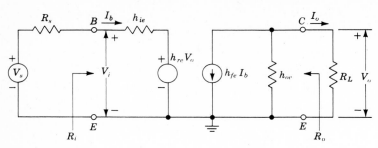

Prob. 10-3

10-4 The amplifier shown uses a transistor whose h parameters are $h_{fe} = 50$, $h_{ie} = 1,100\ \Omega$, and $h_{oe} = h_{re} = 0$.   Calculate

(a) $A_I \equiv \dfrac{I_o}{I_i}$

(b) $R_i = \dfrac{V_i}{I_i}$

(c) $A_V = \dfrac{V_o}{V_i}$

(d) $A_{Vs} = \dfrac{V_o}{V_s}$

(e) R_o and R_o'

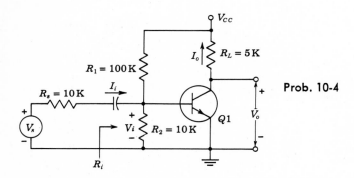

Prob. 10-4

10-5 (a) Draw the equivalent circuit for the CE and CC configurations using the approximate model of Fig. 10-5 subject to the restriction that $R_L = 0$.   Show that the input resistances of the two circuits are identical.
(b) Draw the circuits for the CE and CC configurations subject to the restriction that the input is open-circuited.   Show that the output resistances of the two circuits are identical.   Prove that $R_o' = R_L$.

10-6 For any CE transistor amplifier prove that

$$R_i = \frac{h_{ie}}{1 - h_{re}A_V}$$

HINT: Use the formulas given in Prob. 10-3.

10-7 (a) For a CE configuration, what is the maximum value of R_L for which R_i differs by no more than 10 percent of its value at $R_L = 0$? Use the transistor parameters given in Sec. 10-2 and the formulas in Prob. 10-3.
(b) What is the maximum value of R_s for which R_o differs by no more than 10 percent of its value for $R_s = 0$?

10-8 For a CB connection derive the simplified expressions given in Table 10-2.

10-9 (a) Consider a CB connection with $R_s = 2$ K and $R_L = 4$ K. Find the approximate values of A_I, A_V, A_{Vs}, R_i, and R_o'. Let $h_{ie} = 1.1$ K, $h_{fe} = 50$.
(b) Repeat part a for the CE connection.
(c) Repeat part a for the CC connection.

10-10 (a) In the circuit shown, find the input resistance R_i in terms of the four CE h parameters, R_L and R_e. HINT: Follow the rules given in Sec. 10-3.
(b) If $R_L = R_e = 1$ K and the h parameters are as given in Sec. 10-2, what is the value of R_i?

Prob. 10-10

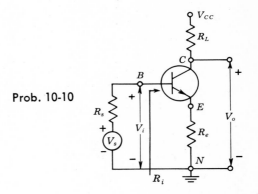

10-11 For the amplifier shown, using a transistor whose parameters are $h_{fe} = 100$, $h_{ie} = 2$ K, and $h_{oe} = h_{re} = 0$, compute $A_I = I_o/I_i$, R_i, A_V, A_{Vs}, R_o, and R_o'.

Prob. 10-11

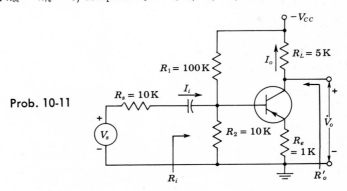

10-12 For the two-stage cascade shown, compute the input and output resistances and the individual and overall voltage and current gains, using the approximate formulas in Table 10-2 with $h_{fe} = 50$, and $h_{ie} = 1.1$ K.
NOTE: Assume each capacitor is arbitrarily large so that it may be considered a short circuit at the signal frequency.

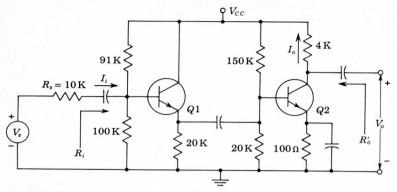

Prob. 10-12

10-13 Compute A_I, A_V, A_{Vs}, R_i, and R'_o for the two-stage cascade shown, using the approximate formulas in Table 10-2 with $h_{fe} = 50$ and $h_{ie} = 1.1$ K.   See note in Prob. 10-12.

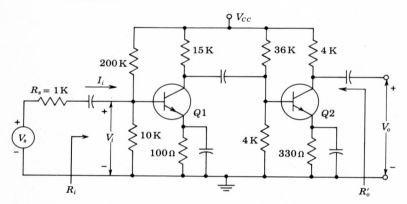

Prob. 10-13

10-14 For the circuit shown, compute A_I, A_V, A_{Vs}, R_i, and R'_o.   Use $h_{fe} = 50$ and $h_{ie} = 1.1$ K.   See note in Prob. 10-12.

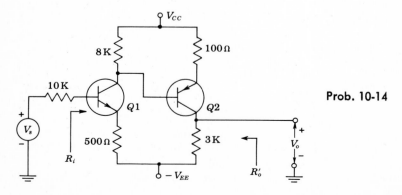

Prob. 10-14

10-15 For the two-stage amplifier shown calculate A_V, A_{Vs}, R_i, and R_o'. Use $h_{fe} = 50$ and $h_{ie} = 1.1$ K. See note in Prob. 10-12.

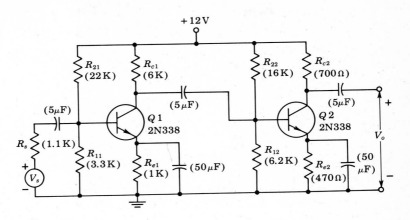

Prob. 10-15

10-16 In the circuit shown, all capacitors are arbitrarily large so that they act as short circuits at the signal frequency. All resistors are in kilohms. For each transistor $h_{fe} = 100$, $h_{ie} = 2$ K, $h_{re} = 0$, and $h_{oe} = 0$. Find I_o/I_i (signal currents).

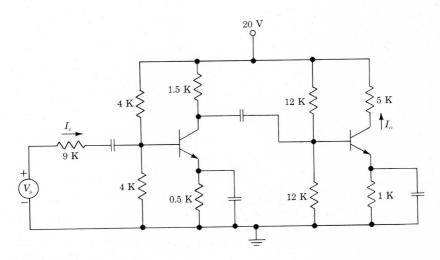

Prob. 10-16

10-17 For the FET whose characteristics are plotted in Fig. 9-3, determine r_d and g_m graphically at the quiescent point $V_{DS} = 10$ V and $V_{GS} = -1.5$ V. Also evaluate μ.

10-18 (a) Verify Eq. (10-42).

(b) Starting with the definitions of g_m and r_d, show that if two identical FETs are connected in parallel, g_m is doubled and r_d is halved. Since $\mu = r_d g_m$, then μ remains unchanged.

(c) If the two FETs are not identical, show that

$$\frac{1}{r_d} = \frac{1}{r_{d1}} + \frac{1}{r_{d2}}$$

and that

$$\mu = \frac{\mu_1 r_{d2} + \mu_2 r_{d1}}{r_{d1} + r_{d2}}$$

10-19 Calculate the voltage gain $A_V = V_o/V_i$ at 1 kHz for the circuit shown. The FET parameters are $g_m = 2$ mA/V and $r_d = 10$ K. Neglect capacitances. See note in Prob. 10-12.

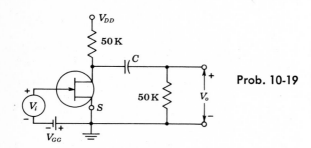

Prob. 10-19

10-20 The circuit shown is called a common-gate amplifier. For this circuit find (a) the voltage gain, (b) the input resistance, (c) the output resistance. Power supplies are omitted for simplicity. Neglect capacitances.

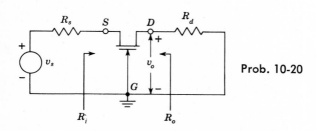

Prob. 10-20

10-21 Find an expression for the signal voltage across R_L. The two FETs are identical, with parameters μ, r_d, and g_m.

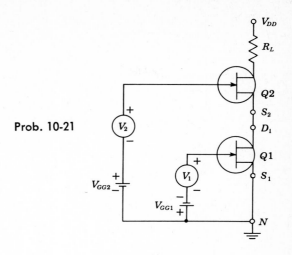

Prob. 10-21

10-22 Each FET shown has the parameters $r_d = 10$ K and $g_m = 2$ mA/V. Find the gain (a) v_o/v_1 if $v_2 = 0$, (b) v_o/v_2 if $v_1 = 0$.

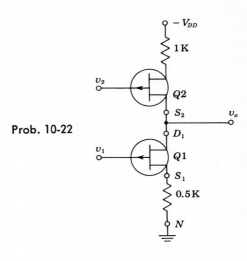

Prob. 10-22

10-23 (a) Prove that the magnitude of the signal current is the same in both FETs provided that

$$r = \frac{1}{g_m} + \frac{2R_L}{\mu}$$

See note in Prob. 10-12.

(b) If r is chosen as in part a, prove that the voltage gain is given by

$$A = \frac{-\mu^2}{\mu + 1} \frac{R_L}{R_L + r_d/2}$$

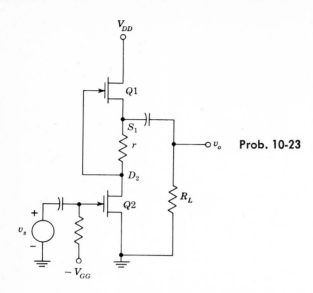

Prob. 10-23

10-24 (a) If $R_1 = R_2 = R$ and the two FETs have identical parameters, verify that the voltage amplification is $V_o/V_s = -\mu/2$ and the output resistance is $\frac{1}{2}[r_d + (\mu + 1)R]$.
(b) Given $r_d = 62$ K, $\mu = 10$, $R_1 = 2$ K, and $R_2 = 1$ K. Find the voltage gain and the output resistance.

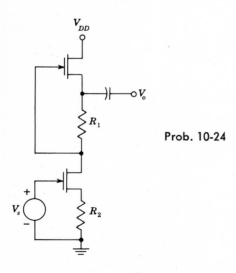

Prob. 10-24

10-25 (a) Determine the quiescent currents and the collector-to-emitter voltage for a silicon transistor with $\beta = 50$ in the self-biasing arrangement of Fig. 10-21. The circuit component values are $V_{CC} = 20$ V, $R_c = 2$ K, $R_e = 0.1$ K, $R_1 = 100$ K, and $R_2 = 5$ K.
(b) Repeat (a) for a germanium transistor.

10-26 A p-n-p germanium transistor is used in the self-biasing arrangement of Fig. 10-21. The circuit component values are $V_{CC} = 4.5$ V, $R_c = 1.5$ K, $R_e = 0.27$ K, $R_2 = 2.7$ K, and $R_1 = 27$ K. If $\beta = 44$, find the quiescent point.

10-27 For the circuit shown
(a) Calculate I_B, I_C, and V_{CE} if a silicon transistor is used with $\beta = 50$.
(b) Specify a value for R_b so that $V_{CE} = 7$ V.

Prob. 10-27

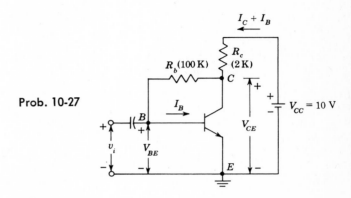

10-28 In the circuit shown, $V_{CC} = 24$ V, $R_c = 10$ K, and $R_e = 270$ Ω. If a silicon transistor is used with $\beta = 45$ and if under quiescent conditions $V_{CE} = 5$ V, determine R.

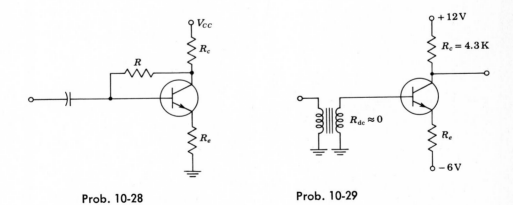

Prob. 10-28

Prob. 10-29

10-29 In the transformer-coupled amplifier stage shown, $V_{BE} = 0.7$ V, $\beta = 50$, and the quiescent voltage is $V_{CE} = 4$ V. Determine R_e.

10-30 In the two-stage circuit shown, assume $\beta = 100$ for each transistor.
(a) Determine R so that the quiescent conditions are $V_{CE1} = -4$ V and $V_{CE2} = -6$ V.

(b) Explain how quiescent-point stabilization is obtained. Assume $V_{BE} = 0.2$ V.

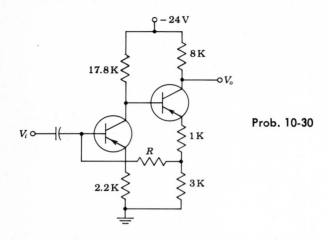

Prob. 10-30

10-31   The CS amplifier stage shown in Fig. 10-23 has the following parameters: $R_d = 12$ K, $R_g = 1$ M, $R_s = 470$ Ω, $V_{DD} = 30$ V, C_s is arbitrarily large, $I_{DSS} = 3$ mA, $V_P = -2.4$ V, and $r_d \gg R_d$. Determine (a) the gate-to-source bias voltage V_{GS}, (b) the drain current I_D, (c) the quiescent voltage V_{DS}, (d) the small-signal voltage gain A_V.

10-32   The amplifier stage shown uses an n-channel FET having $I_{DSS} = 1$ mA, $V_P = -1$ V. If the quiescent drain-to-ground voltage is 10 V, find R_1.

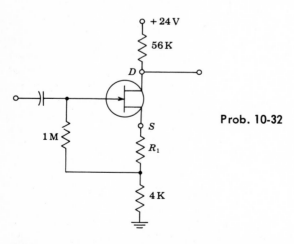

Prob. 10-32

10-33   The FET shown has the following parameters: $I_{DSS} = 5.6$ mA and $V_P = -4$ V.
(a) If $v_i = 0$, find v_o.
(b) If $v_i = 10$ V, find v_o.
(c) If $v_o = 0$, find v_i.
    NOTE: v_i and v_o are constant voltages (and not small-signal voltages).

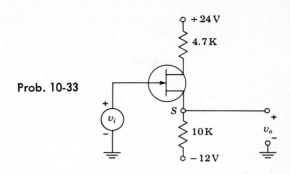

Prob. 10-33

10-34   If $|I_{DSS}| = 4$ mA, $V_P = 4$ V, calculate the quiescent values of I_D, V_{GS}, and V_{DS}.

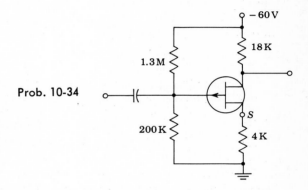

Prob. 10-34

10-35   The drain current in milliamperes of the enhancement-type MOSFET shown is given by

$$I_D = 0.2(V_{GS} - V_P)^2$$

in the region $V_{DS} \geq V_{GS} - V_P$.   If $V_P = +3$ V, calculate the quiescent values I_D, V_{GS}, and V_{DS}.

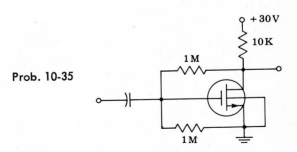

Prob. 10-35

CHAPTER 11

11-1   The input to an amplifier consists of a voltage made up of a fundamental signal and a second-harmonic signal of half the magnitude and in phase with the fundamental.   Plot the resultant.

The output consists of the same magnitude of each component, but with the second harmonic shifted 90° (on the fundamental scale). This corresponds to perfect frequency response but bad phase-shift response. Plot the output and compare it with the input waveshape.

11-2 The bandwidth of an amplifier extends from 20 Hz to 20 kHz. Find the frequency range over which the voltage gain is down less than 1 dB from its midband value. Assume that the low- and high-frequency response is given by Eqs. (11-2) and (11-6) multiplied by a constant A_{Vo}.

11-3 Prove that over the range of frequencies from $10f_L$ to $0.1f_H$ the voltage amplification is constant to within 0.5 percent and the phase shift to within ± 0.1 rad. Make the same assumption as in Prob. 11-2.

11-4 (a) Show that the Bode magnitude plot for a two-pole transfer function is equal to the sum of the magnitude plots of each pole considered separately. (b) Repeat part a for the Bode phase plot.

11-5 Consider a transfer characteristic with two poles such that $f_{p2} = 4f_{p1}$. (a) Plot the Bode magnitude curve. Obtain the actual 3-dB frequency graphically and also analytically. (b) Plot the Bode phase curve.

11-6 Repeat Prob. 11-5 for poles at $f_{p2} = 2f_{p1}$.

11-7 An ideal 1-μs pulse is fed into an amplifier. Plot the output if the bandpass is (a) 10 MHz, (b) 1.0 MHz, (c) 0.1 MHz. Assume $f_L = 0$ and a single-pole amplifier.

11-8 (a) Prove that the response of a two-stage (identical and noninteracting) low-pass amplifier to a unit step is

$$v_o = A_o{}^2[1 - (1 + x)\epsilon^{-x}]$$

where A_o is the midband voltage gain and $x \equiv t/RC$.
(b) For $t \ll RC$, show that the output varies quadratically with time.

11-9 In Prob. 11-8, let the upper 3-dB frequency of a single stage be f_H and the rise time of the two stages in cascade be t_r. Show that $f_H t_r = 0.53$.

11-10 (a) For the transistor CE stage shown with $1/h_{oe} \approx \infty$, calculate the percentage tilt in the output if the input current I is a 100-Hz square wave.

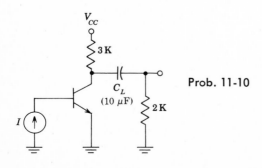

Prob. 11-10

(b) What is the lowest-frequency square wave which will suffer less than 1 percent tilt?

11-11 (a) The transfer function V_o/V_s of an amplifier has n poles $s_1, s_2, \ldots, s_n$ and k zeros $s_{z1}, s_{z2}, \ldots, s_{zk}$, as follows:

$$\frac{V_o}{V_s} = \frac{K(s - s_{z1})(s - s_{z2}) \cdots (s - s_{zk})}{(s - s_1)(s - s_2) \cdots (s - s_n)} = A(s)$$

If the zeros are of much higher frequencies than the poles, show that

$$\left|\frac{A}{A_o}\right|^2_{s=j2\pi f} = \frac{1}{\left[\left(\dfrac{f}{f_1}\right)^2 + 1\right]\left[\left(\dfrac{f}{f_2}\right)^2 + 1\right] + \cdots + \left[\left(\dfrac{f}{f_n}\right)^2 + 1\right]}$$

(b) Show that an approximate expression for the high 3-dB frequency f_H is given by

$$\frac{1}{f_H} \approx \sqrt{\frac{1}{f_1^2} + \frac{1}{f_2^2} + \cdots + \frac{1}{f_n^2}}$$

(c) Show that an expression which gives a more accurate result is

$$\frac{1}{f_H} \approx 1.1\sqrt{\frac{1}{f_1^2} + \frac{1}{f_2^2} + \cdots + \frac{1}{f_n^2}}$$

Verify this, using Eq. (11-15) for the case of
    (i) Two identical poles $f_1 = f_2$.
    (ii) Three identical poles $f_1 = f_2 = f_3$.
Show that the error is within 10 percent.

11-12 Consider a transfer function with poles at 1 MHz and 2 MHz. Assume all other poles and zeros are much larger than 2 MHz. Calculate the high 3-dB frequency. Compare your result with the approximate value obtained from Eq. (11-20).

11-13 If two cascaded single-pole stages have very unequal bandpasses, show that the combined bandwidth is essentially that of the smaller. Assume noninteracting stages.

11-14 Three identical cascaded stages have an overall upper 3-dB frequency of 20 kHz and a lower 3-dB frequency of 20 Hz. What are f_L and f_H of each stage? Assume noninteracting stages.

11-15 It is desired that the voltage gain of an RC-coupled amplifier at 60 Hz should not decrease by more than 10 percent from its midband value. Show that the coupling capacitance C must be at least equal to $5.5/R'$, where $R' = R'_o + R'_i$ is expressed in kilohms, and C in microfarads.

11-16 The parameters of the transistors in the circuit shown are $h_{fe} = 50$, $h_{ie} = 1.1$ K, $h_{re} = h_{oe} = 0$. Find (a) the midband gain, (b) the value of C_b necessary to give a lower 3-dB frequency of 20 Hz. Assume that C_z represents a short circuit at this frequency.   (c) Find the value of C_b necessary to ensure less than 10 percent tilt for a 100-Hz square-wave input.

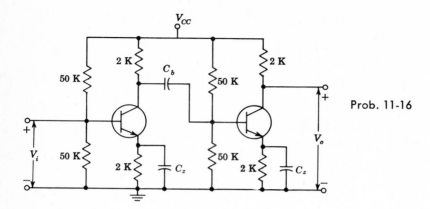

Prob. 11-16

11-17   A two-stage FET RC-coupled amplifier has the following parameters: $g_m = 10$ mA/V, $r_d = 5.5$ K, $R_d = 10$ K, and $R_g = 0.5$ M for each stage.   Assume C_s to be arbitrarily large.   See Fig. 11-8 and Fig. 10-23.
(a) What must be the value of C_b in order that the frequency characteristic of each stage be flat within 1 dB down to 10 Hz?
(b) Repeat part a if the overall gain of both stages is to be down 1 dB at 10 Hz.
(c) What is the overall midband voltage gain?

11-18   A three-stage RC-coupled amplifier uses field-effect transistors with the following parameters: $g_m = 2.6$ mA/V, $r_d = 7.7$ K, $R_d = 10$ K, $R_g = 0.1$ M, $C_b = 0.005$ μF, and $C_s = \infty$.   Evaluate (a) the overall midband voltage gain in decibels, (b) f_L of each individual stage, (c) the overall lower 3-dB frequency.

11-19   (a) Show that the relative voltage gain of an amplifier with an emitter resistor R_e bypassed by a capacitor C_z may be expressed in the form

$$\frac{A_V}{A_o} = \frac{1 + j\omega R_e C_z}{B + j\omega R_e C_z}$$

where $B = 1 + R'/R$, $R' = R_e(1 + h_{fe})$, and $R = R_s + h_{ie}$.
(b) Prove that the lower 3-dB frequency is

$$f_L = \frac{\sqrt{B^2 - 2}}{2\pi R_e C_z}$$

What is the physical meaning of the condition $B < \sqrt{2}$?
(c) If $B^2 \gg 2$, show that $f_L \approx f_p$, the pole frequency as defined in Eq. (11-26).

11-20   In the circuit of Fig. 11-10a, let $R_s = 500$ Ω; $R_1 = R_2 = 50$ K; $R_c = R_e = 2$ K; $h_{ie} = 1.1$ K; $h_{fe} = 50$; $h_{re} = h_{oe} = 0$; $C_b = 5$ μF.
(a) Neglecting the effects of C_z, find f_L for the transistor stage.
(b) Neglecting the effects of C_b, find expressions for f_p and f_o due to C_z alone.

(c) Find a value of C_z for which f_L is virtually unaffected by the presence of the emitter bypass capacitor.

11-21 Find the percentage tilt in the output of a transistor stage caused by a capacitor C_z bypassing an emitter resistor R_e. Use the following method: If V is the magnitude of the input step, then from Fig. 11-10 (and using lowercase letters for instantaneous values),

$$v_o = -h_{fe}i_b R_c = -h_{fe}R_c \frac{V - v_{en}}{R}$$

where $R \equiv R_s + h_{ie}$. Take as a first approximation $v_{en} = 0$. Calculate the corresponding current, and assuming that all the emitter current passes through C_z, calculate v_{en}, and then show that

$$v_o = -\frac{h_{fe}R_c V}{R}\left[1 - \frac{(1 + h_{fe})t}{RC_z}\right]$$

From this result verify Eq. (11-29).

11-22 Show that the low-frequency voltage gain of the FET stage shown with $r_d \gg R_L + R_s$ is given by

(a) $$\frac{A_V}{A_o} = \frac{1}{1 + g_m R_s}\frac{1 + jf/f_o}{1 + jf/f_p}$$

where

$$A_o \equiv -g_m R_L \qquad f_o \equiv \frac{1}{2\pi C_s R_s} \qquad f_p = \frac{1 + g_m R_s}{2\pi C_s R_s}$$

(b) If $g_m R_s \gg 1$ and $g_m = 5$ mA/V, find C_s so that a 50-Hz square-wave input will suffer no more than 10 percent tilt.

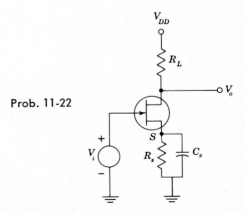

Prob. 11-22

11-23 Show that at low frequencies the hybrid-π model with $r_{b'c}$ and r_{ce} taken as infinite reduces to the approximate CE h-parameter model.

11-24 The following low-frequency parameters are known for a given transistor at $I_C = 10$ mA, $V_{CE} = 10$ V, and at room temperature.

$$h_{ie} = 500 \ \Omega \qquad h_{fe} = 100$$

At the same operating point, $f_T = 50$ MHz and $C_c = 3$ pF, compute the values of all the hybrid-π parameters.

11-25 Given the following transistor measurements made at $I_C = 5$ mA, $V_{CE} = 10$ V, and at room temperature:

$$h_{fe} = 100 \qquad\qquad h_{ie} = 600 \ \Omega$$

$$[A_{ie}] = 10 \text{ at } 10 \text{ MHz} \qquad C_c = 3 \text{ pF}$$

Find f_β, f_T, C_e, $r_{b'e}$, and $r_{bb'}$.

11-26 The hybrid-π parameters of the transistor used in the circuit shown are given in Secs. 11-9 and 11-10. Using Miller's theorem and the approximate analysis, compute
(a) The upper 3-dB frequency of the current gain $A_I = I_L/I_i$.
(b) The magnitude of the voltage gain $A_{Vs} = V_o/V_s$ at the frequency of part a.

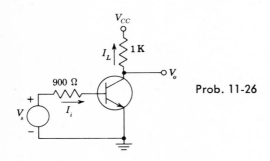

Prob. 11-26

11-27 Consider a single-stage CE transistor amplifier with the load resistor R_L shunted by a capacitance C_L. Apply Miller's theorem (Sec. C-4).
(a) Prove that the internal voltage gain $K = V_{ce}/V_{b'e}$ is

$$K \approx \frac{-g_m R_L}{1 + j\omega(C_c + C_L)R_L}$$

(b) Prove that the 3-dB frequency is given by

$$f_H \approx \frac{1}{2\pi(C_c + C_L)R_L}$$

provided that the following condition is satisfied:

$$g_{b'e}R_L(C_c + C_L) \gg C_e + C_c(1 + g_m R_L)$$

11-28 For a single-stage CE transistor amplifier whose hybrid-π parameters have the average values given in Secs. 11-9 and 11-10, what value of source resistance R_s will give a 3-dB frequency f_H which is (a) half the value for $R_s = 0$, (b) twice the value for $R_s = \infty$? Do these values of R_s depend upon the magnitude of the load R_L? Use Miller's theorem and the approximate analysis.

11-29 A single-stage CE amplifier is measured to have a voltage-gain bandwidth f_H of 5 MHz with $R_L = 500\ \Omega$. Assume $h_{fe} = 100$, $g_m = 100\ \text{mA/V}$, $r_{bb'} = 100\ \Omega$, $C_c = 1\ \text{pF}$, and $f_T = 400\ \text{MHz}$.
(a) Find the value of the source resistance that will give the required bandwidth.
(b) With the value of R_s found in part a, find the midband voltage gain V_o/V_s.
HINT: Use the approximate analysis.

11-30 The hybrid-π parameters of the transistor used in the circuit shown are given in Sec. 11-9 and 11-10. The input to the amplifier is an abrupt current step 0.2 mA in magnitude. Find the output voltage as a function of time (a) if $C_L = 0$. Neglect the output time constant. (b) If $C_L = 0.1\ \mu\text{F}$. Neglect the input time constant. HINT: Use Miller's theorem (Sec. C-4).

Prob. 11-30

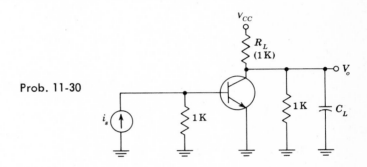

11-31 (a) An MOSFET connected in the CS configuration works into a 100-K resistive load. Calculate the complex voltage gain and the input admittance of the system for frequencies of 100 and 100,000 Hz. Take the interelectrode capacitances into consideration. The MOSFET parameters are $\mu = 100$, $r_d = 40$ K, $g_m = 2.5\ \text{mA/V}$, $C_{gs} = 4.0\ \text{pF}$, $C_{ds} = 0.6\ \text{pF}$, and $C_{gd} = 2.4\ \text{pF}$. Compare these results with those obtained when the interelectrode capacitances are neglected.
(b) Calculate the input resistance and capacitance.

11-32 Calculate the input admittance of an FET at 10^3 and 10^6 Hz when the total drain circuit impedance is (a) a resistance of 50 K, (b) a capacitive reactance of 50 K at each frequency. Take the interelectrode capacitances into consideration. The FET parameters are $\mu = 20$, $r_d = 10$ K, $g_m = 2.0\ \text{mA/V}$, $C_{gs} = 3.0\ \text{pF}$, $C_{ds} = 1.0\ \text{pF}$, and $C_{gd} = 2.0\ \text{pF}$. Express the results in terms of the input resistance and capacitance.

CHAPTER 12

12-1 For the circuit shown, with $R_e = 4$ K, $R_L = 4$ K, $R_b = 20$ K, $R_s = 1$ K, and the transistor parameters given in Sec. 10-2, find
(a) The current gain $I_L/I_s = A_I$.
(b) The voltage gain V_o/V_s, where $V_s \equiv I_s R_s$.

(c) The transconductance $I_L/V_s = G_M$.
(d) The transresistance $V_o/I_s = R_M$.
(e) The input resistance seen by the source.
(f) The output resistance seen by the load.
Make reasonable approximations.   Neglect all capacitive effects.

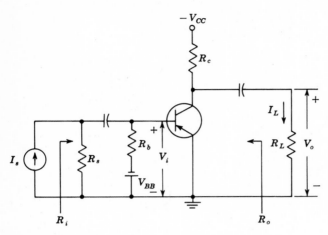

Prob. 12-1

12-2 Repeat Prob. 12-1 for the circuit shown, with $g_m = 5$ mA/V and $r_d = 100$ K.
Note that $V_s \equiv I_s R_s$.

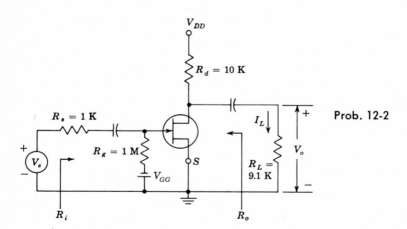

Prob. 12-2

12-3 (a) For the circuit shown, find the ac voltage V_i as a function of V_s and V_f.
Assume that the inverting-amplifier input resistance is infinite, that $A = A_V = -1,000$, $\beta = V_f/V_o = \frac{1}{100}$, $R_s = R_e = R_c = 1$ K, $h_{ie} = 1$ K, $h_{re} = h_{oe} = 0$,
and $h_{fe} = 100$.
(b) Find $A_{Vf} = V_o/V_s = A V_i/V_s$.

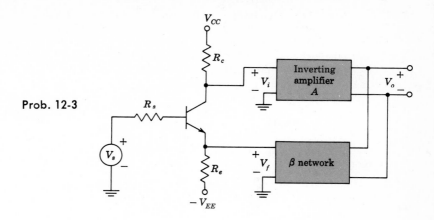

Prob. 12-3

12-4 An amplifier consists of three identical stages connected in cascade. The output voltage is sampled and returned to the input in series opposing. If it is
specified that the relative change dA_f/A_f in the closed-loop voltage gain A_f
must not exceed Ψ_f, show that the minimum value of the open-loop gain A of
the amplifier is given by

$$A = 3A_f \frac{|\Psi_1|}{|\Psi_f|}$$

where $\Psi_1 \equiv dA_1/A_1$ is the relative change in the voltage gain of each stage of
the amplifier.

12-5 An amplifier with open-loop voltage gain $A_V = 1,000 \pm 100$ is available. It
is necessary to have an amplifier whose voltage gain varies by no more than
± 0.1 percent.
(a) Find the reverse transmission factor β of the feedback network used.
(b) Find the gain with feedback.

12-6 The figure shows the transfer characteristic of a nonlinear amplifier. Negative
feedback is applied to this amplifier as shown. Find the new transfer characteristic x_o versus x_s if (a) $\beta = 0.1$, (b) $\beta = 0.05$. Plot the two transfer characteristics on the same figure.

Prob. 12-6

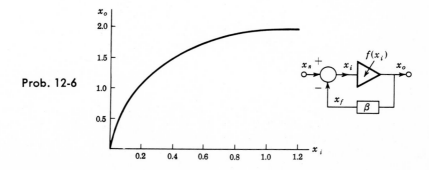

12-7 An amplifier without feedback gives a fundamental output of 36 V with 7 percent second-harmonic distortion when the input is 0.028 V.
(a) If 1.2 percent of the output is fed back into the input in a negative voltage-series feedback circuit, what is the output voltage?
(b) If the fundamental output is maintained at 36 V but the second-harmonic distortion is reduced to 1 percent, what is the input voltage?

12-8 An amplifier with an open-loop voltage gain of 1,000 delivers 10 W of output power at 10 percent second-harmonic distortion when the input signal is 10 mV. If 40-dB negative voltage-series feedback is applied and the output power is to remain at 10 W, determine (a) the required input signal, (b) the percent harmonic distortion.

12-9 A single-stage RC-coupled amplifier with a midband voltage gain of 1,000 is made into a feedback amplifier by feeding 10 percent of its output voltage in series with the input opposing. Assume that the amplifier gain without feedback may be approximated at low frequencies by Eq. (11-2) and at high frequencies by Eq. (12-11).
(a) As the frequency is varied, to what value does the voltage gain of the amplifier without feedback fall before gain of the amplifier with feedback falls 3 dB?
(b) What is the ratio of the half-power frequencies with feedback to those without feedback?
(c) If $f_L = 20$ Hz and $f_H = 50$ kHz for the amplifier without feedback, what are the corresponding values after feedback has been added?

12-10 Assume that the parameters of the circuit are $r_d = 10$ K, $R_g = 1$ M, $R_1 = 40\ \Omega$, $R_d = 50$ K, and $g_m = 6$ mA/V. Neglect the reactances of all capacitors. Find the voltage gain of the circuit at the terminals (a) AN, (b) BN.

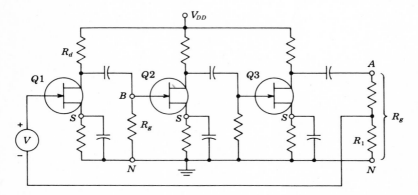

Prob. 12-10

12-11 The transistors in the feedback amplifier shown are identical, and their h parameters are as given in Sec. 10-2. Make reasonable approximations whenever appropriate, and neglect the reactance of each capacitor. Calculate

(a) $A'_{Vf} = V_o/V_i$ and   (b) $A_{Vf} = V_o/V_s$

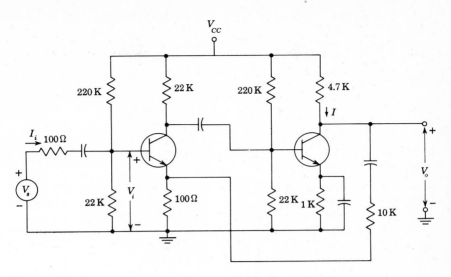

Prob. 12-11

12-12   Consider the transistor stage of Fig. 12-15.   The voltage gain $A_{Vf} = A_f$ is given by Eq. (12-38).   If the relative change dA_f/A_f of the voltage gain A_f must not exceed a specified value Ψ_f due to variations of h_{fe}, show that the minimum required value of the emitter resistor R_e is given by

$$R_e = \frac{R_s + h_{ie}}{h_{fe}} \left(\frac{dh_{fe}/h_{fe}}{\Psi_f} - 1 \right)$$

12-13   In the two-stage feedback amplifier shown, the transistors are identical and have the following parameters: $h_{fe} = 50$, $h_{ie} = 2$ K, $h_{re} = 0$, and $h_{oe} = 0$.

Prob. 12-13

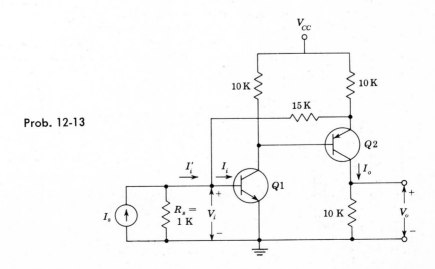

Calculate

(a) $A_{If} = \dfrac{I_o}{I_s}$ (b) $A'_{If} = \dfrac{I_o}{I'_i}$

(c) $A_{Vf} = \dfrac{V_o}{V_s}$ where $V_s = I_s R_s$

(d) Evaluate A_{Vf} from Eq. (12-25) and compare with the result obtained in part c.

12-14 For the circuit shown (and with the h-parameter values given in Prob. 12-13) find

(a) $A_{If} \equiv \dfrac{I_o}{I_s}$

(b) $A_{Vf} \equiv \dfrac{V_o}{V_s}$ where $I_s \equiv \dfrac{V_s}{R_s}$

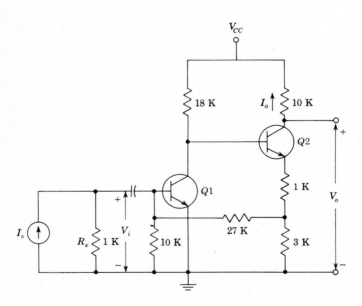

Prob. 12-14

12-15 The transistors in the feedback amplifier shown are identical, and their h parameters are $h_{fe} = 50$ and $h_{ie} = 1.1$ K. Make reasonable approximations where appropriate, and neglect the reactances of the capacitors. Calculate

(a) $A_{If} \equiv \dfrac{I_o}{I_s}$ (b) $A_{Vf} \equiv \dfrac{V_o}{V_s}$ where $V_s \equiv I_s R_s$

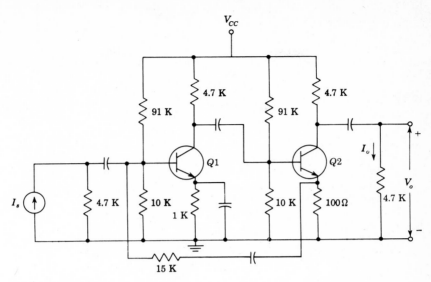

Prob. 12-15

12-16 Let h_{fe} of $Q1$ and $Q2$ of Prob. 12-15 increase to 100. If all other parameters remain constant, repeat Prob. 12-15.

12-17 For the transistor feedback-amplifier stage shown, $h_{fe} = 100$, $h_{ie} = 1$ K, while h_{re} and h_{oe} are negligible. Determine with $R_e = 0$

(a) $R_{Mf} = \dfrac{V_o}{I_s}$ where $I_s \equiv \dfrac{V_s}{R_s}$

(b) $A_{Vf} = \dfrac{V_o}{V_s}$

(c) Repeat the preceding calculations if $R_e = 1$ K.

Prob. 12-17

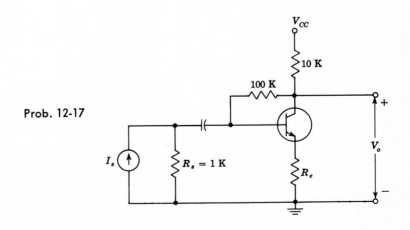

12-18 A three-pole feedback amplifier has a dc gain without feedback of -10^4. All three open-loop poles are at $f = 2$ MHz.
(a) What is the maximum value of β for which the amplifier is stable?
(b) Assume that one of the poles is shifted to $f_1 = 100$ kHz.
Using the value of β found in part a, what is the gain margin of the modified circuit?

12-19 A three-pole amplifier without feedback has a dc gain of -10^3 and poles located at $f_1 = 1$ MHz, $f_2 = 10$ MHz, and $f_3 = 30$ MHz. Dominant-pole compensation is applied to this amplifier.
(a) Find the location of the dominant pole so that the open-loop gain is first constant and then falls to 0 dB at a rate of -20 dB per decade for frequencies $f \le 1$ MHz.
(b) What is the maximum value of β for which this compensated amplifier is stable?

12-20 Pole-zero compensation is used with an amplifier which has -10^3 dc gain and three poles at $f_1 = 1$ MHz, $f_2 = 10$ MHz, and $f_3 = 200$ MHz. The zero of the pole-zero network is selected to cancel the 1-MHz pole of the uncompensated amplifier. Find the pole of the compensating network so that the amplifier is stable with a 45° phase margin at $f = 10$ MHz when $\beta = -0.1$.

12-21 (a) A two-stage FET oscillator uses the phase-shifting network shown. Prove that

$$\frac{V_f'}{V_o} = \frac{1}{3 + j(\omega RC - 1/\omega RC)} = -\beta$$

where V_o is the output voltage of the amplifier and $V_f = -V_f'$ is the feedback voltage.

(b) Show that the frequency of oscillation is $f = 1/2\pi RC$ and that the gain must exceed 3.

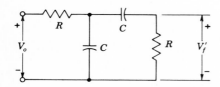

Prob. 12-21

12-22 (a) Find V_f'/V_o for the network shown. The voltages V_o and V_f' are defined in Prob. 12-21.
(b) Sketch the circuit of a phase-shift FET oscillator, using this feedback network. Explain why a two-stage amplifier is needed.
(c) Find the expression for the frequency of oscillation, assuming that the network does not load down the amplifier.
(d) Find the minimum gain required for oscillation.

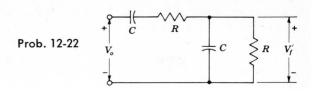

Prob. 12-22

CHAPTER 13

13-1   For the circuit of this problem with $R_i = \infty$, show

$$Y_{of} = \frac{1}{R_{of}} = \frac{1}{R_o}\left(1 - A_v \frac{R}{R + R'}\right) + \frac{1}{R + R'}$$

[HINT: Use the definition in Eq. (10-20) with R_o replaced by R_{of}.]

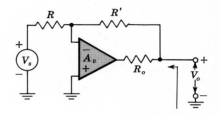

13-2   The circuit shown is a differential amplifier using an ideal OP AMP.
(a) Find the output voltage v_o.
(b) Show that the output corresponding to the common-mode voltage $v_c = \frac{1}{2}(v_1 + v_2)$ is equal to zero if $R'/R = R_1/R_2$.   Find v_o in this case.
(c) Find the common-mode rejection ratio of the amplifier if $R'/R \neq R_1/R_2$.

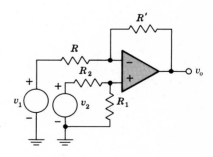

13-3   The emitter volt-ampere characteristic of a transistor in the active region is given by

$$I_E \approx I_S \epsilon^{V_E/V_T}$$

where I_S is a constant.
(a) Verify Eq. (13-12) for the transfer characteristic of the DIFF AMP.
(b) Verify Eq. (13-13) for g_{md}.

13-4 (a) From Eq. (13-12), for the transfer characteristic of the DIFF AMP find the range $\Delta V = \Delta(V_{B1} - V_{B2})$ over which each collector current increases from 0.1 to 0.9 its peak value.

(b) Repeat part a for a collector-current variation from 50 to 99 percent of I_o.

(c) Compare your result with Fig. 13-7.

13-5 (a) Draw the transfer characteristic for the circuit shown, indicating all important voltage values on your sketch. The transistors are silicon having the parameters in Table 4-1. Indicate a reasonable value of Δv_i between clipping levels.

(b) When Q1 is OFF, is Q2 beyond cutoff, in its active region, or in saturation? Justify your answer.

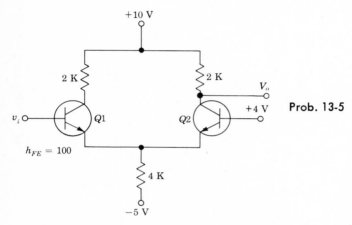

Prob. 13-5

13-6 The figure shows an inverting OP AMP with input resistance R_i and offset voltage V_{io}. Show that V_o is given by

$$V_o = \frac{-A_v R_i (R + R') V_{io}}{RR' + (R_i + R_1)(R + R') - RR_i A_v}$$

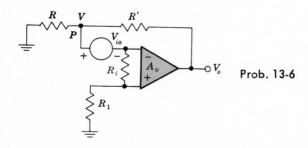

Prob. 13-6

13-7 For the amplifier shown, V_1 and V_2 represent undesirable voltages. Show that, if $R_i = \infty$, $R_o = 0$, and $A_{v1} < 0$ and $A_{v2} < 0$,

$$V_o = A_{v2}[A_{v1}(V' - V_1) - V_2] \qquad \text{where } V' = V_o \frac{R}{R + R'}$$

Show also that, if $A_{v2}A_{v1}R/(R + R') \gg 1$,

$$V_o = -\left(1 + \frac{R'}{R}\right)\left(V_1 + \frac{V_2}{A_{v1}}\right)$$

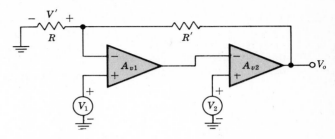

Prob. 13-7

13-8 The input offset voltage V_{io} of an OP AMP is equal to 1 mV at 25°C.   The input offset-voltage drift of this amplifier is equal to 5 μV/°C.   Assume that the open-loop voltage gain is infinite.   Using Prob. 13-6, find the output offset voltage at temperature $T = 100$°C if (i) $R'/R = 1$, (ii) $R'/R = 100$.   At which temperature is the output offset voltage equal to 0.2 V if $R'/R = 200$?

13-9 Design the circuit of Fig. 13-11 so that the output V_o (for a sinusoidal signal) is equal in magnitude to the input V_s and leads the input by 45°.

13-10 Consider the circuit of Fig. 13-11 with $A_V = -100$.   If $Z = R$ and $Z' = -jX_C$ with $R = X_C$ at some specific frequency f, calculate the gain V_o/V_s as a complex number.

13-11 Given the operational amplifier circuit of Fig. 13-11 consisting of R and L in series for Z, and C for Z'.   If the input voltage is a constant $v_s = V$, find the output v_o as a function of time.   Assume an infinite open-loop gain.

13-12 For the given circuit, show that the output voltage is

$$-v_o = \frac{R_2}{R_1}v + \left(R_2C + \frac{L}{R_1}\right)\frac{dv}{dt} + LC\frac{d^2v}{dt^2}$$

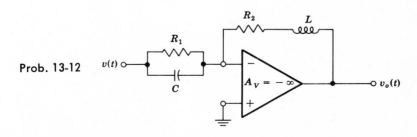

Prob. 13-12

13-13 Consider the operational amplifier circuit of Fig. 13-11 with Z consisting of a resistor R in parallel with a capacitor C, and Z' consisting of a resistor R'.   The input is a sweep voltage $v = \alpha t$. Show that the output voltage v_o is a sweep voltage that starts with an initial step. Thus prove that

$$v_o = -\alpha R'C - \alpha \frac{R'}{R} t$$

Assume infinite open-loop gain.

13-14 Consider the operational amplifier circuit of Fig. 13-11 with Z consisting of a 100-K resistor and a series combination of a 50-K resistance with a 0.001-μF capacitance for Z'. If the capacitor is initially uncharged, and if at $t = 0$ the input voltage $v_s = 10\epsilon^{-t/\tau}$ with $\tau = 5 \times 10^{-4}$ s is applied, find $v_o(t)$.

13-15 In Fig. 13-13b show that i_L is equal to $-v_s/R_2$ if $R_3/R_2 = R'/R_1$.

13-16 The differential input operational amplifier shown consists of a base amplifier of infinite gain. Show that $V_o = (R_2/R_1)(V_2 - V_1)$.

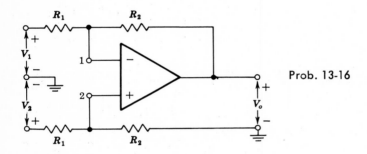

Prob. 13-16

13-17 Repeat Prob. 13-16 for the amplifier shown.

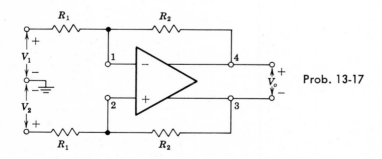

Prob. 13-17

13-18 For the base differential-input amplifier shown, assume infinite input resistance, zero output resistance, and finite differential gain $A_V = V_o/(V_1 - V_2)$.
(a) Obtain an expression for the gain $A_{Vf} = V_o/V_s$.
(b) Show that $\lim A_{Vf} = n + 1$, $A_V \to \infty$.

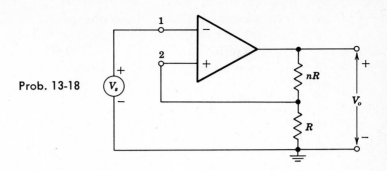

Prob. 13-18

13-19 Verify Eq. (13-20) for the bridge amplifier.

13-20 The circuit shown represents a low-pass dc-coupled amplifier. Assuming an ideal operational amplifier determine (a) the high-frequency 3-dB point f_H; (b) the low-frequency gain $A_V = V_o/V_s$.

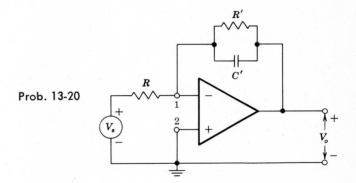

Prob. 13-20

13-21 (a) The input to the operational integrator of Fig. 13-20 is a step voltage of magnitude V. Show that the output is

$$v_o = A_V V(1 - \epsilon^{-t/RC(1-A_V)})$$

(b) Compare this result with the output obtained if the step voltage is impressed upon a simple RC integrating network (without the use of an operational amplifier). Show that for large values of RC, both solutions represent a voltage which varies approximately linearly with time. Verify that if $-A_V \gg 1$, the slope of the ramp output is approximately the same for both circuits. Also prove that the deviation from linearity for the amplifier circuit is $1/(1 - A_V)$ times that of the simple RC circuit.

13-22 (a) The input to an operational differentiator whose open-loop gain $A_V \equiv A$ is infinite is a ramp voltage $v = \alpha t$. Show that the output is

$$v_o = \frac{A}{1 - A} \alpha RC(1 - \epsilon^{-t(1-A)/RC})$$

(b) Compare this result with that obtained if the same input is impressed upon a simple RC differentiating network (without the use of an amplifier).

Show that, approximately, the same final constant output $RC\, dv/dt$ is obtained. Also show that the operational amplifier output reaches this correct value of the differentiated input much more quickly than does the simple RC circuit.

13-23 Given an operational amplifier with Z consisting of R in series with C, and Z' consisting of R' in parallel with C'. The input is a step voltage of magnitude V. (a) Show by qualitative argument that the output voltage must start at zero, reach a maximum, and then again fall to zero. (b) Show that if $R'C' \neq RC$, the output is given by

$$v_o = \frac{R'CV}{R'C' - RC} \left(\epsilon^{-t/RC} - \epsilon^{-t/R'C'}\right)$$

13-24 Sketch an operational amplifier circuit having an input v and an output which is $-5v - 3dv/dt$. Assume an ideal operational amplifier.

13-25 Sketch in block-diagram form a computer, using operational amplifiers, to solve the differential equation

$$\frac{dv}{dt} + 0.5v + 0.1 \sin \omega t = 0$$

An oscillator is available which will provide a signal $\sin \omega t$. Use only resistors and capacitors and assume $v = 4$ V at $t = 0$.

13-26 Set up a computer in block-diagram form, using operational amplifiers, to solve the following differential equation:

$$\frac{d^3y}{dt^3} + 2\frac{d^2y}{dt^2} - 4\frac{dy}{dt} + 2y = x(t)$$

where

$$y(0) = 0 \qquad \frac{dy}{dt}\bigg|_{t=0} = -2 \qquad \text{and} \qquad \frac{d^2y}{dt^2}\bigg|_{t=0} = 3$$

Assume that a generator is available which will provide the signal $x(t)$.

13-27 Sketch in block-diagram form an analog computer, using OP AMPs, to solve the differential equation

$$\frac{d^2y}{dt^2} + 5\frac{dy}{dt} - 2y = \sin \omega t$$

where

$$y(0) = 6 \qquad \text{and} \qquad \frac{dy}{dt}\bigg|_{t=0} = 0$$

13-28 Using Eq. (13-31) and Table 13-2, show that the transfer function of a second-order Butterworth low-pass filter satisfies Eq. (13-29).

13-29 Use the value of $B_2(s)$ from Table 13-2 and verify that

$$B_2(s)B_2(-s)\bigg|_{s=j\omega} = 1 + \omega^4$$

13-30 Use the values of $B_3(s)$ from Table 13-2 and verify that

$$B_3(+s)B_3(-s)\bigg|_{s=j\omega} = 1 + \omega^6$$

13-31 Show that the voltage gain $A_V(s) = V_o/V_s$ in Fig. 13-26a is given by Eq. (13-31).

13-32 Design an active sixth-order Butterworth low-pass filter with a cutoff frequency (or upper 3-dB frequency) of 1 kHz.

13-33 Design an active fifth-order Butterworth low-pass filter with a cutoff frequency of 5 kHz. All capacitors in the system are to have a value of 0.01 μF. Draw the system and specify the values of all resistances. (If some resistance values may be chosen arbitrarily, state which these are.)

13-34 (a) Show that the circuit of the accompanying figure can simulate a grounded inductor if $R_1 > R_2$. In other words, show that the reactive part of the input impedance of this circuit is positive if $R_1 > R_2$.

(b) Find the frequency range in which the $Q = \omega L/R$ of the inductor is greater than unity.

    Assume that the unity gain amplifier has infinite input resistance and zero output resistance.

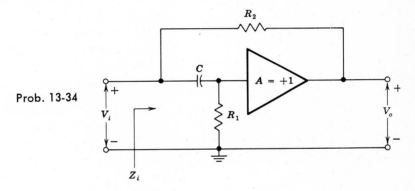

Prob. 13-34

13-35 (a) Show that the circuit of the given figure can simulate a grounded inductor if $A > 1$. In other words, show that the reactive part of Z_i is positive.

(b) Show that the real part of Z_i becomes zero ($Q = \infty$) at the frequency

$$\omega = \frac{1}{R_2C}\sqrt{\frac{R_1 + R_2}{R_1(A - 1)}}$$

    Assume that the input resistance of the amplifier of gain A is infinite.

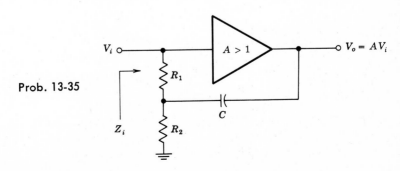

Prob. 13-35

CHAPTER 14

14-1 (a) Verify that the circuit shown gives full-wave rectification provided that $R_2 = 2R_1$.

(b) What is the peak value of the rectified output?

(c) Draw carefully the waveforms $v_i = 10 \sin \omega t$, v_p, and v_o if $R_3 = 2R_1$.

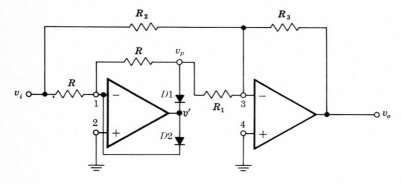

Prob. 14-1

14-2 If a waveform has a positive peak of magnitude V_1 and a negative peak of magnitude V_2, draw a circuit using two peak detectors whose output is equal to the peak-to-peak value $V_1 - V_2$.

14-3 For the feedback circuit shown, the nonlinear feedback network β gives an output proportional to the product of the two inputs to this network, or $V_f = \beta V_2 V_o$. Prove that if $A = \infty$, then $V_o = K V_1 / V_2$, where K is a constant.

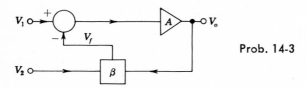

Prob. 14-3

14-4 (a) With the results of Prob. 14-3, draw the block diagram of a system used to obtain the square root of the voltage V_s.

(b) What should be the value of β if it is required that $V_o = \sqrt{V_s}$?

14-5 (a) Verify Eq. (14-4) for the pulse width of a monostable multivibrator.

(b) If $V_z \gg V_1$ and $\beta = \frac{1}{2}$, what is T?

14-6 Verify Eq. (14-10) for the frequency of the triangle waveform.

14-7 The circuit shown consists of two cross-coupled NAND gates. The coupling from the output of N_1 to the input of N_2 is direct (dc), whereas resistance-capacitance (ac) coupling is used from the output of N_2 to the input of N_1. Positive TTL logic is used and the levels are 0 and V_{CC}. Assume that a NAND gate changes state when its input voltage falls below V (≈ 0 V for a TTL gate). Neglect the drop across the clamping diode D. The input v_i is at V_{CC} and at $t = 0$ drops to 0 V for a short time t_p; that is, a negative narrow pulse is applied.

(a) Verify that the circuit behaves as a monostable multivibrator by drawing the waveforms v_R and v_o.

(b) Find the duration T of the output pulse, assuming $T > t_p$.

NOTE: The voltage across a capacitor cannot change instantaneously.

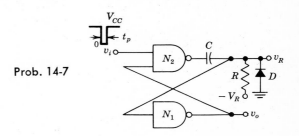

Prob. 14-7

14-8 The Schmitt trigger of Fig. 14-10 is modified to include two clamping Zener diodes across the output as in Fig. 14-8a. If $V_z = 4$ V and $A_v = 5,000$ and if the threshold levels desired are 6 ± 0.5 V, find (a) R_2/R_1, (b) the loop gain, and (c) V_R. (d) Is it possible to set the threshold voltage at a negative value? (e) In part a the ratio of R_2 to R_1 is obtained. What physical conditions determine the choice of the individual resistances?

14-9 The input v_i to a Schmitt trigger is the set of pulses shown. Plot v_o versus time. Assume $V_1 = 3.2$ V, $V_2 = 2.8$ V, and $v_o = +5$ V at $t = 0$.

Prob. 14-9

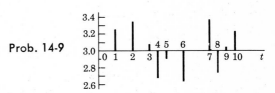

14-10 (a) A regenerative comparator (a Schmitt trigger) is to be designed with -3 V and -2.5 V as the two threshold voltages. Use a DIFF AMP whose gain is 10,000 and whose limiting output voltage is $+5$ V and -5 V. Draw the system and calculate the values of all the parameters (power supplies, resistances, capacitances, etc.) which you add external to the amplifier.

NOTE: The smallest resistance may arbitrarily be taken as 1 K.

(b) Calculate the loop gain.

14-11 (a) Calculate the logic levels at output Y of the ECL gate shown. Assume that $V_{BE,\text{active}} = 0.7$ V. To find the drop across an emitter follower when it behaves as a diode assume a piecewise-linear diode model with $V_\gamma = 0.6$ V and $R_f = 20\ \Omega$. Assume $\beta = 100$.

(b) Find the noise margin when the output Y is at $V(0)$ and also at $V(1)$.

(c) Verify that none of the transistors goes into saturation.

(d) Calculate R so that $Y' = \bar{Y}$.

(e) Find the average power taken from the power source.

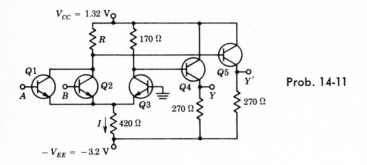

Prob. 14-11

14-12 Verify that, if the outputs of two (or more) ECL gates are tied together as in Fig. 14-14, the OR function is satisfied.

14-13 (a) For the system in Fig. 14-14 obtain an expression for Y which contains three terms.

(b) If in Fig. 14-14 $\bar{Y}_1$ and $\bar{Y}_2$ are tied together, verify that the output is $Y = \bar{A}\bar{B} + \bar{C}\bar{D}$.

(c) If in Fig. 14-14 Y_1 and Y_2 are tied together and if the input to the lower ECL gate is $\bar{C}$ and $\bar{D}$ (instead of C and D), what is Y?

14-14 (a) For the D/A converter of Fig. 14-17 show that when the second most significant bit is 1 and all other bits are zero, the output is $V_o = V_R/4$.

(b) Find V_o if only the third MSB is 1.

(c) Find V_o if only the LSB is 1.

14-15 The figure shows a binary weighted resistor D/A converter.

(a) Show that the output resistance is independent of the digital word and that

$$R_o = \frac{2^{N-1}}{2^N - 1} R$$

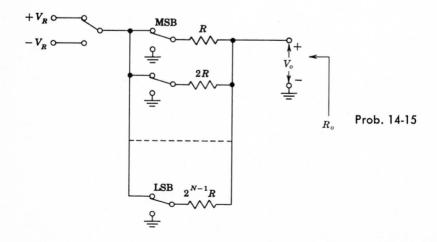

Prob. 14-15

(b) Show that the analog output voltage for the most significant bit is

$$V_o = \frac{2^{N-1}}{2^N - 1} V_R$$

(c) Show that the analog output voltage for the least significant bit is

$$V_o = \frac{1}{2^N - 1} V_R$$

INDEX

INDEX

473